Commercial Satellite Imagery

A tactic in nuclear weapon deterrence

Springer
London
Berlin
Heidelberg
New York
Barcelona
Hong Kong
Milan
Paris
Santa Clara
Singapore
Tokyo

Bhupendra Jasani and Gotthard Stein (editors)

Commercial Satellite Imagery

A tactic in nuclear weapon deterrence

Springer

Published in association with
Praxis Publishing
Chichester, UK

PRAXIS

Professor Bhupendra Jasani
Department of War Studies
King's College London
London
UK

Professor Gotthard Stein
Deputy Director
Programme Group Systems Analysis and Technology Evaluation
Research Centre Juelich
Juelich
Germany

SPRINGER–PRAXIS BOOKS IN GEOPHYSICAL SCIENCES
SUBJECT *ADVISORY EDITOR*: Dr Philippe Blondel, C.Geol., F.G.S., Ph.D., M.Sc., Senior Scientist, Department of Physics, University of Bath, Bath, UK

ISBN 3-54042-643-4 Springer-Verlag Berlin Heidelberg New York

Die Deutsche Bibliothek – CIP-Einheitsaufnahme
Jasani, Bhupendra:
Commercial satellite imagery: a tactic in nuclear weapon deterrence /Bhupendra Jasani; Gotthard Stein. – Berlin; Heidelberg; New York; Barcelona; Hongkong; London; Mailand; Paris; Singapur; Tokio: Springer; Chichester, UK: Praxis Publ., 2002
(Springer praxis books in geophysical sciences)
ISBN 3-540-42643-4

0101 deutsche buecherei0292 deutsche bibliothek

Library of Congress Cataloging-in-Publication Data
A catalogue record for this book is available from the Library of Congress

Printed by MPG Books Ltd, Bodmin, Cornwall, UK

Cover design: Jim Wilkie
Project management: Originator, Gt Yarmouth, Norfolk, UK

Printed on acid-free paper supplied by Precision Publishing Papers Ltd, UK

Contents

Preface

While the suggestion that the International Atomic Energy Agency (IAEA) use commercial satellite imagery was made as long ago as the mid-1980s, the results of the first actual study demonstrating the possibilities was presented in 1994 at the IAEA Symposium on International Safeguards (Jasani, 1994). In the same year the IAEA's Standing Advisory Group on Safeguards Implementation (SAGSI) advised the Director General of the need for an 'assessment of the usefulness, technical feasibility, associated costs and acceptability of the Agency obtaining satellite photographs from commercial sources'. Subsequently, under the UK, German, US, and Canadian Support Programmes to the IAEA, a number of studies were carried out on the use of satellites with panchromatic, multispectral and radar sensors on board. Considerable interest was generated among a number of member states in this area. The European Commission, through its Joint Research Centre at Ispra, has also investigated the usefulness of commercial satellite imagery to the Agency.

Information gathered by the National Technical Means (NTM) of some states has played an important role in uncovering the undeclared nuclear activities of some other states, but access to such satellites is very limited. However, in recent years, the capabilities of commercial observation spacecraft have improved dramatically. This and the non-intrusive nature of satellite-based remote sensing have made its use for the Agency's safeguards procedures attractive. At the same time, it is realised that data gathered from commercial observation satellites is not enough on its own. Collateral information from many sources is needed. With these considerations in mind, the Agency established in 2001 its Satellite Imagery Analysis Laboratory (SIAL), which is now fully operational and part of the safeguards system.

With the help of a number of world-renowned experts in various space-based remote-sensing fields, this book identifies possibilities and challenges associated with the use of satellites by the IAEA. A number of conclusions are drawn and presented.

The Editors would like to take this opportunity to thank the UK, German, Canadian and US Support Programmes to the IAEA for their considerable help and encouragement in establishing the foundation, through a series of studies, for the application of commercial satellite imagery in the Agency's safeguards procedures. We would also like to thank the Department of War Studies, King's College London and the Research Centre Jülich, Germany for providing the infrastructure to carry out various studies demonstrating the usefulness of commercial satellite imagery for the Agency.

Contributors

Phillip J. Baines graduated from the University of Toronto with a BASc in Engineering Science (aerospace option), and a MEng degree in Mechanical Engineering (robotics). He worked at EMS Technologies Canada Ltd (formerly Spar Aerospace Ltd, Space Systems) from 1982 to 2000 in various engineering and project manager assignments. Between 1982 and 1987, he worked on studies investigating the use of civilian remote sensing satellites for arms control verification. During this period he also taught a spacecraft design course at the University of Toronto. Between 1988 and 1997, he worked on development of the gimballed, data relay antenna for the International Space Station programme, large deployable antennas for geostationary mobile communications satellites, and the application of product data management systems on the Canadarm-2 robotic workstation project. Between 1997 and 2000, he worked at the Department of Foreign Affairs and International Trade under a government–industry interchange programme examining new verification technologies for application to existing and future arms control agreements. He is currently employed in the Department of Foreign Affairs and International Trade as a Policy Advisor, Science and Technology, for the Non-Proliferation, Arms Control and Disarmament Division.

Alain Baudoin is responsible for High Resolution Imagery Programmes (SPOT, Pleiades) at the Delegation for Earth Observation, French Space Agency (CNES), which he joined in 1989. He previously worked at the French Geographic Institute (IGN-F) as chief of the Numerical Aerospace Cartographic Service (SCAN) (1985–1986) and then was in charge of the Thesis Department of the National School for Geographical Sciences (ENSG) (1987–1988). He was a student of Ecole Polytechnique (1966–1969) and received a DEA (MSc) in pattern recognition from Paris University (1971). He has held the title of General Geographer Engineer since 1994 and was involved in ISPRS activities as Commission VII secretary (1980–1984), in OEEPE as French delegate (1993–1999) and in the French Society of Photogrammetry and Remote Sensing (SFPT) as General Secretary (1985–1993). He has published several papers on remote sensing and space cartography.

Furio Bellezza graduated in Computer Science. He is co-author of some successful support systems for risk and emergency management in large chemical complexes. Since May 2001, he has worked as Grant-holder at the Non-Proliferation and Safeguards Unit of the Institute for the Protection and Security of the Citizen of the Joint Research Centre, Ispra. He is currently involved in the development of GIS-based support tools for declaration and verification in nuclear safeguards. His main interests are in database design, image processing, GISs and website design.

Morton Canty is a research scientist in the Department of Systems Analysis and Technology Evaluation at the Jülich Research Centre in Germany. He is a Canadian citizen, and received his PhD in Nuclear Physics in 1969 at the University of Manitoba. He attended the University of Bonn for three years on a National Research Council of Canada postdoctoral fellowship, visited the University of Groningen as a guest scientist for one year and worked for two years at the University of Marburg and the Heavy Ion Accelerator Laboratory in Darmstadt. He began his present position in Jülich in 1979, where his principal interest has been the development of statistical and game theoretical models for verification of the Nuclear Weapons Non-Proliferation Treaty. He is the author of three books in German: on non-linear dynamics, statistical classification techniques in remote sensing and algorithmic game theory. He is also co-author of a monograph on mathematical methods for treaty verification. He has written many papers on low energy nuclear physics, nuclear safeguards, applied game theory and remote sensing and has served on numerous advisory bodies to the German federal government and to the IAEA.

Kaluba Chitumbo has a PhD in Physical Chemistry (1978) from the University of Uppsala, Sweden. In 1984 he joined the IAEA s Department of Safeguards and has performed safeguards inspections in various regions. From 1993 to 1996, he was Section Head of one of the Safeguards Operations Sections covering Euratom countries. Since 1996 he has been Director for the Division of Safeguards Information Technology. He has authored and published a number of papers on non-proliferation issues and on safeguards. He was the secretary for Committee II in the 1995 NPT extension conference in New York.

Sergio Contini graduated in Electronic Engineering. He worked for seven years in the field of nuclear safety and probabilistic risk assessment before moving to risk analysis of chemical and petrochemical installations. In 1986 he joined the Joint Research Centre, where he was involved in several projects to support the implementation of the Seveso directives in the European Union. Since early 2000, he has been involved in the development of tools for supporting declaration and verification activities in nuclear safeguards. He has been responsible for the development of several support tools for complex systems reliability, risk analysis, emergency planning and emergency response. He has been a lecturer on risk analysis in several postgraduate courses and has published some 80 papers on system analysis methodologies, risk assessment and management, and GIS-based support tools.

Takashi Hamazaki is a Senior Engineer on the Advanced Land Observing Satellite (ALOS) Project at the National Space Development Agency of Japan (NASDA).

John Hilliard received a B.S. degree from Old Dominion University in 1977. He has worked for seven years for the International Atomic Energy Agency in the Department of Safeguards. Working in the Director's office of the Division of Safeguards Information Technology, he was the focal point for all aspects of the Agency's utilisation of commercial satellite imagery. This included being the R&D project manager and Task officer to the Member States who supported and funded the Departmental effort. He also was the divisional liaison for the three Divisions of Operations and the IAEA Action Team who are the active users of satellite imagery under the IAEA inspection regime. While serving in this capacity he trained inspectors in the use of satellite imagery for inspection purposes as part of a standard Agency satellite imagery awareness course. He also was a fixed presenter for the Department of Safeguards state evaluation course given periodically to Departmental staff. Other activities included acting as Scientific secretary for an international Y2K seminar, technical secretary for various Agency workshops on satellite imagery, and conference co-chairman for a satellite imagery session at a symposium in October 2001 on international Safeguards. Prior to working at the IAEA he owned and operated a small computer software company for 14 years.

Bhupendra Jasani is a Visiting Professor at the Department of War Studies, King's College London, which he joined in 1990. He was previously with the UK Medical Research Council at the University College Hospital Medical School, London (1965–1972), the Stockholm International Peace Research Institute (SIPRI) (1972-1987), and the Royal United Services Institute for Defence Studies (1987–1990). He received a MSc in Nuclear Physics and a PhD in Nuclear Physics and nuclear medicine from the University of London. He has authored, co-authored and edited several books and reports on non-proliferation of nuclear weapons and arms control in outer space. He has published a number of papers on military uses of outer space and verification using commercial observation satellites.

Richard Keeffe has a BSc in Physics and a Master's degree in Material Science. He has been working in the areas of International Safeguards and Nuclear Non-Proliferation for over 25 years. His experiences range from import/export control, non-proliferation, safeguards implementation, and safeguards R&D. He is currently the Head of the Safeguards and Support Section of the Office of International Affairs of the Canadian Nuclear Safety Commission.

R. Krishnan has been with the Indian Space Research Organisation for the last 25 years. He has been closely associated with the Indian Remote Sensing Programme. He graduated from Madras University in Electrical Engineering and holds a PhD from the Indian Institute of Science, Bangalore. His main interests are Image Processing, Remote Sensing and Artificial Intelligence.

Reinhard Loosch retired in 1995 as Deputy Director General for International (and, until German unification, Intra-German) Cooperation at the Federal Ministry for Research and Technology (now 'Education and Research'), where he spent most of his 35 year career in the German civil service. His education – at Tübingen, Princeton and Bonn Universities – was in law and economics, leading to the Second State Examination in Law in 1959. While working with the German Ministry, he served on numerous international councils, commissions and committees, *inter alia* as Governor of the International Atomic Energy Agency from 1972 to 1992 (1987–1988 as Chairman of the Board), member of the European Space Agency (ESA) Council, member and chairman of the OECD Nuclear Energy Agency's Steering Committee and of the Max von Laue–Paul Langevin Institute at Grenoble as well as of several EU bodies and bilateral commissions for science and technology cooperation. Between 1996 and 1997, he was recalled from retirement to serve as the German delegate to the IAEA Board's Committee on Strengthening Safeguards ('Committee 24'). He has authored many articles and papers on various matters (including safeguards) in fields of his official responsibilities.

Irmgard Niemeyer studied Geography, Meteorology and Geophysics at the University of Bonn, Germany. Between 1996 and 1997, she was an Application Manager and Consultant for geoinformation systems (GIS), market and politics research at Dimap GmbH, Bonn, Germany. She did her PhD on 'Change detection and object-oriented, knowledge-based classification of multispectral images for nuclear verification' at the Research Centre Jülich (FZJ), Germany (1997–2000). Since 2000, she has been a Consultant for Remote Sensing and GIS and a Satellite Imagery Analyst within the German Support Programme for the IAEA, Vienna, Austria. She remains a researcher with FZJ in the field of satellite imagery analysis for treaty verification purposes.

Masami Onoda is a Data Policy Specialist in the Satellite Programme and Planning Department at the National Space Development Agency of Japan (NASDA).

J. P. Paquette started his career in the field of imagery interpretation and data analysis in 1981 with the Canadian Forces Intelligence Branch, where he specialised as an Imagery Analyst. As a consultant, he has performed consulting services to government, nongovernment organisations and commercial clients. Examples of his imagery analysis work include: the 1990–1991 Persian Gulf War; Arms Control Verification and NonProliferation Issues; the 1992–1995 Yugoslav/Balkan crisis; refugee monitoring (the 1996 Zaire/Rwandan refugee situation); infrastructure studies; industrial production analysis; industrial pollution monitoring; support to search and rescue efforts; and potential applications of commercial satellite imagery in international safeguards.

Robert L. Pearson is Vice President of CH2MHILL, an international engineering company, headquartered in Denver, Colorado, with offices around the world. He joined CH2MHILL in 2000. He is also an adjunct member of the faculty of the

University of Colorado, Denver, where he teaches a graduate level course on air pollution. From 1992 to 2000, he was a senior scientist with Radian International, an engineering firm. Prior to that, he was an environmental engineer with the Public Service Company of Colorado, an electric and gas utility company in Denver from 1973 to 1992. He received his professional geophysical engineering degree from the Colorado School of Mines in 1968 and his MSc in 1972 and PhD in 1973 from Colorado State University. His dissertation topic was the quantitative remote multispectral sensing of biomass vegetation from aircraft and satellites. More recently, Dr Pearson has conducted work on the use of remote sensing imagery and GIS data in the determination of areas with probable clandestine industrial and military activities.

André Poucet is head of the Non-Proliferation and Nuclear Safeguards Unit at the Institute for Protection and Security of the Citizen of the European Commission's Joint Research Centre (JRC) in Ispra (Italy). After working for the Belgian nuclear industry, he joined JRC-Ispra in 1985 to work on nuclear reactor safety and Probabilistic Risk Assessment. In 1993 he was seconded to the Joint Central Team of the ITER project in San Diego, working on the safety analysis and safety design of this large experimental fusion reactor. In 1997 he joined the safeguards group in Ispra. He received a MSc in Electro-mechanical Engineering and a PhD in Nuclear Engineering from the University of Leuven (Belgium). Since 1991, he has also been a part-time professor in Risk Analysis at the same university. He has published around 100 papers on risk assessment methodology, fusion safety, and the use of information technologies for safeguards.

Bernd Richter works as a scientist at the Jülich Research Centre, which he joined in 1976. He received a Dipl Phys and a DrRerNat in Nuclear Physics from the Rheinische Friedrich-Wilhelms-Universität Bonn. Subsequently, he was a visiting scientist at the Weizmann Institute of Science, Rehovot (1974–1975). He joined the safeguards R&D group within the Jülich Research Centre in 1978 and became the project officer of a number of tasks within the German Support Programme to the IAEA. As of 1985, he has also been cooperating with the German Government in coordinating that programme. Currently, he is a member of the Scientific Council and Coordination Board of the European Safeguards R&D Association (ESARDA), and he chairs the ESARDA Working Group on Containment and Surveillance. He has published papers on various technical issues related to international safeguards.

Stephen W. Robb (BSc) has worked in the Satellite Imagery Analysis Laboratory (SIAL) at the IAEA since October 1999 and is the GIS technical expert and database administrator. Previously, he worked in the IAEA as LAN administrator and Intranet manager.

Gotthard Stein is Deputy Director of the Systems Analysis and Technology Evaluation Programme Group at the Jülich Research Centre and a Visiting Professor at the Department of War Studies, King's College London. He received a Dipl Phys and a DrRerNat in Nuclear Physics from the Rheinische Friedrich-Wilhelms-Universitt Bonn. His current working fields comprise research in global climate change, technology assessment, treaty verification, non-proliferation and international nuclear safeguards. He is a member of the Executive Committee as well as of the Steering Committee of the European Safeguards R&D Association (ESARDA), Head of the German Technology Assessment Group of the Helmholtz-Gemeinschaft and member of the Standing Advisory Group on Safeguards Implementation (SAGSI) of the IAEA.

Gerald Steinberg, director of the Program on Conflict Management and Negotiation, is a member of the BESA Center for Strategic Studies and is a Professor of Political Studies at Bar Ilan University. He specialises in Middle East arms control and strategy, the US–Israeli relationship, and strategic aspects of space technology. He received his PhD at Cornell University (1981), and his doctoral research focused on the interaction between the political and technological dimensions of satellite reconnaissance and arms control verification. He has been a visiting fellow at MIT, the US Institute for Peace, SIPRI, and participates in projects of the United Nations University and Track Two multilateral workshops on Middle East Arms Control and Regional Security. Recent publications include 'Commercial Observation Satellites in the Middle East and Persian Gulf' – a chapter in *Commercial Observation Satellites: At the Leading Edge of Global Transparency*, 2001 – and *Dual Use Aspects of Commercial High Resolution Imaging Satellites*, published by the BESA Center for Strategic Studies, February 1998.

C. Vincent Tao holds the Canada Research Chair Professorship in Geomatics and is Director of Geospatial Information and Communication Technology (Geo-ICT) Lab at York University, Toronto, Canada. His research interests include Internet and wireless GISs, mobile and distributed geocomputing, digital photogrammetry and remote sensing image processing, and LIDAR and SAR data processing. He is a recipient of many international awards and holds several important positions in international working groups and editorial positions in geomatics journals.

Q. S. Bob Truong graduated from Ottawa and McGill Universities and the University of Toronto with BEng, MASc and PhD degrees in Mechanical Engineering. He has been working in the Canadian nuclear industry for 28 years. His expertise and interests cover the following areas: thermohydraulics, aerodynamics, nuclear reactor safety analysis and assessment (specialising in severe accidents), safety research management, and training. His recent work under the Canadian Safeguards Support Programme of the Canadian Nuclear Safety Commission includes training of IAEA inspectors on a variety of safeguards equipment using traditional and computer-based training techniques, and investigation of potential applications of modern science and technology in international safeguards. His urrent effort is directed towards the following topics: applications of airborne (hyperspectral and SAR) remote sensing and multi-sensor commercial satellite imagery in treaty verification; GIS for management of safeguards information;

and tracking of containers carrying nuclear materials. He has actively participated in international meetings and conferences on international safeguards and has published papers on the above topics of interest.

Maurice Ward is with the UK Atomic Energy Authority and is Coordinator of the UK Safeguards R&D Support Programme to the IAEA, which is financed by the UK Department of Trade and Industry. He received a BSc (in Applied Chemistry, specialising in nuclear technology) and an MSc (in Radiological Health and Safety) from the University of Salford and a PhD (in Radiation Chemistry) from the University of Manchester. Before joining the UKAEA (1974), he worked at the Institute of Photo-chemistry and High Energy Radiation of the Italian National Research Council (1970–1973) and was a Research Fellow at Manchester University (1973–1974). He has co-authored a number of papers and reports on the use of commercial satellite imagery to assist international safeguards.

Olga V. Zhdanovich is a Consultant at the Analytical Center for Economics and Management in Aerospace Technologies in Moscow, Russia. She was previously with the International Space University in Strasbourg, France (1998–2000), the Scientific and Research Institute Cometa (1996–1997) and VideoCosmos Inc. (1990–1995) in Moscow, Russia. She received an engineering degree in Cartography from the Moscow Institute of Engineering in Geodesy, Cartography and Aerial Surveying, and an MSc in Environmental Sciences and Policy from the Central European University, Budapest, Hungary. She has published a number of papers on Russian remote sensing programmes, future space activities and sustainable development issues. She was a coordinator of the IAF/ISU/Prospective 2100 Forum on Space Activities in the 21st Century at UNISPACE III.

Figures

Tables

Colour plates (between pages 136–137)

Plate 1 Image processing comparison for hydrological phenomena. *Left side*: A natural colour image of the Bruce Nuclear Generating Facility, acquired by IKONOS on 11 November 1999. The circles indicate the position of discharge points. There is a small surface effect at Bruce B and hints of a plume but it is difficult to see. The pattern of the waves visible in the original data shows clearly that the wind is blowing from the north. *Right side*: The same image after removing the sun glint of the surface waves, and enhancement. A large plume is seen flowing south of Bruce B and curling into the small bays. The slightly less turbid water along the coast between A and B may be a smaller plume originating at Bruce A

Plate 2 An extract from a quarter scene acquired on 3 November 1997 by the US Landsat 5 satellite over the UK Dungeness Nuclear Power Reactor station. The warm water discharged into the sea can be seen in this image which is a combination of the thermal band 6 (red) and bands 5 (green) and 4 (blue). Some heat escaping the switchyard building can also be detected while clearly the reactor buildings are reasonably well insulated. Some of the buildings in the reactor complex can be distinguished as separate buildings

Plate 3 MAD change detection for Space Imaging Landsat-5 August 1998–August 1996, fused to the negative of an IRS-1C panchromatic (5 m) image taken in September 1997. MAD component 1 and 2 exceeding the significance thresholds are in red and green

Plate 4 Extract from a Landsat 5 scene (resolution 28.5 m) acquired over the Oak Ridge facility on 8 September 1984 (lower image). The image is a combination of bands 1 (blue), 2 (green) and 3 (red). The enrichment facility is in the area marked by red rectangle. The top image is from the US early reconnaissance satellite KH4 showing roughly the same area as the rectangle in the Landsat image. The image was acquired on 8 March 1970. Various facilities are identified in the KH4 image, K-25 being the original diffusion plant [source: Landsat/EOSAT and Corona image].

Plate 5 Composite image of the central part of Rome. Colour is taken from the

KFA-1000 image and 2-m resolution is taken from the KVR-1000 image (courtesy of Sovinformsputnik).

Plate 6 Extracts from Landsat images over the K-25 facility on three dates indicated under the images that are combination of bands 6 (red), 5 (green) and 4 (blue). The temperature of the water is indicated by the red colour **(A)**. This was found to be highest on 26 June 1986 and lowest on 10 April 1987 suggesting that the plants were operating in 1984 and 1986 but ceased operating by 1987.

Plate 7 Extracts from two scenes acquired from SPOT (resolution 10 m) on 12 May 1991 (top image) and from a Russian KVR100 sensor (resolution 2 m) acquired on 27 June 1990 (lower image). Perimeter fences can be identified. Also some of the changes that have occurred on the Iraqi reactor site during a period of one year can be identified. The red boxes show new constructions and the blue boxes indicate changes in the site possibly due to bombing during the Gulf War or due to later dismantlement process. This is an example of visual change detection.

Plate 8 A combination of bands 1 (blue), 2 (green) and 6 (red) of the Landsat 5 image acquired on 7 May 1997. The thermal plume from the reactor can be seen at **A** while at **B** possible sedimentation from rivers can be detected.

Plate 9 A combination of three principal components, C4 (red), C1 (green), and C2 (blue), of the Landsat 5 image.

Plate 10 The result of combining the difference between the principal components C2 and C4 (red) with component C2 (green) and component C1 (blue).

Plate 11 A combination of the co-registered ERS-2 image (as red) and the principal components C4–C2 (as green) and the component C1 (blue) as an RGB image. It can be seen that the warm water plumes coincide in all the images.

Plate 12 A USGS Landsat 7 Image from 26 August 1999. False colour composite of the BNPD site using the first, second and third principal components (30 m) fused with panchromatic image (15 m) [courtesy of M. Canty].

Plate 13 Extracts from the Landsat TM images (resolution 30 m) acquired at different times over Gorleben, shown as combinations of the principal components 1 (red), 2 (green) and 3 (blue) (a–c) and the site map superimposed over the image (d). The dates are indicated under each image.

Plate 14 Bi-temporal change detection for the Landsat TM (resolution 30 m) extracts over Gorleben acquired on 24 August 1984 and 31 August 1989 fused with a high-resolution KVR-1000 image. (a) PCA: Shown is the second principal component of the channels 5 (red), 4 (green) and 3 (blue) exceeding three standard deviations. (b) IPCA: Shown is the second iterated principal component of the channels 5 (red), 4 (green) and 3 (blue) exceeding three standard deviations. (c) MAD: Combinations of the MAD Components 6 (red), 5 (green) and 4 (blue) exceeding five standard deviations. (d) Bi-temporal change detection for the Landsat TM extracts over Gorleben acquired on 31 August 1989 and 21 August 1997, fused with a high resolution KVR-1000 image: Combinations of the MAD components 6 (red), 5 (green) and 4 (blue) exceeding five standard deviations.

Plate 15 In this figure the images were combined as band five (red) of 1987 with bands five (green) and four (blue) of 1983. Sites marked **A** and **B** are production reactors and reprocessing facilities. In the image red and green colours represent changed areas. Compare this with the image in Figure 7.26.

Plate 16 This shows a 3 linear combination of the twelve bands of the bi-temporal image in red, green and blue resulting from the analysis using MAD. Twelve infrared bands of Landsat images dated 12 August 1983, 27 July 1986 and 28 June 1987 were used. The changes represented by colours were then superimposed on a single band from the image dated 12 August 1983. Compare this with the image in Figure 7.25.

Plate 17 SAR airborne image showing the discharge at Bruce.

Plate 18 This is an image over south-east India and Sri Lanka acquired by the Ocean Colour Monitor (OCM) on board the IRS-P4 satellite.

Plate 19 Spectral unmixing results – uranium and copper tailings.

Plate 20 Spectral unmixing results – environmental effects in water.

Acronyms and abbreviations

ADEOS	Advanced Earth Observation Satellite
AGR	Advanced Gas-cooled Reactor
ALOS	Advanced Land Observation Satellite
AMSR	Advanced Microwave Scanning Radiometer
AOI	Area Of Interest
API	Application Programming Interface
ARISE	Agricultural Resources Inventory and Survey Experiment
ARSIS	Amelioration de la Resolution Spatiale par Injection de Structures
ASCII	American Standard Code for Information Interchange
ASTER	Advanced Spaceborne Thermal Emission and Reflection Radiometer
AVLIS	Atomic Vapour Laser Enrichment
AVNIR	Advanced Visible and Near Infrared Radiometer
BBC	British Broadcasting Corporation
BNPD	Bruce Nuclear Power Development
BW	Biological Weapons
CAD	Computer Aided Drawing
CASI	Compact Airborne Spectrographic Imager
CCRS	Canadian Centre for Remote Sensing
CD-ROM	Compact Disk-Read-Only Memory
CDSat	Change Detection with Satellite imagery
CEOS	Committee for Earth Observation Satellites
CERES	Clouds and the Earth's Radiation Energy System
CIA	Central Intelligence Agency
CMP	Centre de Mission et de maintien à Poste
CNES	Centre National d'ètudes Spatiales
CNSC	Canadian Nuclear Safety Commission
CSBMs	Confidence and Security Building Measures
CSI	Commercial Satellite Imagery
CSSP	Canadian Safety Support Programme

CST	Core Software Technology
CTBT	Comprehensive Test Ban Treaty
CWC	Chemical Weapons Convention
DARA	Deutsche Agentur für Raumfahrtangelegenheiten
DBMS	Data Base Management Systems
DEM	Digital Elevation Model
DFAIT	Department of Foreign Affairs and International Trade
DLR	Deutsches Zentrum für Luft- und Raumfahrt
DTM	Digital Terrain Model
DVD-ROM	Digital Video Disk-Read-Only Memory or Digital Versatile Disk- Read-Only Memory
ELOP	Electro Optical
EM	Electromagnetic
EMERCOM	Emergencies and the Elimination of the Consequences of Natural Disasters?
EMIS	Electromagnetic Isotope Separation
EMS	Electromagnetic Spectrum
ENVISAT	Environmental Satellite
EO-ICWG	Earth Observations International Co-ordination Working Group
EOC	Earth Observation Center
EORSAT	Electronic Ocean Reconnaissance Satellite
EPS	European Polar System
EROS	Earth Remote Observation System
ERS	European Remote Sensing
ERTS	Earth Resources Technology Satellite
ESA	European Space Agency
ESARDA	European Safeguards Research and Development Association
ETM	Enhanced Thematic Mapper
EU	European Union
EUMETSAT	Europe's Meteorological Satellite Organisation
FAS	Federation of American Scientists
FFT	Fast Fourier Transformation
FMCT	Fissile Materials Production Cut-off Treaty
GB	Giga Byte
GCOM	Global Change Observation Mission
GCP	Ground Control Points
GDTA	Groupement pour le Dévelopment de la Télédétection Aérospatiale
GeoATOMS	Geospatial Atomic Tracking and Organisation Manager System
GHz	Giga Hertz
GIF	Geographics Interchange Format
GIS	Geographic Information System
GLI	Global Imager
GLONASS	Global Navigation System
GMES	Global Monitoring for Environment and Security
GMS	Geostationary Meteorological Satellite

GPS	Global Positioning System
GSD	Ground Sampling Distance
GUI	Graphic User Interface
HRG	High Resolution Geometric instruments
HRPT	High Resolution Picture Transmission
HRS	High Resolution Stereoscopic
HRV	High Resolution Visible
HRVIR	High Resolution Visible and InfraRed
I^2SIP	IAEA Integrated Safeguards Instrumentation Programme
IAEA	International Atomic Energy Agency
IAF	International Astronomical Federation
IAI	Israel Aircraft Industry
IEOS	International Earth Observation Satellite
IFOV	Instantaneous Field Of View
IGN	Institut Géographique National
IGOS	Integrated Global Observation Strategy
IGS	Information Gathering Satellite
IHS	Intensity-Hue-Saturation
INFCIRC	Information Circular
IPCA	Iterative Principal Component Analysis
IR	Infra-red
IRS	Indian Remote Sensing satellite
ISDAS	Imaging Spectrometer Data Analysis System
ISIS	Intelligent Satellite Data Information System
ISMA	International Satellite Monitoring Agency
ISO	Infrared Space Observatory
ISPRS	International Society for Photogrammetry and Remote Sensing
ISRO	Indian Space Research Organisation
ISSSR	International Symposium on Spectral Sensing Research
IT	Information Technology
JASPAS	Japan Support Programme for Agency Safeguards
JERS	Japan Earth Resources Satellite
JMA	Japan Meteorology Agency
JPEG	Joint Photographic Experts Group (Compression Standard)
JRC	Joint Research Centre
KH	Key Hole
LAN	Local Area Network
LEO	Low Earth Orbit
LISS	Linear Imaging Self-scanned Sensor
LOF	Locations Outside Facilities
LoG	Laplacian of Gaussian
Lps	Lines per Second
LUT	LookUp Table
LWRs	Light Water Reactors
MAD	Multivariate Alteration Detection

MB	Mega Byte
MENRT	Ministère de l'Education Nationale, de la Recherche et de la Technologie
METEOSAT	Meteorological Satellite
METI	Ministry of Economy, Trade and Industry
MISR	Multi-angle Imaging Spectro-Radiometer
MITI	Ministry of International Trade and Industry
MODIS	MODerate-resolution Imaging Spectro-radiometer
MOMS	Modular Optoelectronic Multispectral Stereo Scanner
MOPITT	Measurements Of Pollution In The Troposohere
MOS	Marine Observation Satellite
MOX	Mixed Oxide
MRVIR	Medium Resolution Visible and InfraRed
MSMR	Multi-Frequency Scanning Microwave Radiometer
MSS	Multispectral Scanner System
MTM	Multilateral Technical Means
MTSAT	Multifunctional Transport Satellite
MWe	Mega Watt electrical
NASA	National Aeronautical and Space Administration
NASDA	National Space Development Agency of Japan
NEMO	Naval EarthMap Observer
NFCIS	Nuclear Fuel Cycle Information Service
NGO	Non-Governmental Organisation
NIMA	National Imagery and Mapping Agency
NIR	Near Infra-Red
NNWS	Non-Nuclear Weapon States
NOAA	National Oceanic and Atmospheric Administration
NPP	Nuclear Power Plant
NPT	Treaty on the Non-Proliferation of Nuclear Weapons
NRSA	National Remote Sensing Agency (India)
NSCAT	NASA Scatterometer
NTDB	National Topographic Data Base (Canadian)
NTM	National Technical Means
NWS	Nuclear Weapon States
NWT	Northwest Territory, Canada
OCM	Ocean Colour Monitor
OCTS	Ocean Colour and Temperature Scanner
OEEPE	Organisation Européene d'Etudes Photogrammétriques Expérimentales
OPCW	Organisation for the Prohibition of Chemical Weapons
OPS	Optical Sensor
ORNL	Oak Ridge National Laboratory
P	Panchromatic
PAC	Common European Agriculture Policy
PALSAR	Phased Array type L-band Synthetic Aperture Radar

PAS	Priority Acquisition Service
PCA	Principal Component Analysis
POLDER	Polarization and Directionality of the Earth's Reflectance
PPP	Public Private Partnership
PRIS	Power Reactor Information Service
PRISM	Panchromatic Remote Sensing Instruments for Stereo Mapping
PSLV	Polar Satellite Launch Vehicle
PTBT	Partial Test Ban Treaty
PWR	Pressurised Water Reactor
QIS	Qualité des Images SPOT
RADARSAT	Radar Satellite
RAM	Random-Access Memory
RBV	Return-Beam Vidicon
RESTEC	Remote Sensing Technology Center (Japan)
RGB	Red Green Blue
RIMLIS	RIKEN's Molecular Laser Isotope Separation
RMS	Root Mean Square
RORSAT	Radar Ocean Reconnaissance Satellite
RRDB	Research Reactor Data Base
RSAC	Regional System of Accounting
RSC	Russian Space Corporation Energia
RSMA	Regional Satellite Monitoring Agency
SAGSI	Standing Advisory Group on Safeguards Implementation
SAMRO	Satellite Militaire de Reconnaissance Optique
SAR	Synthetic Aperture Radar
SEPT	Société Française de Photogrammétrie et Télédétection
SEU	Single Event Upsets
SFSI	Short wavelength infrared Full Spectrum Imager
SGIT	Safeguards Information Technology
SIAL	Satellite Imagery Analysis Laboratory
SIAM	Safeguards Imagery Archiving and Management system
SILMO	Sèparation des Isotopes par irradiation au Laser de Molècules
SIPRI	Stockholm International Peace Research Institute
SIT	Site Investigation Tool
SOP	Satellite Operating Partners
SPOT	Système Probatoire d'Observation de la Terre
SRD	Stations de Réception Directe
SSAC	State System of Accounting
SSOC	Segment Sol de l'Opérateur Commercial
SSOS	Segment Sol de l'Opérateur Satellite
SSS	Strengthening Safeguards System
SWIR	Short Wave Infra-Red
TIFF	Tagged Imagery File Format
TM	Thematic Mapper
TOMS	Total Ozone Mapping Spectrometer

TRMM	Tropical Rainfall Measuring Mission
TsSKB	Russian Central Specialised Design Bureau
UHR	Ultra High Resolution
UNCOPUOS	UN Committee on the Peaceful Uses of Outer Space
UNISPACE	United Nations conference on the exploration and peaceful uses of outer Space
UNMOVIC	Unite Nations Monitoring, Verification and Inspection Commission
UNSCOM	United Nations Special Commission
UTM	Universal Transverse Mercator
WIFS	Wide Field Sensor
WIS	West Indian Space
WWW	World Weather Watch
XS	Multispectral

1

Introduction

B. Jasani and G. Stein

1.1 BACKGROUND

The need to control the spread of nuclear weapons was already recognised in 1953. In his now famous speech on 'Atoms for Peace', the then US president, Dwight Eisenhower, declared that 'the United States pledges ... its determination to help solve the fearful atomic dilemma – to devote its entire heart and mind to finding the way by which the miraculous inventiveness of man shall not be dedicated to his death, but consecrated to his life' (Eisenhower, 1953). While these aspirations were not realised, nevertheless, during the 1960s, there was a general agreement that the further proliferation of nuclear weapons, both horizontal – spread of nuclear weapons to countries other than the five nuclear weapons states (NWS), namely China, France, Russia, the UK, and the USA – as well as vertical – increase in numbers and quality of nuclear weapons within the NWS – would jeopardise world security. Therefore, non-proliferation was seen to be to the advantage of all states. This was to be achieved through a series of arms control measures. The Treaty on the Non-Proliferation of Nuclear Weapons (NPT) was the first international agreement to control the spread of nuclear weapons. Although the UK, the USA and the former USSR signed the NPT in July 1968, it did not come into force until March 1970. However, by 17 June 1999, 187 states were parties to the NPT including all five NWS.

Under the Treaty, the NWS parties to it 'undertake not to transfer ... nuclear weapons or other nuclear explosive devices or control over such weapons or other explosive devices directly, or indirectly; and not in any way to assist, encourage, or induce any non-nuclear-weapon State to manufacture or otherwise acquire nuclear weapons or other nuclear explosive devices, or control over such weapons or explosive devices' (NPT, 1995, article I). On their part, the non-nuclear-weapon states (NNWS) parties to the NPT 'undertake not to receive ... nuclear weapons or other nuclear explosive devices ...; not to manufacture or otherwise acquire nuclear weapons or other nuclear explosive devices; and not to seek or receive any assistance in the manufacture of nuclear weapons or other nuclear explosive devices' (NPT, 1995, article II).

1.2 IAEA AND ITS SAFEGUARDS

For the purpose of verification of the fulfilment of these obligations, the NNWS parties to the NPT, undertake to accept a safeguards mechanism. This requires the parties to sign an agreement with the International Atomic Energy Agency (IAEA) which would carry out the safeguards procedures (NPT, 1995, article III.1). The mechanism is to deter, by early detection, diversion of significant quantities of nuclear materials from their civil nuclear energy programme to possible weapons production. The safeguards agreement with the IAEA has to be concluded '... not later than eighteen months after the date of initiation of negotiations' (NPT, 1995, article III.4).

Three different types of safeguards agreements are possible:

1. The NNWS party to the NPT, the Tlatelolco Treaty or the Rarotonga Treaty could have an agreement under the INFCIRC/153 (corrected). The basic undertaking a party makes is the acceptance of safeguards on all source or special fissionable material in all peaceful nuclear activities for the exclusive purpose of verifying that such material is not diverted to the production of nuclear weapons or other nuclear explosive devices. Such agreements are known as the full scope safeguards agreements. Inspections of facilities are included (paragraphs 76, 77 and 79 of INFCIRC/153).
2. In the second, the NNWS that do not wish to adhere to the NPT, the Tlatelolco Treaty or the Rarotonga Treaty, usually conclude safeguards agreements with the IAEA under the INFCIRC/66. Under these agreements certain items (materials, facilities and equipment) and in some agreements, technological information, are subject to the signatory undertaking not to use them for the manufacture of nuclear explosive devices or to further any military purposes. The INFCIRC/66 type agreements do not usually cover all nuclear materials in all peaceful nuclear activities but only those which have been imported or for other reasons placed on the inventory.
3. The third applies to the five NWS that have concluded voluntary safeguards agreements with the IAEA and accepted safeguards on nuclear materials in peaceful nuclear activities to check that such materials are not withdrawn for military purposes.

There are four other relevant treaties (United Nations, 1988), parties to which are also required to conclude safeguards agreement with the IAEA. These are all regional treaties:

1. the 1967 Treaty on the Prohibition of Nuclear Weapons in Latin America (Tlatelolco Treaty);
2. the 1986 South Pacific Nuclear-Free Zone Treaty (Rarotonga Treaty);
3. the 1996 African Nuclear-Free Zone Treaty (Pelidaba Treaty); and

4. the 1997 Southeast Asia Nuclear Weapon-Free Zone Treaty (Treaty of Bangkok).

Thus, the importance of the IAEA cannot be overemphasised. While the Agency was created in 1957, the seeds were already sown in 1953 with Eisenhower's 'Atoms for Peace' speech. 'The governments principally involved, to the extent permitted by elementary prudence, should begin now and continue to make joint contributions from their stockpiles of normal uranium and fissionable materials to an international atomic energy agency. We would expect that such an agency would be set up under the aegis of the United Nations.' Over a period of time, this role of a custodian changed to that of a verification agency also. One of the most important tasks for the IAEA was 'to establish and administer safeguards designed to ensure that special fissionable and other materials, services, equipment, facilities, and information made available by the Agency or at its request or under its supervision or control are not used in such a way as to further any military purposes; and to apply safeguards, at the request of the parties, to any bilateral or multilateral arrangement, or at the request of a State, to any of that State's activities in the field of atomic energy' (IAEA Statute, article III.5). Therefore, the Agency provides a legal institutional framework for managing international co-operation as well as control in the use of nuclear energy.

The IAEA carries out its verification activities by making use of national accounting and control system. The Agency applies its safeguards so as to enable it to verify the findings of the national system. There are three basic components of the NPT safeguards: material accountancy, containment and surveillance. Material accountancy is applied to a collection of measurements and other determinations that enable the state and the IAEA to know the location and movement of nuclear material. Containment and surveillance are designed as complementary measures to accountancy. Containment prevents movements of materials whereas surveillance is the detection of movements of nuclear materials.

Surveillance is achieved by using cameras or other devices placed at strategic points so that both the containment measures and the inventory changes are monitored. The Facility Attachment (Facility-Specific Safeguards Agreement), which describes procedures to be applied to a particular plant, is based on information given to the IAEA by the country in control of that plant. The inspection access constraints set forth in the Facility Attachment are observed if surveillance is carried out by IAEA personnel.

In addition to the systems established under the above treaties, two other types of measures have been developed to control commercial nuclear material and technology transfers. One is based on national efforts such as the US Non-Proliferation Act of 1978. Such systems follow international guidelines. Two sets of guidelines have been developed. In 1970, experts from a group of countries, parties to the NPT formed the Zangger Committee,[1] and prepared a set of rules on nuclear exports. It

[1] This consisted of 17 members: Australia, Austria, Canada, the former Czechoslovakia, Denmark, Finland, Germany, Ireland, Japan, Luxembourg, Netherlands, Norway, Poland, the former USSR, Sweden, the UK, and the USA.

supplied the IAEA with a Trigger List consisting of materials, equipment and technologies which, when provided to any NNWS would 'trigger' IAEA safeguards. The results of the Zangger Committee were published in INFCIRC/209 (1974). Over the years this has undergone several modifications which were published under INFCIRC/209/Rev.1.

While the guidelines developed by the Zangger Committee were based on the obligations assumed by states who are party to the NPT (NPT, 1995, article III.2), a similar set of guidelines were developed by a group of countries which exported nuclear materials and technologies. The group is known as the Nuclear Suppliers Group. The guidelines were first published in INFCIRC/254 (1978). The Group began to meet from November 1974 to work out a common approach to their nuclear exports. The first meeting took place in London. On 27 January 1976 (IAEA, 1978), the initial seven members[2] of the London Club endorsed a uniform code for nuclear exports which go beyond the NPT obligations, cover the Zangger Committee's Trigger List and include a common policy to exercise restraint in the transfer of sensitive nuclear technologies. By June 1998, the membership had grown to 35 states (The Zangger Committee, 1998). During their meeting on 3 April 1992, the Nuclear Supplier's Group adopted the guidelines consisting of dual-use items, thereby strengthening it (*Nuclear Fuel*, April 1992).

These two sets of guidelines recognise the IAEA safeguards on all relevant exports but do not require adherence to the NPT by the receiving states; also, it is a pity that instead of strengthening the IAEA system such guidelines were set up outside it, thus weakening the IAEA regime. Moreover, these guidelines do not require the application of IAEA safeguards to all nuclear materials in all peaceful nuclear activities (i.e. full scope safeguards, in a receiving country). In the INFCIRC/254, special restraints are called for in the transfer of sensitive facilities, such as enrichment or reprocessing plants, certain nuclear technologies and weapon usable materials.

There are a number of non-proliferation shortcomings that are also relevant to safeguards. For example, at least six NNWS operate important un-safeguarded civil nuclear facilities. All five NWS operate many un-safeguarded civil and military facilities. Also, until recently, virtually no capability existed within the IAEA to detect undeclared nuclear facilities and activities; this was left to the National Technical Means (NTM) of verification of individual member states and it was hoped that the Agency would be informed of any irregularities detected by them. Moreover, a NNWS party to the NPT is also supposed to inform the Agency of all of its nuclear activities so that the IAEA could carry out its tasks. However, neither seemed to have worked in the case of Iraq or, for that matter, North Korea. Lastly, there always will be some inaccuracies in any measurement made during the safeguards procedures. This will become more important with the increased amounts of the sensitive materials the Agency will encounter during its work in the future.

In 1995, the NPT Review and Extension Conference was convened between 17

2 The initial seven members of the London Club are Canada, France, Germany, Japan, the UK, the USA and the former USSR.

April and 12 May in New York. The parties, without a vote, adopted, as part of three Presidential proposals, indefinite extension of the Treaty. The parties also stated that 'Decisions adopted by its Board of Governors aimed at further strengthening the effectiveness of Agency safeguards should be supported and implemented and the Agency's capability to detect undeclared nuclear activities should be increased.' (NPT Conference Document, 1995).

The Agency's efforts to detect undeclared nuclear activities resulted from the discovery, in the early 1990s, of Iraq's clandestine nuclear programme. In 1993 and 1994, the Director General's Standing Advisory Group on Safeguards Implementation (SAGSI) formulated specific recommendations regarding the detection of undeclared activities. One of the recommendations advised 'assessment of the usefulness, technical feasibility, associated costs and acceptability of the Agency obtaining satellite photographs from commercial sources' (SAGSI, 1994). In June 1993, after discussions in the Board of Governors, these recommendations were translated into the Secretariat's development programme to strengthen safeguards, known as 'Programme 93+2', for cost-effectively strengthening the safeguards procedures (Pellaud, 1994). In 1997, this process led to the negotiation and establishment of the new safeguards protocol INFCIRC/540. This protocol, after implementation, provides the Agency with a powerful tool to detect undeclared materials and activities in a state. Especially, the use of open sources, satellite imagery, further additional information and extended access to nuclear facilities and other locations gives the Agency credible assurance on the absence of undeclared nuclear materials and activities. In the practical implementation of the new protocol, the Agency is now faced with the problem of integrating the old and new safeguards systems to reach especially optimal cost-effectiveness. One element of this problem is related to the reconsideration and relaxation of the former timeliness criteria for the different nuclear material categories, which are important for frequencies with which the routine and *ad hoc* inspections could be carried out. This development could affect, for example, one to two inspections per year for the Light Water Reactors (LWRs) without any further containment/surveillance measures.

1.3 WHY COMMERCIAL SATELLITES?

The attitude of the North Korea was important in 1995 because it may have influenced the thinking of other states in the region on their approach to the 1995 NPT Review Conference. North Korea joined the NPT on 12 December 1985, but it was not until 30 January 1992 that it signed the safeguards agreement with the IAEA. Later that year, on 9 April, North Korea ratified it (IAEA, April 1992) and the safeguards agreement came into force on the following day committing the country to open its nuclear facilities to IAEA inspections within 90 days. On 4 May 1992 North Korea provided the IAEA with a list of its nuclear facilities. The list included uranium mining, refining facilities, a fuel fabrication plant, and three operating reactors, two of which are believed to be plutonium production reactors (IAEA, May 1992).

As part of the safeguards agreement, in order to establish the correctness of the information provided to it by North Korea, the IAEA began its inspection of the country's nuclear facilities in May 1992 (*Nuclear Fuel*, June 1992). As a result of this inspection, and the information acquired by the IAEA from other sources, enough evidence was accumulated by the Agency to declare North Korea in non-compliance with its commitments to the NPT. On 12 March 1993, North Korea announced, under the provision of article X.I of the NPT, its intention to withdraw from the NPT. The reasons for this were given in a statement to the 154 NPT parties and the three depository states, Russia, the UK and the USA. In defence of this action, North Korea stated that, 'It is on the basis of the "intelligence information" fabricated by the United States ... that some officials of the IAEA secretariat are trying to enforce the inspection of our major military installations which are unrelated to nuclear activities' (IAEA, April 1993). It was also stated that the IAEA's insistence on the inspections of some sites was 'an encroachment on the sovereignty of (North Korea), an interference in its internal affairs, and a hostile act aimed at stifling our socialism' (IAEA, April 1993). In spite of several efforts by the Agency to resolve the dispute, North Korea conveyed to the Director General that its decision to withdraw from the NPT would remain unchanged 'until ... IAEA Secretariat returns to its principle of independence and impartiality' (IAEA, April 1993).

A similar battle continues between Iraq and the UN inspection regimes established to monitor Iraq's activities in the field of weapons of mass destruction. For example, it has been reported that Iraq will not allow the newly established UN Monitoring, Verification and Inspection Commission (UNMOVIC) to inspect its territory (BBC News, March 2001). The former inspection team, UN Special Commission (UNSCOM), ended its activities at the end of 1999. In view of this, UNMOVIC has been considering using satellite imagery for its tasks (Devenport, 2000).

This whole saga illustrates the need for the Agency to acquire independently information relevant to its safeguards activities. In order to have confidence in the indefinitely extended NPT, it is essential to consider ways of strengthening the IAEA's safeguards procedures. To this end an examination of the contribution from observations from space becomes vital. For example, in the case of North Korea, the Agency could have used both SPOT and some Russian data to support its need to carry out 'special inspections'.

Had North Korea carried out its threat to leave the NPT, it would have had a serious impact on the 1995 NPT Review Conference and on the treaty itself. For example, with the development and testing (on 29 May 1993) North Korea of its Rodong 1 missile with a range of about 1,000 km, some countries in the region may have become concerned about their security (Imai, 1993). A renunciation of the NPT by North Korea might have triggered other countries in the region to reconsider their attitude to the NPT. Thus, unless better safeguards procedures are developed soon for the IAEA, the success of the NPT may be jeopardised.

Information gathered from space played an important role in uncovering the nuclear activities of, for example, Iraq and North Korea. The non-intrusive nature of observations from space adds to its attractiveness. However, the nature of such sources of information as NTM is such that their use by the IAEA would not be possible. On the other hand, commercial satellite data are generally available without restrictions. The use of data acquired by such satellites for the verification of the NPT was suggested as early as 1987 (Jasani, 1987). Furthermore, the usefulness of multispectral and radar image data for the determination of the operating status of, for example, nuclear reactors, was also demonstrated in 1998 (Jasani *et al.*, 1998 and Jasani, 1999). Therefore, it is important now to examine how feasible, practical and cost-effective such a technique is for the enhancement of the IAEA's safeguards system. It is important, however, to realise that satellite observation by itself is not enough. Collateral information from many sources needs to be collected and examined.

At present China, France, India, Japan, Russia and the United States launch and operate civil remote sensing satellites (see Table 1.1). The European Space Agency (ESA), a 15-nation consortium, followed suit when the ERS-1 satellite was launched on 16 July 1991. Of these nations, only France and the USA have been actively distributing their satellite data on a commercial basis but India, Japan and Russia have now joined them.

Since the launch of the first civil remote sensing satellite (Landsat 1) by the USA in 1972, the quality of images, measured in terms of the resolution of a sensor, from subsequent civil satellites has improved some 100 fold. For example, Russia is marketing images with about 2 m resolution and the USA has been declassifying images with similar resolutions acquired from its early (until about 1972) classified satellites (Ruffner, 1995). Not only this but the USA has launched a satellite, IKONOS (Napier, 1999), with resolution of about 1 m and is considering the launch of a satellite with 0.5 m resolution (Loeb, 2000). Thus, verification, using such satellites, becomes possible.

During the past three to four years some criticism has been levelled against the verification procedures carried out by the IAEA. The main criticism, for example, has been the Agency's failure to detect until quite late much of the Iraqi nuclear weapons programme. By and large this has been because of the lack of adequate information available to the IAEA on the undeclared nuclear activities in such regions. This has been recognised by the Director General of the Agency who has stated that '... satellite imagery can be very helpful to inform inspectors where there is reason to look for evidence ... it does not seem reasonable that the Agency should refrain from utilising a type of information that is available to the whole world even commercially – and that has long been used for confidence building in the field of arms control. ... it would be preferable for an international organisation to obtain such imagery from several sources or from an international satellite agency' (Blix, 1995).

Also, it should be noted here that the IAEA's safeguards would be triggered when sales of nuclear materials or facilities occur and are reported to the Agency by the exporters and or by the recipients (article III.1 and 2 of the NPT, paragraph 34, INFCIRC/153, and paragraph 15, INFCIRC/66/Rev.2). If a facility is built or nuclear material is acquired indigenously, then it is the responsibility of the NNWS party to the Treaty to inform the IAEA so that it can apply safeguards. If

Table 1.1. Some current and future commercial remote sensing satellites belonging to various countries are summarised below.

		Resolution in pixel size (m)		
Country Satellite	Date of launch of first satellite	Panchromatic	Multispectral	Thermal/IR
OPTICAL				
Brazil/China		20	–	–
Zi Yuan CBERS	1999–2001	5	10	40–80
I & II CBERS III & IV	–			
France				
SPOT-4	1998	10	20	
SPOT-5	2002	2.5		
India				
IRS-1C,-1D	1995, 1997	5.8	25	
IRS-P5	1999–2000	2.5		
Israel				
EROS-A	1999	1		
Japan				
ALOS	2003	2.5	10	
Russia				
Resurs-F series	1989–98	2		
USA				
KH-1 to 4	June 1959–December 1963	7.6		
KH-4A	August 1963–October 1969	2.7		
KH-4B	September 1967–May 1972	1.8		
KH-6	March 1963–July 1963	1.8	30	120
Landsat-5	1984	–	28.5	60
Landsat-7	1999	15	3–5	
IKONOS-1	attempted, April 1999	0.8–1	3–5	
IKONOS-2	September 1999	0.8–1		
QuickBird-1	October 2001	0.61	2.44	
Orbview-3	2001	1–2	4	
Orbview-4	expected 15/3/2001	1–2	4 & 8 hyper-spectral	
Earlybird-1	2000	3	15	
ASTER	18 December 1999		15(bands 1–3), 30(bands 4–9), 90(bands 10–14)	
EO-1	21 November 2000	10	30(9 bands) 30(220 bands)	
RADAR				
ESA				
ERS-1, & -2	1991 & 1995	25		
Japan				
JERS-1	1992	18		
Canada				
Radarsat	1995	9–100		
Russia				
Almaz-1B				
USA				
SIR-C	1994	8–30		

this does not occur, as seems to have been the case with Iraq, the Agency, at present, has no possibility to know when and where to carry out its inspections unless a member state with better sources of information informs the Agency of any undeclared facilities of another state party to the Treaty. Such a source could be observations made from satellites.

Here is the importance of observation satellites for monitoring nuclear activities of parties to the NPT as well as of those to the other regional non-proliferation treaties. The use of satellites for monitoring was first recognised publicly when US spacecraft detected the undeclared nuclear facilities of Iraq. The USA gave the IAEA the relevant information enabling the Agency to conduct inspections. This method worked in the case of Iraq because it followed the 1991 Gulf War. On the other hand, the use of such a technique could cause considerable political difficulties, highlighting the general problems that the IAEA faces and illustrating the need for the IAEA to acquire its own information. At present, one possible source is information gathered by civil commercial satellites.

The intrusive nature of any verification procedure complicates the control of the spread of weapons and the IAEA's safeguards procedure is no exception. It is important to devise a non-intrusive monitoring method for the Agency in order to enhance its safeguards without creating political difficulties. Observations from space could help. This is particularly true for the detection of possible undeclared nuclear sites. In the following chapters, various methods and techniques used in remote sensing from space that could be used by the IAEA are examined in detail. It must be realised that data acquired by commercial observation satellites range from panchromatic, multispectral, hyperspectral to even radar. Radar has been shown, for the first time, to be very important and useful for safeguards purposes since it has a day and night and all-weather capability (Jasani, 1999).

1.4 REFERENCES

BBC News (2001) Iraq rejects new UN weapons team. BBC News: World: Middle East, 2 March,
http://news.bbc.co.uk/hi/english/world/middle_east/newsid_1197000/1197850.stm

Blix, H. (1995) Statement to the Review and Extension Conference of the Parties to the Treaty on the Non-Proliferation of Nuclear Weapons, IAEA, C22, 17 April, New York.

Devenport, M. (2000) UN considers spy satellites over Iraq, BBC News Online: World: Middle East, 23 August,
http://news.bbc.co.uk/low/english/world/middle_east/newsid_890000/890483.stm

Eisenhower, D. (1953) Address by President Eisenhower to the 470th Plenary Meeting of the United Nations General Assembly, 8 December,
http://www.iaea.org/worldatom/About/atoms.html

Imai, R. (1993) Asian ambitions, rising tensions. *The Bulletin of Atomic Scientist*, June, Vol. 49, No. 5, pp. 33–36.

IAEA (1978) INFCIRC/254, February.

IAEA Press Release (1992) PR 92/20, 10 April.

IAEA Press Release (1992) PR 92/24, 5 May.

IAEA (1993) INFCIRC/419. Statement of the Government of the Democratic People's Republic of Korea, 12 March 1993. Annex 7, 8 April.

Jasani, B. (1987) 'Satellite monitoring – programs and prospects'. In: Jasani, B. and Sakata, T. (eds) *Satellite for arms control and crisis monitoring*, pp. 1–8, Oxford University Press, Oxford.

Jasani, B., Canty, M., Richter, B., Stein, G., Niemeyer, I. and Ward, M. D. (1998) 'Commercial Satellite Imagery and Safeguards: Some case studies using multispectral and radar data.' Report to the IAEA, SRDP-R256, JOPAG/11.98-PRG-284, Rask JNT D00988.

Jasani, B. (1999) Civil Radar Observation Satellites for IAEA Safeguards. *Institute of Nuclear Materials Management*, Vol. XXVII, No. 2, Winter, pp. 29–34.

Loeb, V. (2000) 'U.S. is relaxing rules on sale of satellite photos', *Washington Post*, 16 December, p. A03.

Napier, G. (1999) IKONOS satellite launches into space. *Space Imaging*, 24 September, http://www.spaceimaging.com/newsroom/releases/1999/inorbit.html

NPT Conference Document (1995) NPT/Conf.1995/L5, the second principle.

Nuclear Fuel (1992) 13 April, p. 406.

Nuclear Fuel (1992) 22 June, pp. 15–16.

Pellaud, B. (1994) Safeguards in transition: Status, challenges, and opportunities. *IAEA Bulletin*, September, Vol. 36, No. 3, pp. 2–7.

Ruffner, K. C. (ed.) (1995) CORONA: America's first satellite program. Center for the Study of Intelligence, Central Intelligence Agency, Washington, D.C., p. xiv.

SAGSI (1994) SAR-17, Report to the Director General on the 38th Series of SAGSI meeting, 21–24 March.

United Nations, Status of multilateral arms regulation and disarmament agreements, 3rd edition, 1987, (United Nations, New York, 1988), pp. 71–76.

The Zangger Committee (1998) http://www.usun-vienna.usia.co.at/zangger.htm

2

The new safeguards protocol – the need for new verification technologies

B. Richter and G. Stein

2.1 INTRODUCTION

The discovery in the early 1990s of a clandestine nuclear weapons development programme in Iraq, a NPT member state, led to the consensus that the IAEA's capabilities should be extended to the detection of undeclared nuclear activities as well as the verification of completeness and correctness of states' declarations. Thus, the international safeguards are in a state of transition. The new INFCIRC/540-based (IAEA, 1997) safeguards system has been designed to overcome shortcomings of the currently applied system which was established in the 1970s. Under INFCIRC/153 safeguards, the IAEA is entitled to verify all declared nuclear material in NNWS. The new, so-called integrated safeguards system will combine INFCIRC/153- and INFCIRC/540-based safeguards, in order to provide the IAEA with the capability to gain credible assurance on the absence of undeclared nuclear materials and activities in NNWS. This is obviously a task which will have an unprecedented impact on national sovereignty. Therefore, states will sign and ratify the additional safeguards protocol under the condition that there will be a significant reduction of safeguards effort in the combined new system. At the very least this should be achieved under the condition of cost neutrality (Häckel and Stein, 2000).

This chapter deals with impacts of integrating the INFCIRC/153 Model Agreement and the INFCIRC/540 Model Protocol in connection with improving not only the effectiveness but also the efficiency of safeguards verification measures. In this context, it discusses the consequences for the implementation and, in particular, for the development of new verification technologies. For a better understanding, the evolution of safeguards is briefly outlined, followed by a description of the structure of the new safeguards systems and further trends.

2.2 EVOLUTION OF SAFEGUARDS

International safeguards are characterised by a strong dynamical element. The international safeguards system started with the design and implementation of the INFCIRC/66-based safeguards system. As 'Pre-NPT Safeguards' this agreement is exclusively related to those nuclear items which are designated for safeguards. Therefore, in principle, it is possible that under this system national nuclear activities lie outside IAEA safeguards. The system itself has a broad scope, so that not only nuclear material and activities but also (e.g. heavy water) can be safeguarded. The next step in the evolution of the safeguards system was the negotiation and implementation of INFCIRC/153 as the safeguards system of the NPT. On the one hand, NPT safeguards in NNWS require the verification of all nuclear materials as well as activities; on the other hand, the application of safeguards is limited to nuclear material. In addition, the inspectors' access in facilities is restricted to so-called strategic points at which nuclear material is accessible for measurement or can be monitored.

In this evolution one can identify two main directions: the first is related to extending the scope of safeguards; the second concerns the element of universality. To begin with one has to analyse the structure of INFCIRC/153. The objective of this safeguards system is the timely detection of the diversion of significant quantities of nuclear materials. This safeguards goal implies a quantitative resolution in time and material quantity so that appropriate conclusions can be drawn by the IAEA. The main elements of the system are accountancy and measurements which allow the Agency to perform these quantitative calculations. The general objective of INFCIRC/153 to verify the correctness of states' declarations is achieved not only by deploying the above mentioned quantitative elements but also by making use of some qualitative features such as those which involve design information verification. The aspect of universal application to all was, in the 'pre-540' time, covered by voluntary offers on the part of the NWS and itemrelated safeguards (INFCIRC/66) by states still outside the NPT.

The new INFCIRC/540 Model Protocol opens the door wider in both directions: scope and universality. As regards the universality issue it has been recognised during the discussions in the 'Committee 24' of the IAEA Board of Governors that all states must be concerned about undeclared nuclear activities and that globalisation does not stop at the nuclear market. In consequence, there are now different INFCIRC/540 Model Protocols for both NNWS and INFCIRC/66-states which take into account their different boundary conditions. So, one can conclude that, in principle, a new vision of universality has been reached.

As regards the scope of INFCIRC/540 the following can be stated. Controlling nuclear materials and activities to reach credible assurance on the absence of undeclared nuclear materials and activities implies a qualitative safeguards system, since these conclusions can only be drawn on a qualitative basis.

All this has to be understood under the inherent realisation that an absolute assurance on the absence of undeclared nuclear material and activities can never been reached. The factual situation leads to the conclusion that new tools and technologies for safeguards are needed to cope with the requirements, especially with regard to the IAEA achieving better assurance.

2.3 STRUCTURE OF THE NEW SAFEGUARDS SYSTEMS AND FURTHER TRENDS

Within the given legal framework, strengthening of the safeguards system comprises the following elements:

- early provision of design information and an extended reporting scheme;
- increased co-operation between the IAEA and the national/regional safeguards authorities (i.e. in the European Union, the Euratom Safeguards Office where this co-operation is known as the New Partnership Approach);
- use of environmental sampling at specific locations;
- use of remote monitoring and satellite imagery, and increased access to states' nuclear activities.

The IAEA is interested in implementing these measures in consultation with its member states. For the complete implementation of the INFCIRC/540 Model Protocol additional legal authority is required in order to allow for the request of the provision of additional information by the states (i.e. expanded declaration). Also important in this connection is the use of widearea environmental sampling by the IAEA, and the IAEA's right to complementary access to declared sites and to locations the states have declared or informed about or to those selected by the IAEA.

According to the expanded declaration approach, each member state will be obliged to provide additional information about nuclear-related activities. The Agency will have the possibility of checking this against information acquired independently from open sources and that from Agency activities. These open sources and IAEA activities basically include:

- literature and media reports;
- satellite remote sensing data;
- environmental monitoring; and
- routine inspections in the declared sector.

If inconsistencies or problems should result from a comparison of the states' declarations and the above information sources, then the IAEA would be permitted to undertake on-site verification activities at locations not containing nuclear material. However, the IAEA will have no mandate to verify the expanded declarations in a mechanistic manner (i.e. binding the inspection effort to the number of facilities and amount of nuclear material in a state). The effect would be that the majority of IAEA resources – as in the conventional safeguards system – would be directed towards countries with a low proliferation risk.

The role of open sources can be summarised as follows:

- consistency check of declared information;
- preparation of inspections;
- accumulation of evidence on the absence of undeclared activities;
- early-warning of possible undeclared activities;
- information gathering on technological developments; and
- information tracking.

Strengthening the safeguards system through enlarging its scope and reaching universal application relates to improving the effectiveness of the international safeguards system. In addition, there is the need to improve the cost-effectiveness and efficiency of safeguards. This complex problem will be influenced by a number of factors and driving forces. For example, on the one hand there is a mainstream development towards saving resources and budgets around the globe which will not stop at UN institutions, and, therefore, also not at the door to IAEA safeguards; diplomatic circles are asked to discuss and implement such budgetary reductions. On the other hand, integration of INFCIRC/153 and INFCIRC/540 means combining old and new safeguards. This task should be performed with the expectation of reducing overall inspection effort, as already mentioned. Derived from this discussion, it is anticipated that there are possible savings in the future safeguards system.

In contrast, additional IAEA safeguards resources may be required due to the following. The global energy system is still in a process of growing quantitatively as well as qualitatively. There is no evidence that global population increase and energy consumption will be de-coupled. As a consequence of this, nuclear energy has a strong perspective, especially under the concept of sustainable development. The increase of national nuclear programmes leads to the need for additional inspection resources for the safeguarding of nuclear materials and facilities by the IAEA.

Future activities in nuclear disarmament or cutting off fissile material production also lead to substantial additional safeguards efforts for the IAEA. In the so-called Trilateral Initiative it is foreseen that the IAEA will verify weapon-origin and other fissile material released from defence programmes in Russia and in the USA. In this initiative, the USA and Russia determine which fissile materials they will submit to IAEA verification. The IAEA should have the right and obligation to verify that fissile materials remain separate from nuclear weapons production.

In order for the Fissile Materials Production Cut-off Treaty (FMCT) to become effective and efficient, four requirements are of utmost importance (Stein, 2001).

2.3.1 Universality

The FMCT shall be universally applied on a global basis. This requires accession to both the future fissile material production cut-off and the comprehensive test ban treaties, not only by the present NWS but also by the threshold states India, Israel, and Pakistan as well as by the present NNWS. It is anticipated that India, Israel, and

Pakistan play a key role as regards the entry into force of the FMCT. Without these states, the FMCT should not be considered universal.

2.3.2 Non-discrimination

All rights and obligations arising from the FMCT have to be equally applied to all parties to the treaty.

2.3.3 Irreversibility

Fissile material inventories once declared for civilian uses must not be devoted to military purposes, thereafter.

2.3.4 Transparency

The compliance with both the prohibition of fissile material production for military purposes and non-transfer of civilian inventories into military uses must be reliably verified. The kind and scope of the remaining military inventories have to be as transparent as possible.

Also, the following factors are important for the further structuring of safeguards.

2.3.5 Political and institutional aspects

One prominent problem is the question as to whether the IAEA, in carrying out its verification activities, should be allowed to differentiate between states while, at the same time, not violating the principle of non-discrimination. There is a challenge by the new protocol INFCIRC/540 towards such a differentiation, since hard and soft factors can influence the practical safeguards implementation. What are the factors that are of relevance in a state? Here are some examples: fuels cycle specific elements, the role/independence of the State System/Regional System of Accounting and Control (SSAC/RSAC), and transparency.

A top agenda item and question of primary importance is to what extent the IAEA should relegate verification activities to the SSAC or RSAC. In this context, it is also of interest to know how far the IAEA is able to implement the tool of quality control of regional or SSAC safeguards.

2.3.6 Safeguards approaches

It has already been mentioned that integrating INFCIRC/153 and INFCIRC/540 aims at combining qualitative and quantitative control structures on an equal level. This is a new task for the IAEA and has also to be carefully considered in formulating the evaluations and conclusions in its annual Safeguards Implementation Report.

Another important agenda item concerns timeliness and detection probabilities. Since the integrated safeguards system presently offers a possibility to obtain credible assurance on the absence of undeclared material and activities, one can consider a reduction of the classical timeliness and detection goals (e.g. from three months to one year for spent fuel and from one month to three months for fresh MOX fuel).

Last but not least, another topic on the agenda is to re-visit other classical safeguards criteria. This with the intention of adapting them to the new conditions of the integrated safeguards system and also taking into account inspection schemes with random or unannounced character.

2.4 IMPACTS ON THE DEVELOPMENT OF NEW VERIFICATION TECHNOLOGIES

The IAEA Strengthened Safeguards System (SSS) will have to be based on improved and cost-effective verification techniques taking into account new facility projects in advanced nuclear fuel cycles. To this end the IAEA had been advised to embark on, what has become known as, the IAEA Integrated Safeguards Instrumentation Programme (I^2SIP). The objective is to enhance the IAEA's inventory of mutually compatible instruments by introducing the concepts of standardisation and modularity, in order to facilitate the design of customised integrated verification and monitoring systems which can operate unattendedly with remote data transmission. This concept will also facilitate implementation procedures, servicing and training for technicians and inspectors (Remagen, Richter and Stein, 2001).

In general, advanced safeguards techniques comprise measurement sensors, optical surveillance units, seals, and components for the collection of data. As far as safeguards specific requirements are not addressed by the commercial market, safeguards specific development efforts need to take place.

In the measurement sector, instrumentation is required with specific sensor heads, (i.e. radiation detectors with high neutron counting efficiency at low gamma sensitivity, gamma detectors operating in a wide dynamic dose rate range, high resolution gamma detectors operating at room temperature, and small size detector systems with high sensitivity). In a first step, commercially available detector equipment is being screened and tested under realistic conditions for its safeguards appropriateness. In the nuclear electronics sector, digital signal processing electronics and data acquisition modules are being developed.

Combining measurement equipment with optical surveillance techniques will yield integrated safeguards systems with new capabilities. However, standardised modules for radiation monitoring systems capable of networking and remote data transmission are lacking. There is an urgent need for the development of modular, miniaturised, digital spectrometer components with data authentication and encryption, which will readily lend themselves to integration with digital image surveillance techniques.

The new digital video techniques allow for the configuration of small to very complex image surveillance systems as well as the integration with other monitoring equipment in unattended systems and implementation of remote data transmission. The issue is whether such systems will have the capability of reducing onsite inspection effort. This is being analysed in a number of remote data transmission field trials performed in various countries under different conditions. Prerequisites are the integration capability of devices as well as data authentication and encryption. Test criteria are reliability in connection with lossless data acquisition and user-friendliness with regard to system operation and data review. A meaningful data review also requires powerful data processing by automatically correlating (e.g., image and measuring) information.

In addition to the objective of reducing the inspectors' routine on-site efforts, the use of complex unattended integrated measuring and monitoring systems will allow for decreasing the inspectors' exposure to radiation and, last but not least, reduction of escorting requirements on the plant operators' side.

Development efforts should also focus on low power consumption components, in order to ensure that mains power outages can be bridged using commercially available power backup modules in order to ensure uninterrupted data collection. As a matter of principle, equipment development and maintenance costs should be kept to a minimum by identifying, adapting, testing, and using, as far as possible, commercial off-the-shelf hardware and software.

As the safeguards market is extremely small, the industry is not able to sponsor safeguards specific equipment development. In addition, the IAEA's R&D budget is limited. Therefore, most of the dedicated technical and methodological development is being sponsored by IAEA member states. At present, there are 15R&D programmes in support of the IAEA (i.e. member states' Support Programmes) including the one sponsored by the Euratom Safeguards Office. The IAEA provides management and evaluation support and co-operates on the expert level as the future end user of the development results.

In order to improve its ability to detect undeclared nuclear facilities and activities, the Agency is interested in the systematic collection, review and evaluation of open source information on states' nuclear activities in addition to all information available to it through its safeguards activities. To this end, the IAEA needs support in improving available databases and implementing methodologies of information review and analysis. There is a wide variety of open sources which can be categorised as follows.

Human sources may include inspectors, industry staff, analysts, consultants, university staff, laboratory staff, equipment suppliers, trade association professionals, as well as private contacts at meetings and conferences. Published literature comprises periodicals and scientific journals, reports, books, brochures, patents, news, and media reports. 'Grey' literature refers to company reports, meeting notes, conference materials, and white papers. Electronic media include commercial databases, the internet, and information services. Organisational contacts can be specified as professional societies, lobbying groups, specialised interest groups (e.g. Non-Government Organizations – NGOs), technology or equipment vendors, and government sources. Remote and local sensing refers to satellite imagery and environmental sampling.

The IAEA's task is not only to acquire the information but also to arrive at a knowledge base. To this end the Agency has to solve the practical problems of collecting, classifying, structuring, filtering, categorising, correlating and analysing the data. This requires the application of appropriate tools which are provided on the software market or have to be adapted in consultation with experts.

Important features in coping with the open source data are credibility, diversity, relevance of user needs, no data overload, problem of missing key data, and the necessity of assessing the relevance and quality of the data.

One particular information source is commercial satellite imagery, which, by itself, is not expected to be capable of detecting undeclared nuclear activities. However, it has a great potential to indicate anomalies which the IAEA will be able to resolve by other means such as special inspections and complementary or managed access. Satellite sensors provide a variety of multispectral information (i.e. images in the visible, infrared, and radar ranges). This information is to be analysed, (e.g. to monitor ground activities and changes around nuclear facilities as well as to determine operational characteristics and temperature variations in cooling reservoirs). Development efforts aim at providing software tools for change detection and site characterisation using new mathematical algorithms. It should be stated that satellite imagery has been identified to have the widest range of application in verification treaties (i.e. for NPT, Chemical Weapons Convention, Comprehensive Test Ban Treaty, future Fissile Materials Production Cutoff Treaty, and future Biological Weapons Convention).

Another method to detect undeclared nuclear activities is environmental sampling (e.g. swipe sampling inside declared enrichment plants and hot cell facilities, to detect undeclared high enrichment of uranium and plutonium separation). Development efforts are needed to adapt the analytical techniques to the facilities under consideration, provide reference materials, and organise a network of qualified analytical laboratories.

In addition, INFCIRC/540 foresees environmental sampling of air, water, vegetation, soil, and smears, in order to enable the IAEA to draw conclusions about the absence of undeclared nuclear activities over wide areas. Efforts are required to characterise source terms, model atmospheric as well as waterborne transport of nuclear signatures, and to develop screening methodologies, sampling procedures and analytical methods.

An advantage in the use of satellite imagery and environmental monitoring is intrusiveness: this is not only of paramount importance for nuclear safeguards but also for other verification regimes. The other positive aspect of these technologies is the huge technological future potential and therefore the overwhelming perspectives, (e.g. through improving resolution of multispectral satellite imagery or through improved particle analysis for environmental monitoring).

2.5 RESEARCH AND DEVELOPMENT

In the future of IAEA safeguards, R&D will have an increasing role. This will be necessary in order to provide new methodologies and to contribute to the introduction of advanced technologies, so as to increase the efficiency and effectiveness of the safeguards. But, clearly, the new technologies will have to be adapted to the framework of the basic safeguards principles and agreements. The perspective of an universal safeguards system to be applied to all states, including the nuclear weapons states, has to be taken into account.

The safeguards needs can be categorised into the following areas of R&D activities:

- facility specific verification concepts;
- monitoring techniques including global data communication links;
- measurement techniques (i.e. sensor technology including data generation);
- data and information treatment, computing technology, data collection and organisation in a datarich environment;
- mathematical and statistical tools as well as decision support;
- socio-political aspects of safeguards; and
- interlinkage with nonsafeguards inspection regimes.

Although facility verification concepts will still be needed, in principle it can be stated that R&D strategies have been changed from facility-oriented to more general approaches. Examples for general R&D areas are (Foggi and Petraglia, 1998):

- evaluation of satellite imagery;
- remote monitoring;
- improved data transfer and treatment; and
- analysis of inspection strategies.

Additional research and development needs arise from the impacts of integrating the old INFCIRC/153 safeguards with the new INFCIRC/540 system. These areas are related to the following.

2.5.1 Protection of sensitive information

The additional access which is expanded in scope and space leads to the need to perform managed access and to protect sensitive information. Centrifuge enrichment technology is one example. A second related example evolves from the Trilateral Initiative. Here, the protection of sensitive weapons materials-related information is essential to further implementation of an effective safeguards system.

2.5.2 Authentication and tamper-resistance

The preceding discussion addressed the need for unattended and remote safeguards equipment to strengthen the efficiency of safeguards through reduction of on-site inspection effort. An essential precondition for the applicability of this technology is the availability of an improved authentication and tamper-resistance (i.e. the IAEA must have the assurance that it collects authentic information and detects any falsification of the data as well as any tampering with the safeguards equipment).

Experience has shown that commercial off-the-shelf systems are not applicable in any case. Normally, sensor heads and data generators have to be newly developed.

2.5.3 Environmental testing of equipment

With the implementation of modern low-power-consuming digital image surveillance systems in control areas a new problem has recently been discovered: Single Event Upsets (SEU) induced through neutrons (e.g. from the reactor environment) (Rosenstock, *et al.* 2001). These events have caused a general reconsideration of the structure of environmental qualification of safeguards equipment. These effects have to be taken into account in the planning and performing of development tasks for safeguards equipment.

2.5.4 Social and institutional transparency

As already mentioned, there is a need to take into account socio-political aspects in safeguards. One research area could reflect the necessary flow of information and budgets in a country associated with the clandestine development of a weapons programme. Indicators for the detection of such a programme could be gained through open information which could be released (e.g. by institutions which are concerned with the programme and have a constitutionally independent function in the state). These aspects could complement the IAEA's so-called 'physical model', which gives technical indications for the development or production of materials and technologies which are needed for a nuclear weapons programme in a state (Fischer, 2000).

2.5.5 Modeling

Modeling of inspection strategies (e.g. the impact of unannounced inspections or the connection and interaction of different factors to determine the effectiveness and efficiency of safeguards) will become more important.

2.5.6 Software development

Further software development is needed, especially for satellite imagery. This software could support the application and implementation of tools (e.g. for change detection in the structure of buildings or infrastructure), which are of relevance for undeclared activities.

2.6 SUMMARY

The new INFCIRC/540-based safeguards system has opened a complete new dimension for the development and implementation of new safeguards technologies. Not only the IAEA's new task of detecting undeclared nuclear materials and activities but also the opening of the scope to control both NNWS and NWS has a strong impact on structure and density of IAEA safeguards.

Safeguards technologies which support the quantitative element of safeguards are still of importance and their standard has to be kept on a high level. Safeguards technologies which have a qualitative character are new and under development. They will gain more importance in the future. Combined digital measuring and surveillance systems with remote data transmission, environmental monitoring, and satellite imagery, are technologies which have a huge future potential. But they need further research and development so that the IAEA and also other international verification agencies can benefit from their successful implementation.

2.7 REFERENCES

Fischer, W. (2000) Measuring 'Specific Transparency' – a Sensible Concept? Proceedings of the 22nd Annual ESARDA Meeting, 9–11 May, Dresden, pp. 127–139.

Foggi, C. and E. Petraglia (1998) 2nd Workshop on Science and Modern Technology for Safeguards, EUR 19059, Albuquerque.

Häckel, E. and Stein, G. (eds) (2000) *Tightening the Reins – Towards a Strengthened International Nuclear Safeguards System*, Springer-Verlag, Berlin.

IAEA (1997) INFCIRC/540. Model Protocol Additional to the Agreement(s) between State(s) and the International Atomic Energy Agency for the Application of Safeguards, September, Vienna.

Remagen, H. H., Richter, B. and Stein, G. (2001) The German Support Programme to the IAEA, submitted for publication in the proceedings of the 42nd Annual INMM Meeting, 15–19 July, Indian Wells.

Rosenstock, W., Köble, T., Metzger, S., Lennartz, W., Henschel, H., Schoop, K., Neumann, G., Lange, S., Richter, B. and Arlt, R. (2001) Neutron Induced Soft Errors in Digital Surveillance Technology, submitted for publication in the proceedings of the 23rd Annual ESARDA Meeting, 8–10 May, Bruges.

Stein, G. (2001) Perspectives for International Safeguards – What is on the Agenda? Nonproliferation Conference, March, Tokyo.

3

Use of commercial satellite imagery in strengthening IAEA safeguards

K. Chitumbo, S. Robb, J. Hilliard

3.1 INTRODUCTION

Over the last several years, the IAEA has been engaged in a programme to strengthen the effectiveness and improve the efficiency of Agency safeguards. In that programme, one of the measures identified was that of the improved analysis of information available to the Agency, including that available through open sources.

The group of outside experts established by the Director General of the IAEA to advise on safeguards matters, SAGSI, recommended in 1994 to include as part of the safeguards programme an assessment of the usefulness, technical feasibility, associated costs and acceptability to member states, of the Agency obtaining satellite images from commercial sources. The Department of Safeguards developed a Work Plan for such an assessment, consisting, *inter alia*, of commercial imagery availability, potential safeguards applications, resource implications, and legal and policy issues.

Pursuant to this Work Plan, the Agency requested certain member states, through the Member State Support Programme, to advise it on the subject matter. The findings in that pilot study report were presented to the Agency in November 1996 (Jasani *et al.*, 1996). In September 1997, the use of satellite imagery for verification purposes was discussed in a seminar held by the Secretariat on the application of open source information (IAEA, 1998). A follow-on technical workshop, devoted solely to satellite imagery, was held in September 1998. Two papers were tabled at that meeting: an earlier draft of this chapter and a discussion paper on legal issues.

In 2000 the Agency established an operational unit with limited imagery analysis capability. Subsequently, in 2001 the Agency's Satellite Imagery Analysis Laboratory (SIAL) became fully operational and has been successfully integrated into the safeguards system.

Objective

The primary objective of this chapter and of the other elements completed under the Work Plan is to assess, from the Agency's perspective, the potential contribution to safeguards offered by satellite imagery. In seeking to meet this primary objective it is necessary to address some of the associated issues such as sources of supply, availability and costs. This chapter, along with other supporting documents, such as specific case studies, is intended to provide the basis for the introduction of commercially available satellite imagery into the information review process of the IAEA's SSS.

3.2 POTENTIAL SAFEGUARDS APPLICATIONS OF COMMERCIAL SATELLITE IMAGERY

This section examines the potential safeguards applications of commercial satellite imagery from several perspectives. Satellite imagery can be viewed from the roles it might play in general areas of information use. Another approach is to examine the usefulness of satellite imagery as a means of detecting elements of the nuclear fuel cycle and related activities, and the extent to which specific features can be identified. It is also useful to assess the relative worth of satellite imagery with respect to the type of safeguards obligations that the member state has assumed. Finally, an important part of the assessment is to weigh the contribution of satellite imagery to the process of verifying the completeness and correctness of a state's declaration under an Additional Protocol.

Current high-resolution, 1 m Ground Sampling Distance (GSD) panchromatic and 4 m GSD multispectral commercial satellite imagery products are proving particularly useful when used to verify state declared information. Recent experiences of SIAL have shown these products to be of value in many tasks. For example, panchromatic imagery can identify physical features of a site such as small buildings, security fencing, cooling towers, power lines, and construction and dismantling activities. Multispectral imagery can be used to further enhance analysis when fused with panchromatic data of 1 m GSD. Further, member state support programmes have shown that analysis of data collected from sensors utilising the thermal bands can indicate the operational status of facilities

3.2.1 Safeguards applications: general

As an initial step in a systematic examination of the potential contributions of satellite imagery to the Agency's safeguards programme, the use of this new data product in some of the general applications will be considered.

3.2.1.1 As reference information

Satellite imagery could be viewed as a reference resource, much like a map or some other geographic form of information. For example, satellite imagery might be used:

- to help orientate inspectors visiting a site for the first time;
- to provide a common point of reference when discussing the site with a member state;
- as an aid to inspection planning, particularly when considering large sites with many inspection areas; and
- to help interpret other information received about a site, such as maps, diagrams, photos and observations.

3.2.1.2 As confirmation of Agency acquired or generated information

Satellite imagery can be used as a complementary source of information in confirming or giving added credibility to the correctness and completeness of information available to the Agency from Agency acquired or generated data, such as technical co-operation projects, expert trip reports and fuel cycle databases (Power Reactor Information Service (PRIS), Research Reactor Data Base (RRDB) and Nuclear Fuel Cycle Information Service (NFCIS)). Most importantly, this category would include the use of imagery for verifying data supplied by member states under an Additional Protocol. This special application of satellite imagery is examined in greater detail later in this section.

3.2.1.3 Change detection and ongoing monitoring

Satellite imagery can be used to detect changes and to conduct ongoing monitoring at declared sites provided there is observable evidence. Typical examples would be expansion of existing utilities or structures, changes in general or specific activity levels, or the construction of new facilities. If detected, changes might trigger further investigation (e.g. on-site inspection or other safeguards-related activities such as environmental sampling). In this role, satellite imagery might be seen as a primary source of information to be complemented by other information sources.

3.2.1.4 Assessing open source information available to the Agency

For information acquired from open sources (e.g. from media reporting, wire services, and from organisations such as the Monterey Institute's Center for Non-proliferation Studies), it is often difficult to assess the credibility of the information. Only on few occasions is there multiple and independent reporting of the information. Commercial satellite imagery, in some instances, may be able to establish the correctness of the open source reporting.

3.2.1.5 Detecting undeclared activities and undeclared sites

Indiscriminate countrywide searches for undeclared activities and sites would be prohibitively expensive both in terms of human resources and in imagery costs. The first indication of a possible undeclared activity or site will most likely come from sources of information other than the Agency's satellite imagery. Then, as a complementary tool, satellite imagery would be a valuable source of information. Additionally, there may be some value in using imagery to search for undeclared activities or facilities on declared sites. In part this will be covered by change detection, but it also embraces the search for undeclared buildings, evidence of activities not compatible with declared functions, and evidence of concealment activities.

3.2.2 Satellite imagery and the fuel cycle

3.2.2.1 Assessment of fuel cycle and related activities

The nuclear fuel cycle and related activities, such as heavy water production and nuclear research centres, present a diverse set of physical features to an imagery analyst. Not only are there major differences in the characteristics of each type of activity with respect to others (e.g., enrichment versus reactors versus reprocessing), but the results are highly dependent upon whether the activity is being assessed at the laboratory, pilot or commercial scale. Further, for some fuel cycle activities there are different technologies that could be associated with a single type of activity. Each of the technologies may present entirely different opportunities for the use of satellite imagery. An examination of the potential contribution of satellite imagery in the detection of nuclear fuel cycle and related activities is contained in Appendix 3.1.

3.2.2.2 Results of the assessment

The assessment of satellite imagery as a tool for detecting fuel cycle and related activities and for analysing such factors as the operational status, features and level of production of activities once identified provides a range of results. There are few nuclear fuel cycle activities that can be confidently identified from satellite imagery absent from any other information. On the other hand, the contribution of satellite imagery increases dramatically when used in conjunction with other information. Further, when the use of the satellite imagery is not only to identify the type of activity but also to gain the maximum amount of information about the activity, then it is clear that the higher the quality of the imagery, the greater the return.

Based on the results of the assessment the following observations can be made about the applicability of satellite imagery to the identification of the various fuel cycle and related activities.

- An important general attribute of satellite imagery, when the facility location and type are already known, is the ability in most cases to determine with confidence whether the facility is active or not. This application of satellite imagery offers the potential for reducing the need for a physical visit to an activity in order to determine its operational status.
- Uranium mining and milling operations have signatures, such as piles of ore and tailings, large ore trucks and the mill itself. However, being able to distinguish a uranium mine from other mining operations would depend upon having complementary information.

- Acquisition of natural uranium by *in situ* leaching is a form of uranium extraction that has unique signatures visible through satellite imagery. Both the pattern and the spacing of wells for this process, as well as the evaporation ponds offer the opportunity for identification, even without complementary information.
- Thorium extraction would not be detectable by satellite imagery alone.
- Conversion activities, without complementary information, are generally not distinguishable from other chemical plants by satellite imagery. Canisters for nuclear material could be a strong indicator if able to identify them on site. If a conversion plant were known to be collocated with another nuclear facility, such as an enrichment plant, the possibility of identification would be enhanced.
- Identification of enrichment activities by satellite imagery is highly dependent upon the type of process being used.
- For uranium enrichment activities using uranium hexafluoride as feed material, a common signature would be the cylinders of depleted uranium stored in the open. The high power demand of facilities using the gaseous diffusion and aerodynamic processes would create signatures such as the presence of a nearby power plant and power lines and switch yards. Heat generated in these plants could be detected by infrared imagery.
- Of the enrichment processes that use UCl_4 as a feed material, the electromagnetic isotope separation process would be the most detectable by satellite imagery because of the high power demand. The other two processes, chemical and ion exchange, have the appearance of plants in the chemical industry.
- The enrichment processes that use uranium metal are the most difficult to identify from satellite imagery. Both methods, atomic vapour laser enrichment and plasma enrichment have not been deployed in commercial size plants, which adds to the uncertainty of the interpretation.
- There are no signatures from nuclear fuel fabrication that would allow the identification by satellite imagery with any confidence, unless there was complementary information.
- Power reactors are perhaps the most easily distinguishable of the fuel cycle activities, but also the activity whose presence is least likely to not already be known.
- Non-power reactors have many of the signatures of power reactors. However, the size and location, along with the absence of other signatures such as power lines or cooling towers, make the identification more difficult.
- Reprocessing plants of a commercial scale have several signatures that enhance the probability of detection by satellite imagery. Laboratory scale reprocessing in hot cells, by contrast, would be more difficult to detect.
- Heavy water production, like enrichment, has many different kinds of processes. Identification is difficult because of the similarity between most heavy water production facilities and chemical plants. One process, the ammonium-hydrogen exchange process, is usually in a facility collocated with fertiliser plants, which aids in the identification process.

- Detection of research centres in most cases will be the result of the identification of a research reactor and tall 'smoke' stacks for the dispersal of off-gases.

3.2.3 Satellite imagery and INFCIRC/153

3.2.3.1 Basis of safeguards applications

States entering into an INFCIRC/153 type agreement with the Agency undertake to accept safeguards on all source and special fissionable material in all peaceful nuclear activities within its territory, jurisdiction or control. The Agency, within the terms of the agreement, is charged with the responsibility of verifying that such material is not diverted to nuclear weapons or other nuclear explosive devices.

The potential uses of satellite imagery by the Agency in carrying out its verification responsibilities in states with comprehensive safeguards agreements are far greater than in non-nuclear weapons states not party to a comprehensive safeguards agreement and in the nuclear weapon states.

3.2.3.2 Potential uses of satellite imagery

While clearly the greatest return from satellite imagery is in the case of a state which has submitted an Expanded Declaration under an Additional Protocol, the return in increased effectiveness from satellite imagery remains near that level for countries with only INFCIRC/153 type agreements in force.

3.2.4 Satellite imagery and INFCIRC/66

3.2.4.1 Basis of safeguards applications

For the application of safeguards in states with INFCIRC/66 agreements there is a much more limited objective. For these agreements the task of the Agency is to ensure that materials, services, equipment, facilities or information subject to the agreement are used only in accordance with the agreements. The Agency neither has the authority nor the responsibility to seek out information on those elements of the state's nuclear programme that is not under the purview of the INFCIRC/66 agreement.

3.2.4.2 Potential uses of satellite imagery

There are limited, but useful opportunities for satellite imagery to make a contribution to the implementation of safeguards in states with INFCIRC/66 agreements. These include:

- Assistance in design review. The Agency is required to perform design reviews of principal nuclear facilities in order to be satisfied that each such facility will permit the effective application of safeguards. Satellite imagery would enable the Agency to closely monitor the progress of a facility under construction in a non-intrusive way.

- Confirmation of special reports. Under INFCIRC/66 agreements, a state is required to provide special reports to the Agency in the event that an unusual incident occurs involving actual or potential loss or destruction of, or damage to, any safeguarded nuclear material or principal nuclear facility. The use of satellite imagery would prove valuable in the early assessment of Chernobyl-type accidents and natural disasters such as earthquakes or floods.
- Determination of operational status. Satellite imagery would serve a useful complementary role in the case of new facilities to ensure that inspections are carried out prior to the start of operations at a new facility, and to determine when a facility has been shut down, such as a light water power reactor for refuelling.
- Verification of sealed storage. In those instances where a member state chooses to stockpile source material in a sealed storage facility, the Agency may be able to complement the routine annual inspection of the site with monitoring by satellite imagery.

3.2.5 Satellite imagery and voluntary offers of nuclear weapons states

The potential uses of satellite imagery for safeguards implementation in the nuclear weapon states are useful but less extensive than for the other states. The limitations in usefulness are primarily due to the limited nature of the voluntary offers by the nuclear weapon states that restricts Agency access to only a portion of the state's nuclear programme.

3.2.6 Satellite imagery and the Additional Protocol

The potential contribution that satellite imagery would make to the evaluation of information supplied by states under an Additional Protocol was examined for each of the submissions required by Article 2. It was found that satellite imagery could make a contribution to the verification of the correctness and completeness of the information supplied through an Additional Protocol.

For many of the reporting requirements, satellite imagery could be used by the Agency to verify the accuracy of the locations reported. In this regard, satellite imagery could further be used in the planning of site visits. This is especially important since many of the sites reported under an Additional Protocol are not sites which the Agency has visited in the past. For these categories of use the Agency should investigate the possibility of using archived imagery as a cost saving means.

Some potential uses of satellite imagery in connection with the verification of information provided under an Additional Protocol would best be served by having high resolution imagery. These applications include being able to detect specific details at a site, such as observing spent fuel cask transfers, nuclear fuel receipts and maintenance activities.

The application of satellite imagery to detect changes over time at the reported facilities also falls into this category of applications. There are uses of satellite imagery in connection with an Additional Protocol that offer the possibility of tangible cost savings by either reducing the number of visits or eliminating the necessity of a visit. These uses include the monitoring of remote and unmanned sites such as those containing high level waste, and to verify the operational status of a reported site, such as a uranium mine.

Appendix 3.2 contains more specific information on the results of the examination of the contributions of satellite imagery to verifying the information provided by an Additional Protocol.

3.3 RESOURCE IMPLICATIONS

This section identifies the five principal areas where the Agency will incur costs in the use of commercial satellite imagery as an additional source of information. Each of the areas (imagery, hardware, software, training and personnel) is discussed. This is followed by a discussion of how the Agency would monitor the use of satellite imagery from a cost/benefit perspective.

3.3.1 Imagery costs

Costs of satellite imagery today vary considerably depending, *inter alia*, on the system, the resolution and the area of coverage. IKONOS imagery products typically cost between US$3,000 and US$5,000, SPOT imagery costs approximately US$1,500 per frame, IRS-1C images between US$900 and US$3,900 per frame, and Russian archive data such as KVR-1000 imagery costs between US$900 and US$2,500. As more commercial satellites become available, prices are expected to fall, partly because of increased competition, and partly because the improvements in capabilities will increase demand which means that overhead costs can be spread over a higher sales volume. Notwithstanding these long-term positive factors, enhanced resolution imagery as it becomes available over the next five or six years is unlikely to vary appreciably from the cost range shown above. As a very rough indication of the order of magnitude of the costs, and using the current prices as a guide, the total cost of purchasing imagery for very selective applications would probably be measured in tens of thousands of dollars per year. By contrast, the total cost for images of every declared facility/site would likely be measured in hundreds of thousands of dollars.

The Agency has achieved significant cost savings by entering into contractual agreements with commercial satellite imagery vendors. In addition, the Agency has utilised other avenues to effect cost savings through its use of less expensive archived imagery, which although of lesser quality in some instances, has nonetheless met the requirements of some tasks. The use of aerial photography where available may also offer the possibility of cost savings. It has already been determined that the photo archives of the Agency contain numerous aerial photos of nuclear facilities. These may be adequate for such tasks as site orientation for inspectors, or for use with current imagery to see what changes have occurred at a particular site over time.

3.3.2 Hardware resources

Traditionally, specialist photo-interpretation light tables were necessary for viewing and interpretation of film containing satellite imagery. Today, the imagery analyst requires a well-equipped computer workstation; imagery is acquired in electronic form, and high-resolution video monitors are replacing light tables.

A representative hardware setup for an imagery analyst would consist of a personal computer with the following capabilities:

- minimum 1.5 GHz Pentium 4 or equivalent processor;
- minimum of 1 GB of RAM memory;
- minimum hard drive capacity of 36 GB;
- high-performance video display card with 64 MB or more of video RAM;
- dual high-quality video monitors with 21″ screen displays;
- DVD-ROM drive; and
- 48 speed CD-ROM drive.

Peripheral equipment, which could be shared by two or more persons, would include a colour printer, a high-resolution flatbed scanner and large capacity storage devices for system backup.

3.3.3 Software support

Commercial software to aid in the interpretation of imagery has been available for some time and, in software terms, is at a very mature state. The Agency has evaluated and is currently using the ENVI, ERDAS Imagine and PCI imagery analysis software packages. These software packages; they are summarised as follows:

- ENVI version 3.1 by Research Systems Inc. The strengths of the ENVI applications are in the user interface for accessing the processing and image enhancement modules. These modules permit the user to browse and enhance the imagery through highly-developed algorithms and techniques without altering the original data set.
- The weakness of the ENVI suite is in its lack of rapid access to multiple data sets. ERDAS Imagine version 8.3 by ERDAS Inc. The strengths of the Imagine application are in its user interface, which allows the user to quickly learn the basic imagery browsing and enhancement tools. Its weakness is in the use of its advanced features that are costly add-ons and require extensive training.
- ImageWorks version 7 by PCI Geomatics Group. The strengths of the ImageWorks suite are in the capability of accessing and processing all the remote sensing industries data set formats. It permits quick access to imagery with ease. The modularity of the suite permits rapid access to multiple enhancement algorithms and tools. Its weaknesses are in its user interface in that it is not an intuitive user interface.

Software packages under consideration require additional review in order to determine which tools best meet the needs of the Agency. The capabilities sought from software ultimately selected include the following:

- image archiving, storage and retrieval, necessary for managing large imagery databases;
- change detection through 'side by side' feature comparison;
- creation of line drawings and cartographic quality maps from imagery;
- generation of digital elevation models;
- identification of new features using radar data; and
- image enhancement.

3.3.4 Training

Training is a necessary element of any plan by the Agency to incorporate satellite imagery into the information review process. Workshops held by the Agency that addressed satellite imagery and the assistance provided thus far by the member state support programmes have shown that there are well qualified resources to draw upon. Indeed, the Agency is grateful to both the Canadian and Swedish support programmes for their assistance in developing and providing training in the use of satellite imagery in the review process.

3.3.5 Personnel

Trained and experienced staff will be needed to use commercial satellite imagery on almost any scale. The numbers will depend in the long term, of course, upon the extent of the Agency's use of imagery. By early 2002, SIAL will have in place two imagery analysts and a geospatial database administrator at the professional level, and one general staff support person. Additionally, persons from each of the operations divisions and from the Division of Safeguards Concepts and Plans have received training and familiarisation to enable them to interact effectively with the specialists. Training courses will continue to expand the basic working knowledge of staff with regard to satellite imagery's uses and capabilities.

3.4 COST/BENEFIT ANALYSIS

3.4.1 Costs

The costs associated with the use of satellite imagery by the Agency are relatively easy to calculate in the case of hardware, software, personnel and training. At the present time the greatest cost uncertainty is that for imagery. Although the Agency has negotiated a bulk order satellite imagery contract, this is scalable and based on overall UN imagery purchases, which means it is difficult to reasonably estimate future costs.

3.4.2 Benefits

The calculation of benefits from satellite imagery is, of course, dependent upon the safeguards applications selected. From analyses reflected in this paper and in the case studies a clearer picture begins to emerge as to which applications provide the greatest contribution. All benefits from the use of satellite imagery are not quantitative. Some are qualitative. Moreover, some benefits can only be quantified using empirical data.

Benefits can be further categorised as:

- those that contribute to more *effective* safeguards; effectiveness, in this case, is the increased confidence that the Agency can attach to its evaluations or to its chances of detecting undeclared activities; and
- those that contribute to more *efficient* safeguards; efficiency, in this case, is the ability of the Agency to achieve its safeguards objectives in less time or at less cost.

3.4.3 Measuring the benefits

The most fundamental measurement of benefits from new or improved techniques is in the impact on inspection activities. This requires a determination of how the use of satellite imagery reduces the number of, and/or the frequency of, on-site inspection activities. Potential applications of satellite imagery that may result in a reduction of inspections include the following:

- confirmation of the correctness and completeness of the site descriptions and other information provided in an Additional Protocol;
- checking of locations and operational status of activities; and
- detection of changes and ongoing monitoring at declared sites.

Other benefits from potential applications of satellite imagery are less quantifiable but, nonetheless, are real in terms of increased effectiveness and efficiency. These include:

- increased confidence in the correctness and completeness of the information provided by the member state or generated by the Agency;
- increased ability to detect undeclared activities and sites;
- increased ability to check the credibility of open source reporting;
- better prepared inspectors; and
- better inspection planning.

From a broader perspective there is a benefit from the increased confidence of the world community in the effectiveness of the Agency that comes from being seen as using the most technologically advanced tools. Relatedly, the use of satellite imagery will add to the deterrent value of the SSS in dissuading potential proliferators from engaging in undeclared activities.

While this cost/benefit discussion falls short of being able to quantify the expected benefits for a given level of spending, it does indicate a high level of confidence that benefits will be derived from the use of satellite imagery.

3.5 IMPLEMENTATION

3.5.1 IAEA and SIAL

At the time of writing, late 2001, SIAL is operating with a limited capability, both in regard to laboratory hardware and staffing. Its resources are limited to one imagery analyst and one geospatial database administrator – however, it is receiving substantial support from the Member State Support Programme that has allowed for the provision of expert imagery analytical support. This has resulted in several consultant-based visits from each member state that has served as a vehicle for the transfer of experience and knowledge that has permitted SIAL to rapidly and successfully become integrated into the safeguards system.

By early 2002, SIAL will increase its staffing levels to include a second imagery analyst and a second IT support staff member. The laboratory infrastructure will be in place (Figure 3.1) and a greater variety and volume of products will result.

The laboratory in SIAL will consist of three high-end workstations, a UNIX based geospatial database server and peripheral devices such as plotter, large format colour laser printer and scanning device. The laboratory will also have an area that

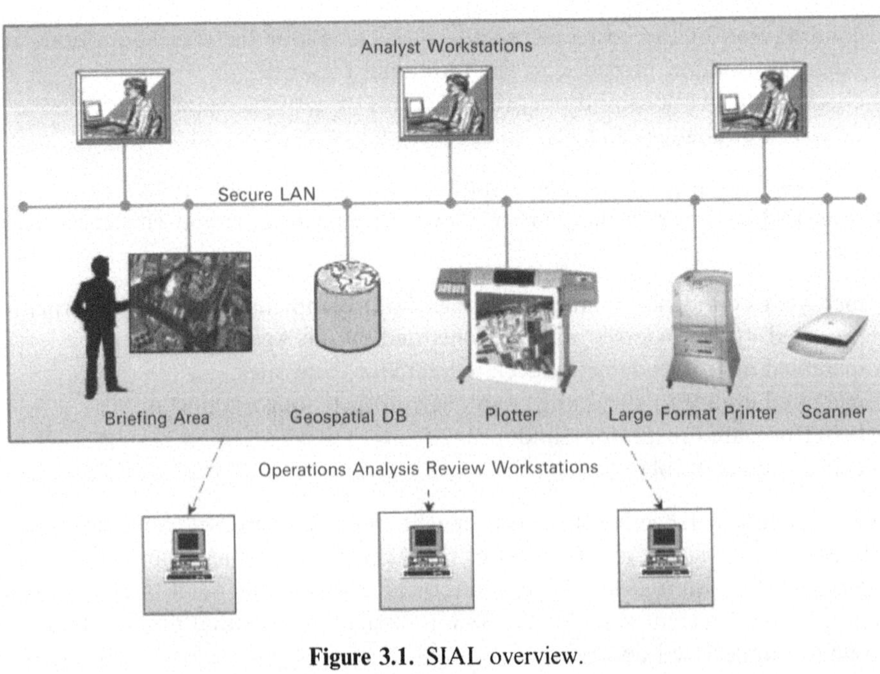

Figure 3.1. SIAL overview.

will have a high-resolution plasma display for briefings. All devices will be connected by a local area network (LAN) that is isolated and thus inaccessible from outside the laboratory. In addition, all devices connect to the secure network with interface cards that have built in encoders for on-the-fly data encryption. The geospatial database itself is based on CARTERRA Analyst Professional software which will be the analysts' interface to all imagery and datasets upon which analysis will be performed.

Products created in SIAL are delivered to operations' areas and can be in the form of hard-copy reports, GIS database projects and data. Each of the operations areas have adopted the same secure/isolated infrastructure as the SIAL. This permits them easily to review products of analysis created in the SIAL. The integration of GIS and other safeguards computerised systems is a near-term aim of the division of Safeguards Information Technology (SGIT). This integration would take the form of the GIS being utilised as a data provider for new unique safeguards applications as well as the ultimate graphical user interface (GUI) for the safeguards inspectorate.

3.5.2 SIAL tasking process

Although both SIAL staff and inspectors share the same goal they are viewing it from very different perspectives. The imagery analyst uses one particular skill-set while the inspector utilises another to equal effect. This is where the tasking process begins. The definition of the task by the inspector is the first step in the tasking process and forms the basis of subsequent decision making and development of the task.

After submission of the task form, the task is subjected to an evaluation process (Figure 3.2). The first step is to consult the imagery analyst on the feasibility of the task – is the task suitable for the application of imagery analysis? Typically, satellite

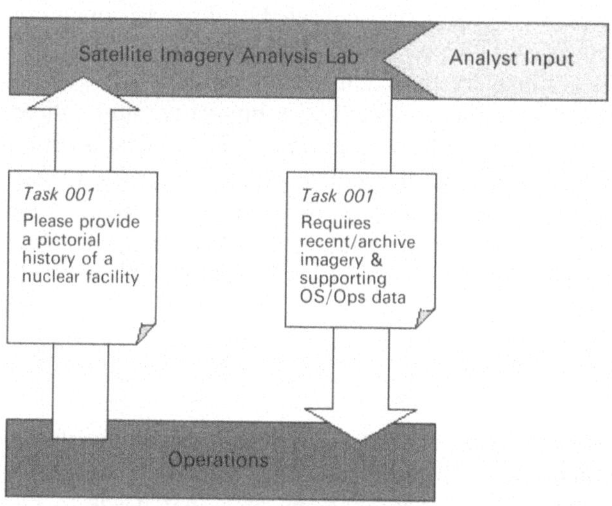

Figure 3.2. SIAL task evaluation process.

imagery will provide an overview of a visited area of interest (AOI) that is sometimes difficult for the inspector to achieve. However, on occasion, satellite imagery may be requested as it offers exclusive access to an AOI, and this must be considered by the analyst in his or her recommendations for imagery acquisition and application of available technology. For example, if the task was the verification of a state's submitted Additional Protocol data, the analyst may recommend purchasing a single, recent, high-resolution, 1 m GSD, image of the AOI which may be directly compared to the submitted site plan. If the task were the verification of the operational status of an uranium mine, the analyst may specify an image collection strategy of one image per quarter for the next twelve months, further stipulating that the image should not be collected on weekends or known state national holidays. In addition, the analyst could task SGIT's information analysts with researching open source information that would supplement the analysis process. Once the analyst has reviewed the task and made the necessary proposals, the inspector is informed of the task's dependencies and is asked to agree on a task completion deadline where a finished report will be formally presented by the SIAL to the responsible operations director.

The first action performed on imagery acquired by SIAL is ingesting that imagery into the GIS. This means that the imagery is indexed by the geospatial database and is now available for easy and rapid analysis and manipulation. Every image will be subjected to first level analysis. This requires that the AOI is identified and its borders and buildings are digitally created and saved as vector layers that are then directly linked to the image from which they were derived. If further datasets, which could be imagery, table or vector line data, exist because they were identified by the analyst and have been obtained or are present as a result of past analysis, it is at this stage that they are linked to the imagery being exploited (see Figure 3.3). It is also at this stage that supplementary data is compiled for the analyst. This dossier could contain either or all of State supplied information, in the case of Additional Protocol, data collected by inspector/operations division, and/or data collected by SGIT Open Source analysts.

The synthesis of imagery analyst, inspector, datasets and technology represents a multitiered mechanism (Figure 3.4), that is innovative and will remain unique to SSS. The interaction between inspector and imagery analyst is continuous and open, the outcome of which is a set of products and cutting edge tools that are proving essential for operations in the role of decision management. These products and tools themselves are becoming the fundamental components of subsequent analytical study that will further enhance the efficiency and effectiveness of the inspectorate.

3.6 FUTURE

The biggest challenge facing SIAL is the tremendous volume of work it faces trying to maintain global focus. The cost associated with purchasing satellite imagery is prohibitive, but imagery itself is only one area that needs to be addressed. The flexibility SIAL requires to independently and confidentially specify when imagery

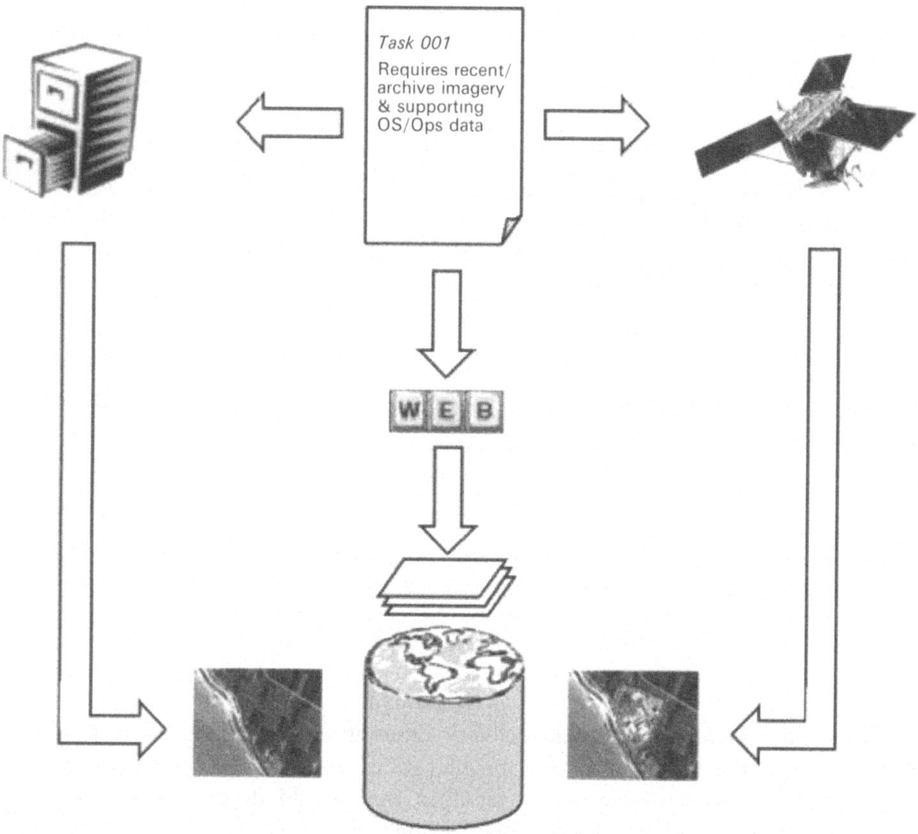

Figure 3.3. Data collection process.

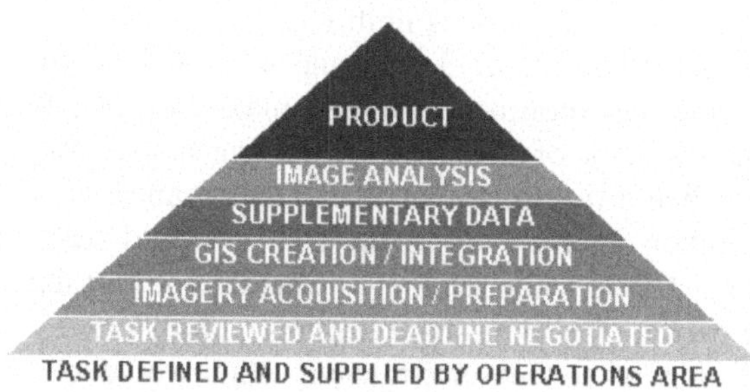

Figure 3.4. Task process and analysis construction.

acquisition takes place can only be realised with the introduction of new technology supported by a sophisticated spatial data and high-speed connectivity infrastructure. At present, should SIAL request a satellite image in its need to perform a monitoring role, respond to a nuclear safety or security and physical protection emergency, in an area that may be inaccessible or unobservable through the traditional inspection

Table 3.1. UN umbrella agreement estimates expenditure.

Objective of Expenditure	Total
Fee for fibre-optic access to UN HQ, 512K Port/256 CIR including installation and 3 years access	US$191,289
Purchase of 20 minutes per month satellite time for 3 years (UN umbrella agreement)	US$8,640,000
Cost to Department of Safeguards/IAEA of 25% capacity totalling 5 minutes per month for 3 years	US$2,160,000
Training one week (10 persons)	US$20,000
On-cell imagery analysis support costs:	
GIS Application Specialist (80 hours)	US$7705.60
Senior GIS Application Specialist (80 hours)	US$10,150.40
Senior Imagery Expert (80 hours)	US$12,315.20
Department of Safeguards/IAEA Total	US$2,401,460
Future costs relating to purchase of a ground station for UN Headquarters, New York, USA	Approx. US$20m

process, its satellite tasking request is scheduled at the discretion of the satellite vendor. This can result in image delivery times in excess of those acceptable by SIAL. To be able to respond to a proliferation crisis or nuclear emergency, the department of SSS has identified a solution that would dramatically increase its ability to maintain global focus and, at the same time, provide the flexibility desired to respond to emergencies within a 24-hour time frame.

SIAL's proposal is that the Department of Safeguards purchases a percentage of satellite time under a United Nations umbrella agreement. The umbrella agreement (cost estimate presented in Table 3.1) would see the UN purchase a total of 20 minutes satellite time per month returning a maximum of 240 images per month. SSS would purchase 25% of this time producing a maximum of 60 images per month. The result being that the UN, other UN organisations, the Department of Safeguards, and other departments of the IAEA, would realise an estimated data cost saving of at least 50% while retaining ownership of that data. Further, purchasing satellite time permits the 'time share' owners to fully control satellite operations, allowing the creation of collection strategies. This would offer SIAL an effective monitoring mechanism that at a minimum would specify where and when an imagery acquisition should take place, and in response to crisis, SIAL could realise image tasking, collection and delivery within vastly superior time periods.

The fact that other UN organisations will have access to the satellite sensor and will specify imagery acquisitions that would populate an UN imagery archive means that the Department of Safeguards will have access to a wealth of imagery data that affords the global coverage it requires. The establishment of a UN imagery archive, a spatial data infrastructure, would result in imagery data being shared between UN

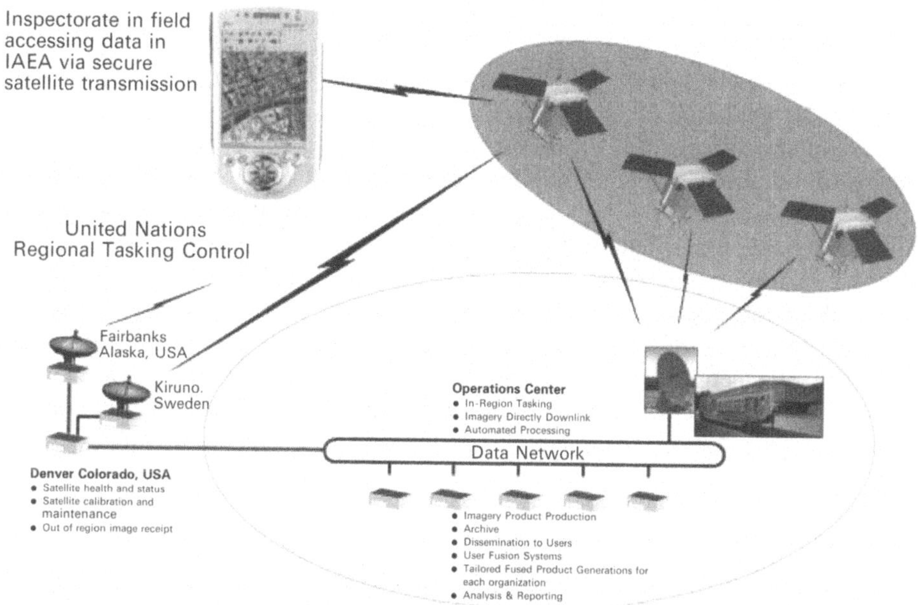

Figure 3.5. United Nations spatial data and high-speed connectivity infrastructure.

organisations and the IAEA without further cost or licensing agreement complications. In addition, there exists the possibility that imagery analysis performed in other UN organisations on an area of interest not monitored by the IAEA, may reveal potentially vital information that the IAEA could be made aware of and exploit to their advantage.

The system would be supported by a high-speed connectivity infrastructure that would allow rapid transfer of data, after image collection by the satellite sensor was complete, between IAEA Headquarters, Vienna, Austria and UN Headquarters, New York, USA. This system will also scale to other vendor satellite ground stations allowing unprecedented access to multi-vendor imagery data, both current and archive. Because satellite technology is an evolutionary process, it is also conceivable that an overall solution would be extended to include the provision of a satellite receiving ground station (see Figure 3.5). This would be used to directly receive satellite data and transmit satellite tasking information.

3.7 SUMMARY CONCLUSIONS

The examination of potential applications of satellite imagery has led to the view that this publicly available source of information can contribute to the effectiveness and the efficiency of the Strengthened Safeguards System. This view is based on the results of an in-depth internal examination of potential applications of satellite imagery and inputs from experts in the field through technical papers and discussions with IAEA staff.

It has been determined that the Agency has the legal authority to purchase commercially available satellite imagery. Notwithstanding the potential contributions of satellite imagery to the implementation of safeguards, the Agency must be fully aware of the policy considerations that might affect its own use of satellite imagery. The availability of high-resolution imagery from open sources in the near future will greatly contribute to the usefulness of this data product in the Agency's information review process. It is anticipated that high resolution will be available from multiple suppliers in several states.

The experiences of other users of satellite imagery have contributed to the Agency's assessment. In particular, the organisational lessons of relevance from the work of the Western European Union Satellite Centre and from the experiences of the Agency's Action Team and UNSCOM. In that connection the Agency's approach will be to acquire imagery specific to the needs of the individual task. Indiscriminate countrywide searches for undeclared activities and sites would be prohibitively expensive and is not contemplated.

The applications for satellite imagery identified of interest are primarily in those areas in which the imagery complements other information sources. The benefits of satellite imagery to the safeguards programme are both qualitative and quantitative. Qualitative because the value is in adding confidence and accuracy to the information review; in this category there are the broader benefits from being a deterrent against undeclared activities and for raising the credibility of the Agency in general. Quantitative because there are also potential benefits to be derived; in some cases the duration of an inspection trip might be reduced due to a better informed and prepared inspector. In other cases, visits to sites may not be necessary if the location and status of a site can be determined from satellite imagery.

The costs of the components of a programme to include satellite imagery into the information review process are both predictable and manageable. Initially, the cost of the images will be one of the most costly elements. The task-by-task approach to satellite imagery use will allow the Agency to control this part of the programme costs. By entering into an implementation plan that does not require a major initial investment and one that is able to assess the cost benefits in a deliberate and step-wise manner, the Agency will be able to arrive at a level of use in which the returns are commensurate with the costs.

3.8 THE NUCLEAR FUEL CYCLE AND RELATED ACTIVITIES: WHAT CAN BE SEEN FROM SATELLITES

The use of satellite imagery as a tool for detection is examined on a step-by-step basis, through the nuclear fuel cycle, from mining and milling through reprocessing, and then for the related activities of heavy water production and nuclear research centres. For purposes of this assessment the activity is assumed to be at a mature state of deployment. In an operational setting, however, the imagery analyst would have to give consideration to whether the activity is in the R&D phase, laboratory or pilot plant stage of operation. Whether the activity is in production, under construction, being expanded, operating, shutdown or decommissioned are also important attributes to be assessed by the analyst. It should be kept in mind that the value of satellite imagery in each of these cases would increase significantly when used in conjunction with other information sources.

3.8.1 Source material acquisition

There are various means of obtaining uranium and thorium, and there are few common characteristics. The principal categories include the following.

Uranium from mining operations

Mining operations to extract uranium ore can be conducted by open-cut or by underground mining. Moreover, these operations may be exclusively for the extraction of uranium or they may be in conjunction with copper or gold mining. Signatures related to these operations that would be visible to a satellite include the presence of piles of ore and tailings, large ore trucks and the uranium mill, except in the unlikely case of an underground mill. In the case of uranium extraction in conjunction with other mining operations, the copper and gold mines themselves become visible indicators of possible uranium mining. The principal difficulty in this case would not be in seeing the mining operation but, rather, in being able to conclude that the mining was for uranium.

However, in conjunction with other reporting, including the state's declaration, satellite imagery would be a valuable complementary means of adding to the Agency's knowledge about the level of activity. With respect to whether the operation is active or inactive, satellite imagery can be a cost-effective means of determining operational versus shutdown status because of the usually remote locations of mining and milling operations.

Uranium by in situ *leaching*

This form of uranium extraction does not involve the digging of ore (as in the previous case) and usually consists of a number of well fields and a central processing plant. The process involves the injection of a leaching agent into the ore zone through injection wells and the recovery of the resultant uranium solution through a production well.

High-resolution satellite imagery would be able to detect two important features of this means of uranium extraction. First, there is a common pattern of wells consisting of a grid with four injection wells at the corners and a production well in the centre; spacing generally ranges from 15 m to 30 m between the injection wells and the production well. Secondly, evaporation ponds to dispose of wastes are a distinctive feature of the *in-situ* leaching process. These two features would allow identification of uranium mining using the *in situ* leaching process. They may also be a way of monitoring the operational status of the mine, but the utility of both indicators is relatively short lived. Most of the uranium is reportedly recovered during the first six months of operation.

Uranium from phosphate

Since most phosphate ores contain some uranium, this process has been a means of obtaining uranium in several places around the world. Whether or not uranium is being produced as a by-product, however, would not be discernible from satellite imagery. One potential benefit of satellite imagery is observed where uranium extraction is known to occur when the phosphoric acid plant is operating. Imagery could be used to determine the operational status of the phosphoric acid plant and thereby indirectly provide information on the status of uranium extraction.

Uranium from seawater

This method is only included for completeness since, on cost grounds, it has not been attempted on a commercial scale. The most likely scenario, should uranium ever be obtained by this process, would be in conjunction with a desalination plant. It would require a lot of equipment that is common to other methods of uranium extraction. If detected at a desalination plant, the presence of the equipment would raise questions.

Thorium from monazite

Thorium is primarily obtained as a by-product of monazite processing for rare earths. Thus far there has been a limited demand for thorium relative to the demand for rare earths and this has led to an oversupply of thorium compounds and residues.

A typical means of mining the monazite is by dredging sand dunes or a riverbed. It may also be acquired dry by using heavy earth-moving equipment. The material is then taken to a wet mill for processing. Whether or not thorium is being separated in the process could easily be detected on the ground but would not be discernible from satellite imagery. The utility of satellite imagery is at best an indirect one, whereby with collateral information it would be possible to observe the monazite mining.

Thorium from uranium ore

If a uranium extraction operation were also extracting thorium it would not be detectable from satellite imagery.

3.8.2 Conversion

Conversion processes are of two types: (1) Conversion 1: those processes that are transformations of natural nuclear material in order to provide feed material for isotope separation or reactor fuel fabrication. (2) Conversion 2: those transformations subsequent to enrichment or reprocessing, sometimes known as re-conversion. There are transformations under both Conversion 1 and Conversion 2 that result in different end-products. The presence on site of canisters used for nuclear materials could be a strong indicator. The primary obstacle to conversion plant identification is the similarity between conversion plants and many chemical plants. Identification would be enhanced if it were known that a conversion facility was collocated with a known facility, such as an enrichment plant.

3.8.3 Enrichment

Nine methods of separating the isotopes of uranium were examined. These methods require feed materials in a form compatible with the process. The feed materials are either uranium hexafluoride (UF_6) or uranium chloride (UCl_4) or uranium metal. The value of satellite imagery will be examined based on the feed materials used in the enrichment processes.

Enrichment of UF_6

A common feature of facilities engaged in uranium enrichment using UF_6 feed material is the depleted uranium produced by the processes. Often the large containers of depleted UF_6 are stored in the open on large concrete rafts and would be visible from satellite imagery.

Gas centrifuge enrichment

Facilities used in this process would have few signatures applicable to satellite imagery. Although the operation of a plant requires electrical power and water for cooling, these requirements are low compared to other processes.

Molecular laser enrichment

A process known in the USA as Molecular Laser Isotope Separation (MLIS), in France as Séparation des Isotopes par irradication au Laser de Molécules (SILMO), and in Japan as Riken's Molecular Laser Isotope Separation (RIMLIS). It would be similar to gas centrifuge enrichment in low-power requirements. This process, which has not yet been deployed in an operational plant, would also not have significant distinguishing features applicable to detection by satellite imagery.

Gaseous diffusion enrichment and Aerodynamic enrichment

Two enrichment processes with characteristics of interest from a satellite imagery perspective. The high-power demand of both would likely result in a need for the construction of a power plant in the vicinity. In any event, there will be a number of high-voltage supply lines into the facility and large size switchyards at the plant. Both processes require cooling water that results in a need for a large number of cooling towers. The heat generated in the operation of these plants could also be detectable by infrared satellite imagery.

Another common feature of these two enrichment methods is the need for large buildings to accommodate the many stages used in the process. This signature, of course, is dependent on the intended output of the plant. Use of smaller facilities for the clandestine operation of these processes cannot be ruled out.

Enrichment of UCl_4

Anhydrous UCl_4 is the feed material for Electromagnetic Isotope Separation (EMIS) enrichment. In its solution form it is also the feed material for Chemical Exchange enrichment and Ion Exchange enrichment. Of the three processes, EMIS enrichment offers the best possibility for the use of satellite imagery for safeguards purposes. In that regard, the experiences in Iraq have been of value.

Plant structure for the Ion Exchange enrichment process would be modular. Buildings would be less than 10 m high, built to withstand explosion rather than earthquakes. The plants for both Chemical Exchange and Ion Exchange enrichment would have the appearance of a plant in the chemical industry.

The facilities for EMIS enrichment can be housed in comparatively small buildings with few, if any, signatures for detection by satellite imagery. However, for this method the electric power demand is high with several tens of megawatts being required for even a small facility. High-voltage incoming power lines with local transformer stations should be observable.

Enrichment of uranium metal

Uranium metal is used as feed material in Atomic Vapour Laser Enrichment (AVLIS or SILVA) and Plasma Separation enrichment. Both methods have undergone research and development but no known commercial size plants have been constructed.

Atomic vapour laser enrichment

Plant characteristics associated with AVLIS remain largely supposition. However, such a facility should feature two types of buildings: a laser building and a separator building, both designed to seismic standards. The process requires little energy, more than 20 times less than gaseous diffusion or aerodynamic enrichment processes.

Plasma separation enrichment

For this process, a plant would consist of a main building housing the separator and adjacent workshops for refurbishing the separator internals, a liquid nitrogen facility and a small switchyard. This process has low energy and cooling requirements.

Satellite imagery would make little or no contribution to the application of safeguards or in the detection of facilities engaged in the enrichment of uranium metal.

3.8.4 Fuel fabrication

There are no observables that would indicate the value of satellite imagery to the safeguards process or to the identification of undeclared fuel fabrication activities.

Moreover, significant indicators would be unlikely even during construction of these facilities.

3.8.5 Reactors

There is a wide range of reactor types. They can be categorised by their purpose: to produce electricity, for research, for plutonium production, or for propulsion. They can alternately be categorised according to the coolant used, the moderator, the fuel type, or the power levels.

There are several signatures related to satellite imagery identified for reactors, including containment buildings, and the presence of high-voltage power lines and cooling towers. The absence of coal and large oil storage tanks are also important indicators for identifying nuclear power plants. This category of nuclear activity is the most easily distinguishable from satellite imagery.

3.8.6 Reprocessing

A commercial scale reprocessing site has several signatures that would help in the identification process using satellite imagery. These include road and rail access for bringing in the spent fuel; high-capacity water supply; large-scale water treatment; cooling towers; heavy lifting equipment; electrical power source; and a building with a heavy duty overhead crane. The presence of transport canisters at the location is also a possibility. These canisters may be from about 2 m to over 6 m long and weigh from 50–100 tonnes.

3.8.7 Heavy water production

The ability to identify a heavy water production plant from satellite imagery is dependent upon the type of technology used. For the most common technologies, water-hydrogen sulphide exchange process and the ammonium-hydrogen exchange process, there are indicators, such as tall exchange towers, to assist in the detection. The collocation of fertiliser plants in the case of the ammonium-hydrogen exchange process is a significant indicator. The similarity between heavy water production plants and chemical plants adds to the difficulty of identification.

3.8.8 Research centres

Nuclear research centres in most cases can be identified by the presence of a research reactor as evidenced by its containment building dome. Tall 'smoke' stacks for the dispersal of off-gases are also common indicators.

3.9 AN ASSESSMENT OF SATELLITE IMAGERY AND THE ADDITIONAL PROTOCOL

Each of the provisions of Article 2 of the Additional Protocol are examined to determine what contribution satellite imagery might make to the process of verifying the declared information.

Article 2a(i)

This provision calls for a general description of and information about nuclear fuel cycle related R&D activities which do not involve nuclear material.

Satellite imagery would be a further way of verifying the accuracy of the locations reported. As reference information, satellite imagery could be used in the planning of site visits by inspectors.

Article 2a(ii)

This provision provides an opportunity for the Agency to request information additional to the other requirements of the Additional Protocol on operational activities of safeguards relevance at facilities and Locations Outside Facilities (LOFs) where nuclear material is customarily present.

The unlimited nature of the types of information that might be requested makes assessing the relevance of satellite imagery case-specific. The Article 2 Guidelines give examples, some of which (e.g. spent fuel cask transfers, nuclear material receipts and the schedule of maintenance activities) could have indicators observable by satellite imagery.

Article 2a(iii)

This provision calls for a general description of each building on each site, including its use and, if not apparent from that description, its contents. A map of the site is to be included.

This information could be checked against satellite imagery of the site, both current and historical, for omissions and inconsistencies.

Article 2a(iv)

This provision calls for a description of the scale of operations for each location engaged in the activities specified in Annex I of the Additional Protocol.

Few, if any, of the activities listed in Annex I can be identified from satellite imagery, particularly with respect to the scale of operations. However, given that this type of information has not been a part of the safeguards programmes in states, satellite imagery can provide an initial authentication of the locations reported and in the establishment of an Agency database for these types of activities.

Article 2a(v)

This provision calls for information specifying the location, operational status and the estimated annual production capacity of mines and concentration plants associated with the production of uranium and thorium.

Satellite imagery could be an effective tool for checking operational status and location, and for ongoing monitoring of status.

Article 2a(vi)

This provision calls for information on source material which is not yet subject to the application of safeguards.

Satellite images could contribute to the Agency's knowledge of the location and physical features of these sites, and would be useful complementary information in the preparation for visits by Agency inspectors.

Article 2a(vii)

This provision calls for information regarding the quantities, uses and locations of nuclear material exempted from safeguards according to the relevant provisions of INFCIRC/153.

Satellite imagery has little or no value in verifying this information, except to check location.

Article 2a(viii)

This provision calls for information regarding the location or further processing of intermediate or high-level waste containing plutonium, high enriched uranium or uranium-233.

As noted by the Agency this provision currently applies to relatively small amounts of nuclear material. However, in the future, the number of such sites and the contained special fissile material could increase significantly. Satellite imagery could be used to assist in monitoring remote unmanned and sealed sites. The ability to detect change at these sites by satellite imagery would be of particular value.

Article 2a(ix)

This provision calls for information on exports of items in Annex II of the Additional Protocol, and upon request, information on imports of these items.

Satellite imagery would have limited value, except as an aid in the preparation of inspectors for visits to the locations of the imported items. In many cases such locations would not have been previously visited.

Article 2a(x)

This provision calls for the member state's general plans relevant to the development of the fuel cycle for the succeeding ten-year period.

The value of satellite imagery would be limited to providing an early point of reference for potential use later for detection of changes.

Article 2b(i)

This provision is the same as Article 2a(i), except in this case the reporting is limited to R & D in enrichment and reprocessing by entities not sponsored by the government. As in Article 2a(i), satellite imagery would not contribute much to the verification of the declared information itself, but it would be a further way of verifying the accuracy of the locations reported. As reference information, satellite imagery could be used in the planning of site visits by inspectors.

Article 2b(ii)

This provision calls for information, to be provided upon request of the Agency, about activities at locations specified by the Agency and not included in the member state's declaration.

The relevance of satellite imagery to checking the member state's response to such a request will be case-specific. Imagery could, however, be part of the process by which locations are identified which raise questions. In other situations, the use of satellite imagery in connection with other sources of information may provide answers to questions without the need to visit the site.

Article 2c

This provision calls upon the member state to provide clarifications and amplifications upon request of the Agency on any information it has provided. The purpose of this provision is to facilitate implementation of the Additional Protocol and to help ensure the correct understanding by the Agency of the information provided by the member state.

If satellite imagery can make a contribution to this objective, especially if it allows the Agency to avoid recourse to complementary access, it will be valuable. Satellite imagery could be used to help monitor the operational status of facilities declared under the Additional Protocol.

3.10 REFERENCES

Jasani, B., Richter, B., Stein, G., Ward, M. D., and Killinger, M. (1996) *Enhancing IAEA Safeguards Using Commercial Satellite Imagery: A pilot study*, November 1996, Report SRDP-R240, JOPAG/12.95-PRG-265, Task A.16/JNT D882.

Safeguards: Sources and Applications of Open Source Information, IAEA Report and Proceedings of a Seminar held in Vienna, 15–19 September 1997. IAEA, Vienna, 1998.

4

The role of satellite imagery for monitoring international treaties

B. Jasani

4.1 INTRODUCTION

A by-product of the defence observation satellites developed and deployed by the USA and the former USSR was their use for monitoring the compliance with the terms of their bilateral arms control treaties. Thus, monitoring from space formed a vital element of their NTM of verification of bilateral and multilateral agreements. However, not all nations have access to such capabilities even though they are parties to several important multilateral arms control treaties. Moreover, today, neither the USA nor Russia are willing to share widely either the technology or the information obtained by their NTMs. This is beginning to give an impetus to the development of a multilateral technical means (MTM) of verification. In this process, the commercial remote sensing satellite could play an important role.

Arms control, disarmament and confidence building measures are some of the important elements of security. If a multilateral arms control agreement is to be credible, it needs to be 'effectively' verified. A number of such treaties have provisions for on-site inspections for their verification (Table 4.1). However, these usually are carried out in a very limited way and only at declared sites because the inspected state is generally reluctant to make itself too transparent. Therefore, non-intrusive ways need to be examined. It was suggested that civil earth-orbiting observation satellites could be used to monitor multilateral agreements such as the 1970 NPT (Jasani, 1990). This and the Chemical Weapons Convention is an example of an international measure that requires verification on a multilateral basis (i.e. MTM of verification). There are a number of other already existing treaties the effectiveness of which could be enhanced by such verification procedures (see Table 4.1). A number of countries now have the capability to build and launch civil observation satellites. Moreover, the advent of high-performance commercial remote sensing satellites now

Table 4.1. Some multilateral treaties that could be verified by commercial observation satellites.

Treaty	Activities limited and prohibited	Comments
Antarctic Treaty (1961)	Establishment of military bases, military manoeuvres and testing of weapons (e.g. nuclear explosions or any type of weapons)	On-site inspections and aerial observations allowed (Article VII); exchange of data on activities; some of the activities prohibited can be observed by current civil satellites
Partial Test Ban Treaty: PTBT (1963)	Any nuclear weapons test in the atmosphere, in outer space, and under water	No verification provisions in the Treaty but civil satellites could be used
Outer Space Treaty (1967)	Placing in orbit of nuclear weapons and other weapons of mass destruction; establishment of military installations and fortifications and military manoeuvres on celestial bodies	On-site inspections at launch and some data exchange is possible (Articles X and XI)
The NPT (1970)	Transfer of nuclear weapons by NWS; receipt, manufacture or other means of acquisition of nuclear weapons by NNWS	Main verification mechanism has been reporting to the IAEA; satellites have now been introduced
Sea-Bed Treaty (1972)	Emplacement of nuclear and other weapons of mass destruction on the sea-bed, the ocean floor and in the subsoil thereof	On-site inspection allowed (Article III)
BW Convention (1975)	Prohibits the development, production, stockpiling or acquisition of bacteriological and toxin weapons; destruction of existing stocks	Consultations could be carried out for verification of compliance (Article V)
CWC (1997)	Prohibits development, production of, otherwise acquire, stockpile or retain or transfer, directly or indirectly chemical weapons; use of CW; destroy all chemical weapons and their production facilities	Exchange of information and consultations and OPCW to monitor activities of the Parties; commercial observation satellites could be used
CTBT (not in force)	Prohibits any nuclear explosion in any environment whether for weapons or peaceful purposes	Mainly seismology; on-site inspections; monitoring radionuclide particulates and gases; hydro-acoustic and infra-sound monitors; commercial observation satellites could be used

make it possible for all countries to use this capability for verifying multilateral treaties as the images can be purchased by anyone.

In view of this, commercial satellite capabilities are examined in this chapter. Various image processing techniques designed to extract as much useful information as possible are also discussed briefly.

4.2 SATELLITE ORBITS

Generally, the orbital characteristics of an earth-orbiting spacecraft reveal its functions because they are optimised for a particular mission. Spacecraft are usually placed in orbit by large multi-stage rockets that apply accelerating forces over a period of several minutes. In the initial phase of the flight, the spacecraft accelerates through the atmosphere in a vertical direction lasting for about 10 seconds. The first stage of the rocket takes the spacecraft to the upper atmosphere at an altitude of about 60 km. The second and the third stages then gradually bring the spacecraft to its orbital height. Once this is achieved, the spacecraft is turned into horizontal position and given sufficiently high velocity in this direction so that it will not fall back to earth. Instead, it will continue in an elliptical or circular path around the earth (Figure 4.1) with the centre of the earth at one of the foci of the ellipse. The satellite is then said to be in orbit.

The motion of a satellite around the earth is influenced by the effects on it of the rotation and the shape of the earth and of other planets. However, these effects can be used to advantage. For example, the earth rotates around its axis approximately once every 1,440 minutes (i.e. every 24 hours). If the period (the time the satellite takes to go around the earth once) of a satellite is chosen as a multiple of 24 hours then the ground tracks[1] will repeat each day. For example, the ground tracks will repeat over a particular area at every sixteenth orbit if the period is 90 minutes (1,440/90 = 16). On the other hand, a different period will result in a gradual shift of the ground tracks each day so that there is complete coverage of a large area. The other parameters that cause shifts in the ground tracks of a satellite are called the right Ascension of the ascending node (Ω), and the argument of perigee (ω). These are two of the six basic parameters, called the orbital elements, which completely describe the size, shape and orientation of an orbit of a satellite in space. The position of the orbital plane in space is usually given in a terrestrial sidereal rectangular co-ordinate system, the axes of which are shown in Figure 4.2.

The origin, **O**, of the co-ordinate system is the centre of the earth; the z-axis is oriented towards the North Pole and the earth's equatorial plane is in the xy-plane. The x-axis is oriented towards the Vernal equinox or the first point of Aries.[2] The

[1] This is the projected path traced out by a satellite over the surface of the earth.

[2] The orbital plane of the earth is called the plane of the ecliptic. The angle between the earth's equatorial plane and the ecliptic plane is 23° 7′. The line of intersection of these two planes defines an important direction in space, known as the point of Aries. At any particular time this direction is exactly known and all celestial directions can conveniently be referred to it.

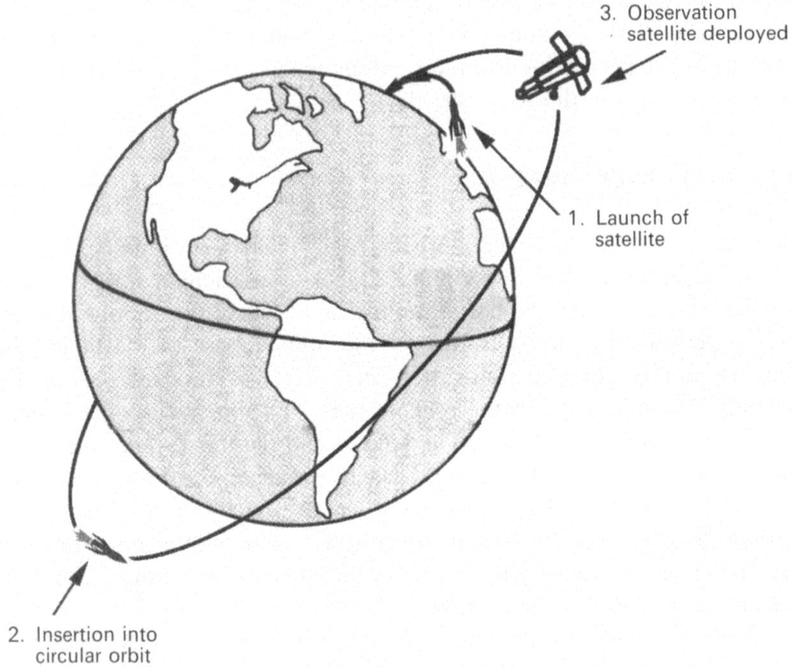

Figure 4.1. The launch of a satellite into a circular orbit.

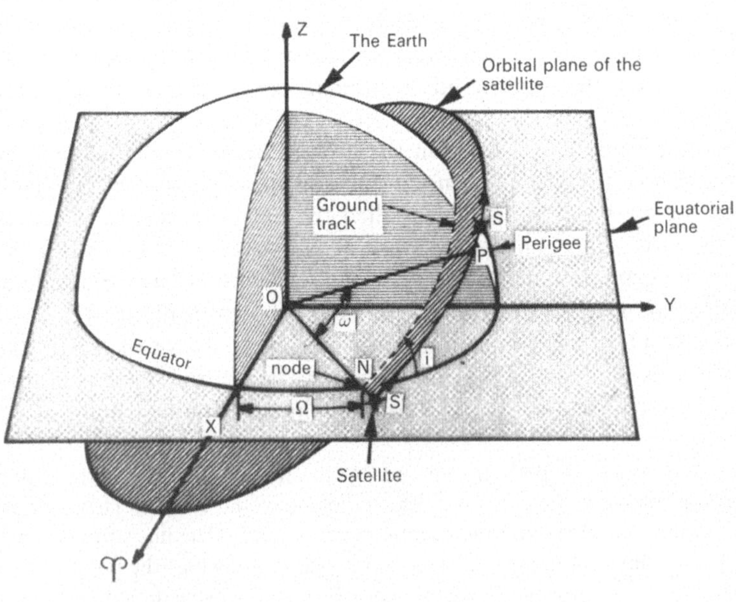

Figure 4.2. Geometry of a satellite orbit.

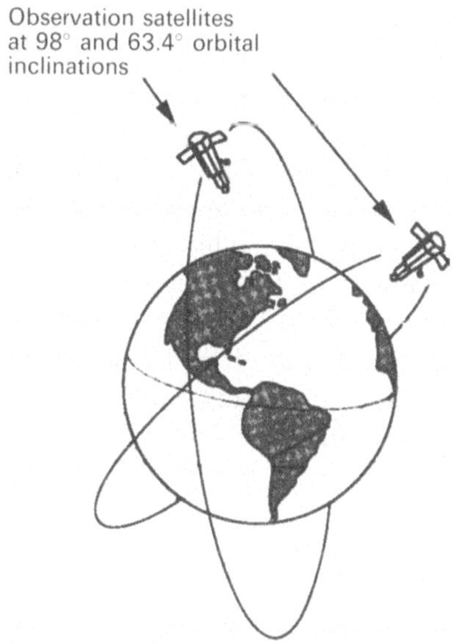

Figure 4.3. Typical orbits for observation satellites.

point of intersection of the satellite ground track with the equator is known as the ascending node (**N**). The angle Ω between the x-axis and line **ON** defines the right Ascension of the ascending node. During the lifetime of a satellite, Ω does not remain fixed; this is because the orbital plane turns round the earth's axis while keeping the orbital inclination[3] constant. The change in Ω is known as precession. This occurs because the earth is not spherical so that its centre of gravity is not at the same point as its centre of mass. Moreover, the mass is not uniformly distributed. For low-altitude orbits with small orbital inclinations, **i**, the rate of change of Ω can be as much as 9° per day. However, for an orbital inclination of 90°, this is zero.

Generally, observation satellites are in circular orbits with inclinations of about 90°. Such an orbit is called a polar orbit (see Figure 4.3). For observation satellites the change in Ω is used so that they pass over a region of interest on the earth at the same time of the day throughout the active lifetime of the satellite. Therefore, observations are always referred to the same local time so that changes in activity in the region can be compared from day to day. This is achieved by placing a satellite in a sun-synchronous orbit in which the satellite is orbited with its orbital inclination of almost 90°. The plane of the satellite orbit contains both the earth and the Sun. In this type of orbit the satellite crosses, for example, the equator, at just about local noon on the sunlit side of the earth and local midnight on the dark side. The earth rotates under the satellite orbit, which is fixed in space, at a rate of 15° per hour, so

[3] This is the angle between the orbital plane of the satellite and the equatorial plane of the earth.

that the equatorial and temperate zones of the earth can be photographed with the sun always high in the sky.

The orientation of the ellipse within the orbital plane of a satellite is given by ω, the second of the two orbital elements considered. It is the angle between the line **ON** (see Figure 4.2) in the equatorial plane and the line **OP**, where **P** is the perigee. The line **OP** always lies along the major axis of the ellipse. The oblateness of the earth also affects ω, which for all except one value of orbital inclination is not constant. This means that the major axis of the orbit rotates in the orbital plane in the direction of motion of the satellite. The rate of change of ω is zero when the orbital inclination is 63.4° (see Figure 4.3). Thus, for a satellite launched at this orbital inclination, its perigee will remain stable along any chosen latitude. For other values of orbital inclination, ω could vary by up to 5° per day.

4.3 SATELLITE CAPABILITIES

The characteristics of an imaging sensor may be described by various technical parameters. Two of the important ones are the spatial (or photographic) and spectral resolutions. We consider the spatial resolution first.

Spatial resolution

The spatial resolving power or resolution of an imaging system is the ability of the sensor to record small details in the scene. The quantification of the smallest detail or size of object that can be seen in an image is, however, a complex issue and depends both upon the spatial resolution of the system and the characteristics of the scene being observed. These include lighting conditions, configuration, size and shape of the objects being observed and atmospheric effects on the light travelling to and from the object being observed. The resolving power of a sensor may be defined as the minimum distance between two identical small objects when they can still be distinguished as two separate objects. While such a definition holds true for a photographic film camera system, for scanning devices a new concept, the instantaneous field of view (IFOV), has to be considered.

The IFOV is defined as the size of the spot on the ground 'seen' by one particular point in the image or, in the case of an electro-optical device, seen by a scanning sensor at the instant of observation. A scanner is made of a series of small sensors, each of which is called a pixel or picture element. This represents the point in the image or a square (see Figure 4.4).

Both photographic products as well as images from electro-optical devices are available commercially. It is, therefore, useful to examine the relationship between the pixel size and the photographic resolution. In Figure 4.5, the response under ideal conditions of an electro-optical device with pixel sizes ranging from 10 m to 1 m to observations of a B-52 bomber aircraft is depicted. It can be seen that a sensor with a pixel resolution of 10 m, such as the SPOT satellite, can detect the B-52 as an aircraft without necessarily identifying it as a B-52 aircraft. However, with a

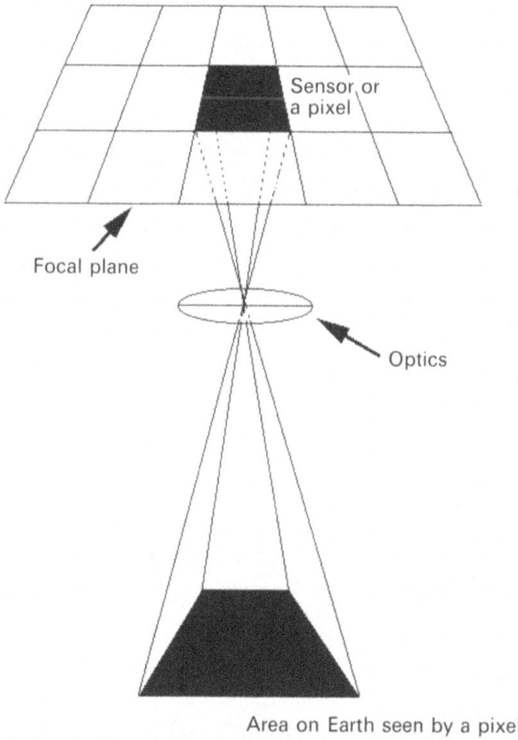

Figure 4.4. The concept of pixel resolution.

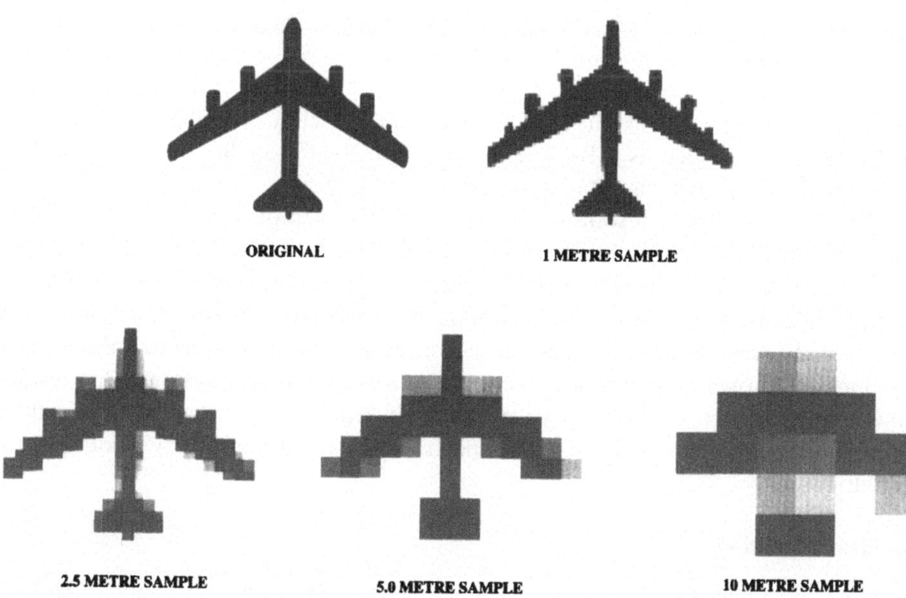

Figure 4.5. The effect of resolution on the satellite image of a B-52 bomber aircraft.

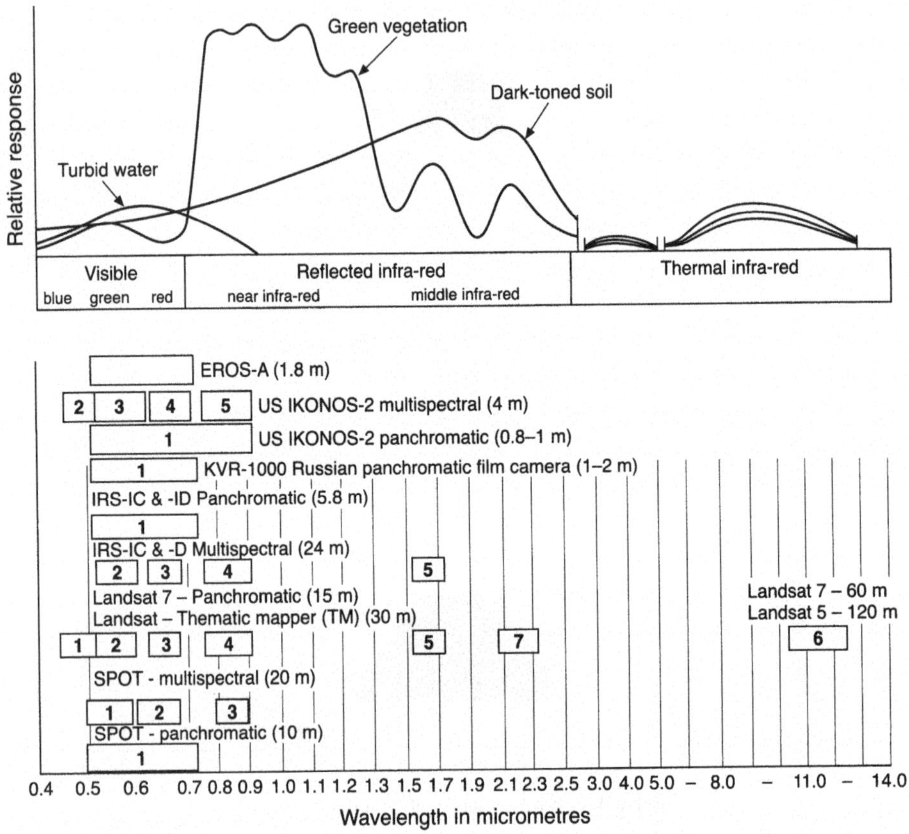

Figure 4.6. Comparison of spatial resolutions and spectral responses of various commercial observation satellites.

resolution of 1 m, such as that of the IKONOS satellite, a B-52 can clearly be identified and described as a B-52. The future IKONOS satellite, with a resolution of 0.5 m, should provide a detailed description of a B-52.

Thus, in this book, resolution is defined as the size of a pixel (Figure 4.4) in metres or the photographic resolution (or ground resolution) as two to 2.4 times the pixel resolution. This is because the photographic resolution depends on the contrast ratio; for a contrast ratio of 2 : 1, it is about two times the pixel resolution. Comparison of spatial resolutions and spectral responses of various commercial observation satellites are shown in Figure 4.6. Here the ratio between the photographic resolution and pixel size of about two is used. It can be seen that most satellites with multispectral sensors have a few spectral bands and hence very poor spectral resolutions.

Spectral resolution

A human eye can see light in wavelengths from 0.38 (deep blue colour) to 0.71 micrometres (μm) (red colour). Wavelengths beyond 0.71 μm extend into the

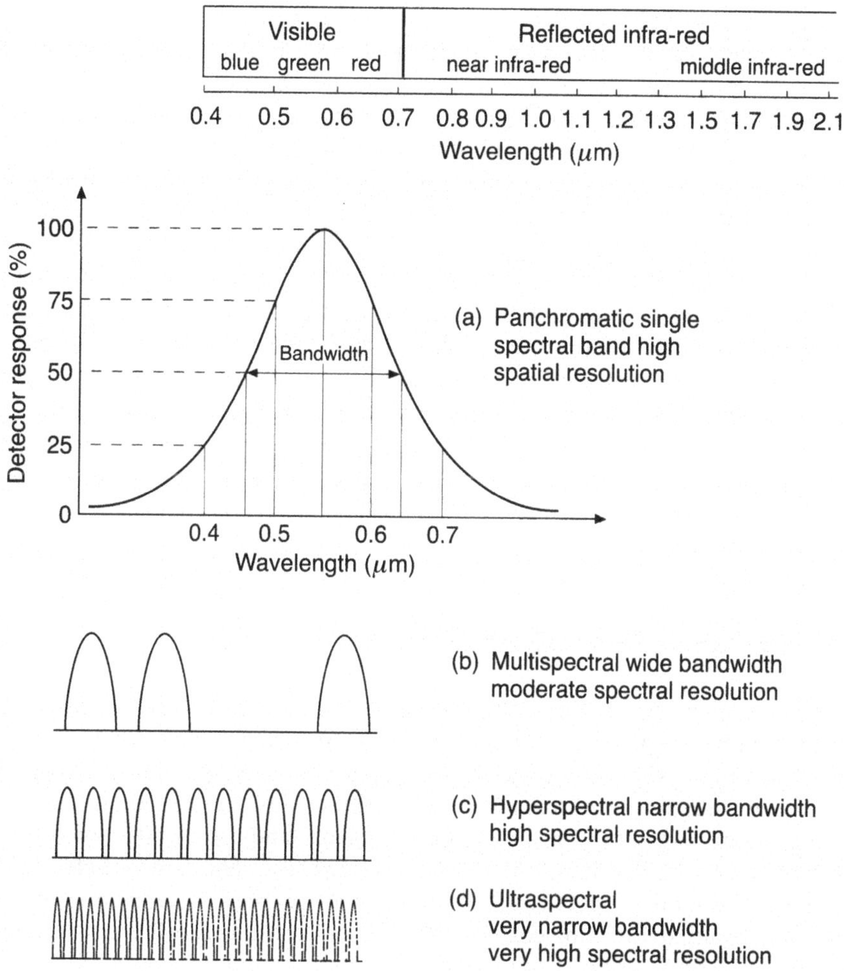

Figure 4.7. This shows different types of spectral data and illustrates the concept of spectral resolution.

infrared (IR) which is invisible to the human eye. Reflected IR extends to 2 μm and beyond this is longer wavelength IR energy sensed as heat. This spectrum extends from 3 μm to about 14 μm (see Figure 4.6). The spectral resolution is the wavelength interval that is detected by a detector. Usually, the response of a detector to light is not sharply defined but as the wavelength increases, the sensor response increases to a maximum and then decreases (Figure 4.7). Spectral resolution or bandwidth is then defined as the wavelength interval recorded at 50% of the peak response of a detector. From Figure 4.7a, for example, it would be 0.10 μm, the width of the curve at 50% of the peak response. Therefore, the narrower the distribution curve or bandwidths the better the resolution of a sensor.

Optical images can be divided into panchromatic, multispectral, hyperspectral or ultraspectral data. Bandwidth and the number of spectral bands distinguish between these. For example, a panchromatic image is one in which data is acquired over a wide range of wavelengths in a single spectral band (see Figure 4.7a). In a multispectral sensor, data are collected simultaneously in three or more but less than 10 spectral regions of the electromagnetic (EM) spectrum with broad bandwidth (see Figure 4.7b). Examples of these are the US Landsat, French SPOT and Indian IRS-1C satellites (see Figure 4.6). A hyperspectral sensor has a narrow bandwidth and several hundred spectral bands (see Figure 4.7c). The first such system was the US LewisSat, orbited on 22 August 1997 with 384 spectral bands in the range of 0.4 to 2.5 micro-metres and a spectral resolution of 5 to 6.25 nano-metres. Unfortunately, by 27 August, contact with the spacecraft had been lost. However, a US company, Orbimage, is planning to launch a satellite with a hyperspectral sensor with a 1 m panchromatic, 4 m multispectral and 4 m hyperspectral sensors with 200 channels having a spectral range of 0.4 to 2.5 μm. Ultra-spectral sensors (see Figure 4.7d) are still at the R&D stage. These are expected to have a very narrow bandwidth and thousands of spectral bands.

4.4 SPACE-BASED REMOTE SENSING

The UN resolution on the 'Principles relating to remote sensing of the Earth from outer space', was adopted on 3 December 1986. According to its article I, remote sensing is 'sensing of the Earth's surface from space, by making use of the properties of electromagnetic waves emitted, reflected or diffracted by the sensed object, for the purpose of improving natural resources management, land use and the protection of the environment.' (United Nations, 1986). Although this definition includes neither surveillance of military activities nor weapons and weapons systems, a number of bilateral and multilateral arms control treaties have legitimised all types of remote sensing of the earth's surface from space.

Astronauts and cosmonauts made the first remote sensing observations of the earth's surface from space in the 1960s using conventional hand-held photographic cameras. Since then, however, major technical advances have been made in sensors for use on board unmanned satellites. Present sophisticated systems include multispectral electro-optical scanners and radar sensors which could transmit data in real time to ground facilities. Radar can give day and night and all-weather capabilities. The USA and the former USSR developed such techniques initially to observe each other's military arsenal and activities. The technology was later adopted to collect information useful for such civil applications as town planning, mapping and for the exploration of natural resources. While the USA, in 1972, made the transition from purely military application to civilian use of space-based remote sensing, in the former USSR no such transition is discernible. Russia sells imageries degraded from its military reconnaissance satellites.

The use of civil remote sensing satellites for monitoring security related activities in general and nuclear activities – such as monitoring nuclear related arms control treaties in particular – is a new application. Although the method is in its infancy, it is possible to extract considerable amounts of useful information.

In the following sections, satellite capabilities in terms of image resolutions, and some of the widely used image processing and enhancement techniques which are now available commercially, are discussed very briefly.

4.5 IMAGE ANALYSIS TECHNIQUES

Normally, information gathered from space is either recorded on an on-board tape recorder to be played back when the satellite is over a ground receiving station, or the data are transmitted via communication satellites to the main ground station. The latter depends on whether a communications spacecraft is capable of handling the data rates resulting from observation satellites. For example, the US Landsat-5 transmits the thematic mapper (TM) data at a rate of 83 MB/s. While the French SPOT satellite has a tape recorder on board, its data transmission rate is 50 MB/s and that of the European ERS-1 is 100 MB/s. The commercial communications satellites are probably not capable of handling such large data rates at present. Therefore, such data are either transmitted via a defence communications satellite or transmitted when the satellite is in line of sight of a ground station.

The data thus collected need to be analysed and interpreted. A wide range of computer-based image enhancement techniques is available for this. In general, image processing can be summarised as image restoration, rectification and registration, enhancement and classification. Some of these techniques are briefly reviewed below.

4.5.1 Contrast enhancement

A digital image consists typically of many pixels whose densities are byte-encoded with dark to light values from 0 to 255. Figure 4.8 shows a typical histogram of pixel levels obtained from a panchromatic image (also shown in the figure) containing one spectral band, ranging in this case from 17 to 79, with the maximum number of pixels having pixel level at 25. These are the characteristics of a SPOT panchromatic image (Figure 4.8) over the UK nuclear reactor facility at Sellafield. The image appears very dark. The spread of the histogram determines the contrast of the image. Increasing its width by 'stretching' the histogram over the complete dynamic range will increase the contrast making features more visible. Figure 4.9 shows the image over Sellafield facility after contrast stretching was applied. This procedure is referred to as linear contrast enhancement.

4.5.2 Edge enhancement

For many earth science applications, it is important to be able to recognise land features. A relatively poor resolution can be compensated somewhat by using mathematical techniques to highlight specific aspects. For detection of nuclear facilities and

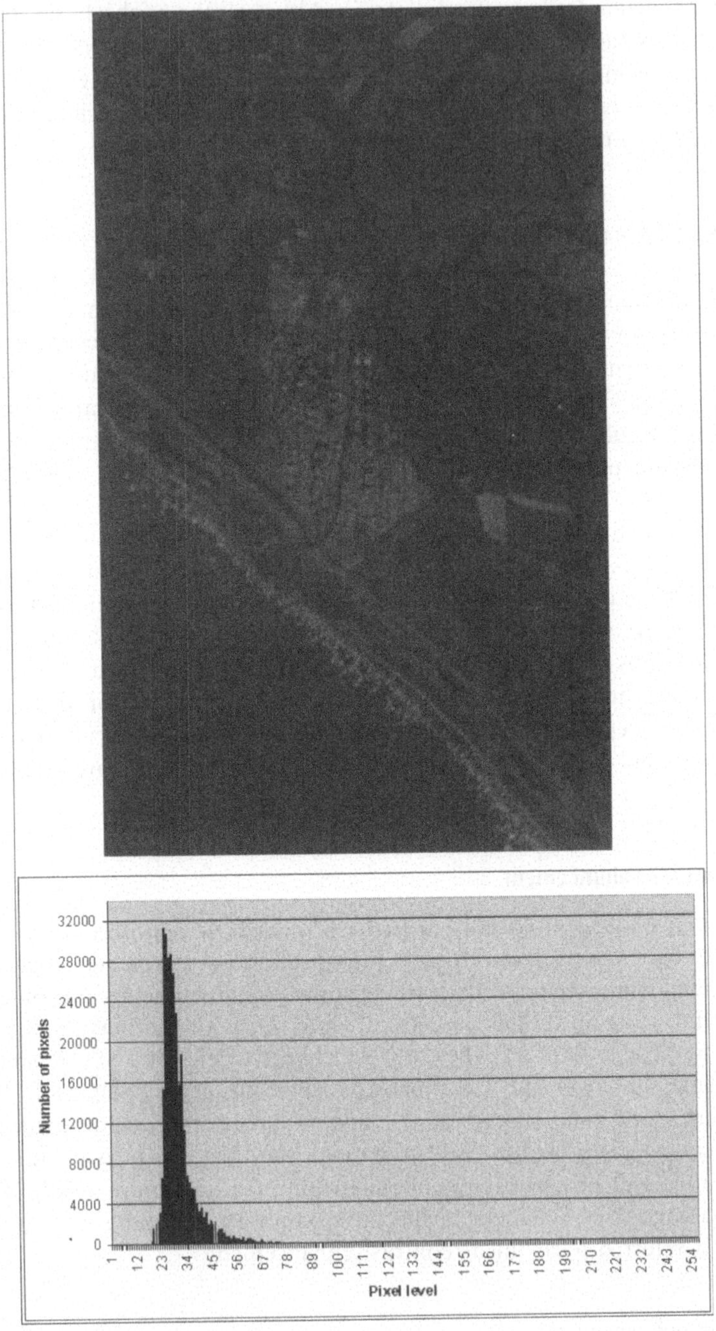

Figure 4.8. A section of a full SPOT scene (10 m resolution) over the former UK plutonium production reactor at Sellafield (top) and the distribution of the number of pixels as a function of the pixel levels (below). The image was acquired on 19 March 1990 [source: CNES/SPOT].

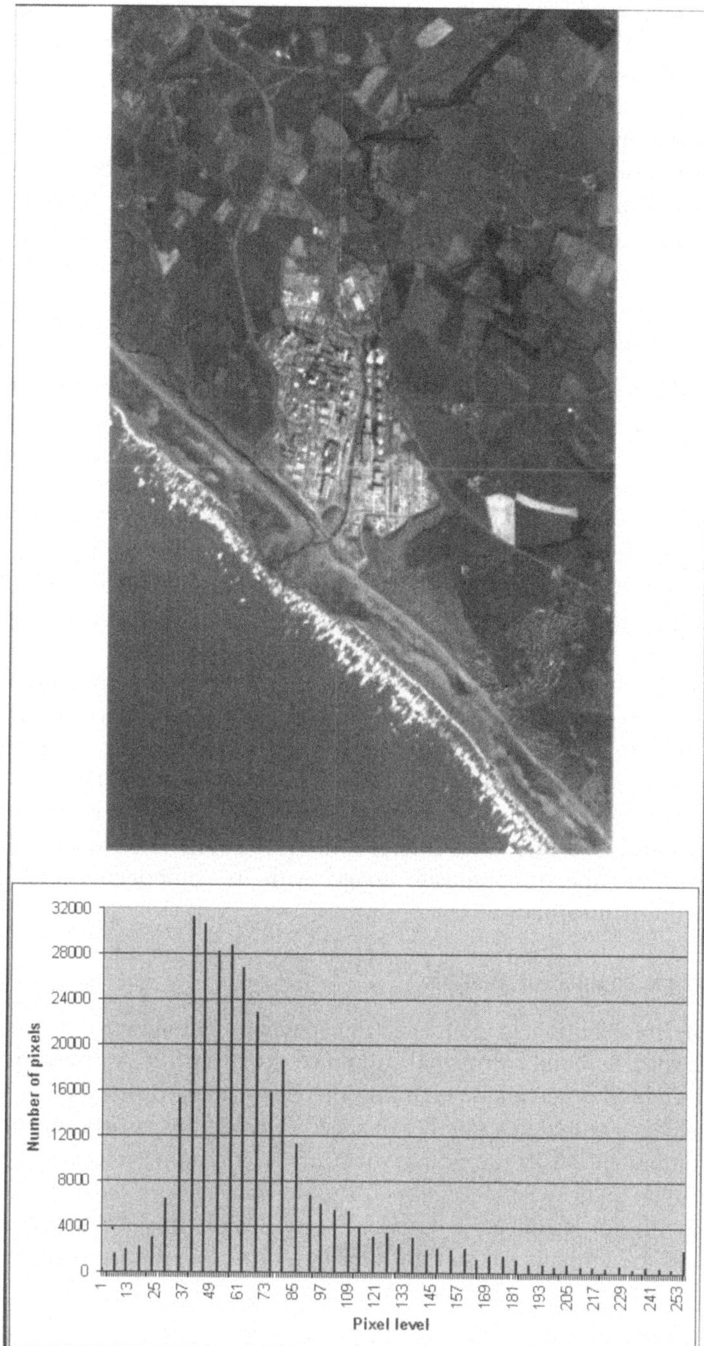

Figure 4.9. A contrast stretch as applied to the image in Figure 4.8.

weapons systems and related infrastructures, geometric shapes of, for example, buildings and the general site layout are obviously of particular relevance. Edge enhancement is one relevant technique, usually implemented by passing a simple template across the image or by filtering its two-dimensional Fourier transform. Abrupt changes in pixel values characteristic of edges and borders are enhanced, making them easier to interpret. Figure 4.10 is an edge-enhanced image over the Sellafield facility in which a sharpening filter was used. Such a filter increases the contrast between adjacent pixels. Buildings and roads are better defined compared with those in Figure 4.9.

4.5.3 Image manipulation

Performing simple arithmetic operations on combinations of spectral components can often generate useful images. Basic operations such as subtraction or multiplication can be used to detect changes in a scene. Computing image ratios in various combinations of infrared and visible spectral bands can be used to distinguish vegetation covers. Advantage is taken of the different reflectance of vegetation at different wavelengths (see Figure 4.11). In the near infrared, healthy vegetation reflects some 40–50% of incident energy while unhealthy or dead vegetation reflects about 30%. This might be used, for example, to differentiate camouflaged targets from background terrain.

Different building materials will also have different reflectance values, and the combination of spectral components can be useful in discriminating subtle variations in a scene not apparent in individual images or in false colour composites. Ratio images can also be used to generate false colour composites. This has a twofold advantage: it enables combination of data from more than two bands into a single colour image and also facilitates interpretation of subtle reflectance differences. It should be realised that a composite image, with its components least correlated, contains the most information.

4.5.5 Principle component analysis

Much of the information in multispectral images is redundant due to correlation between the spectral bands. Principal component analysis (PCA) is a very useful and robust data reduction technique enabling the use of most of the available information in the scene acquired in a number of spectral bands. By associating the first three principal components with the primary colours of a graphics display one can often generate a single image which accounts for more than 95% of the information in the seven bands of, for example, a Landsat scene. With this technique often man-made features are emphasised so that interpretation of an image becomes relatively easier.

4.5.6 Unsupervised cluster analysis

Many techniques and algorithms exist for finding land-cover clusters in satellite images. Unsupervised classification is primarily a computer process of selecting

Figure 4.10. The effect of applying a sharpening filter to the image in Figure 4.9.

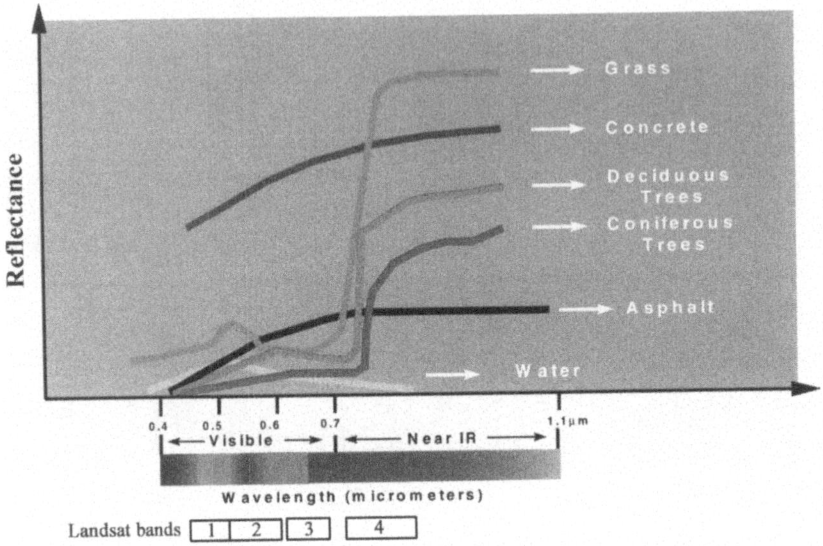

Figure 4.11. Spectral reflectance of different materials [source: *Multispectral Imagery Reference Guide* (1997), LOGICON Geodynamics, Inc., USA].

natural groupings or patterns of pixels with similar characteristics. In this process the analyst has minimum input. To uncover statistical patterns that are inherent in the image data, the computer is programmed to use a specified set of parameters. This method is usually used when less is known about the data. After such classification, the analyst gives meaning to the resulting classes.

4.5.7 Supervised land-cover classification

Supervised classification presumes that the analyst has ground-truth information. Pixels from 'training regions' of the scene, whose land-cover categories are known *a priori*, are used to train a mathematical model such as a neural network to recognise these categories (Bishop, 1995). After the model has been trained, it can be used to classify the entire scene, provided all possible categories have been 'learned'. If a particular installation were to be 'under observation' over a long period of time, the method might be used to document automatically the changes in land use over time.

4.5.8 Pattern recognition

This is a technique of finding known patterns in an image that can be extracted through spectral classification or by geometrical shapes. By spatially and spectrally enhancing an image, pattern recognition can be performed by humans or automatically by a computer system; spectral recognition can be performed automatically. Also if, for example, some common shapes of nuclear facilities are determined, it is possible to train a computer to seek out such patterns in an image. Automatic pattern recognition is by no means an easy task. However, descriptions of weapons systems and deployment facilities are often known. Armed with such information as the dimensions of objects, for example, the use of digital elevation model (DEM) techniques could be a very useful tool, as it would enable creation of three-dimensional pictures of sites and objects to be identified and therefore, enhance interpretability. The site diagrams would also make the automatic pattern recognition task more feasible.

4.5.9 Geographic information system

The geographic information system (GIS) is yet another powerful tool that enhances the interpretability of images acquired from space. Normally, image interpretation is carried out in isolation without actually integrating with a satellite-acquired image, any other external information. An important starting point is to geocode the image by re-sampling each data set to a uniform pixel size that is registered to a geographic co-ordinate system. The GIS is designed for collecting, storing, organising, integrating, manipulating and analysing and displaying a set of geographic data. These could be digital maps or non-spatial information such as tabulated demographic information.

4.6 SUMMARY

Thus, it can be seen that civil remote sensing satellites are becoming capable for the monitoring of arms control treaties. In addition, there are several such satellites in orbit operated by various countries, which not only provides adequate coverage on the earth but also overcomes the problem of data authentication. Moreover, simple contrast stretching and edge enhancement techniques are useful in bringing out details in an image enabling recognition of features of interest. Using principal component analysis a multispectral image could be enhanced further. Supervised and unsupervised cluster analysis, and a detailed examination of multispectral images taken at different times over a particular site can also be a useful technique in, for example, detecting changes in a scene without having recourse to ground-truth information.

As the sophistication of image processing software and techniques increases and as they become available, extraction of more information from civil remote sensing data becomes possible. In this context neural network and pattern recognition techniques are important methods to be investigated. The use of GIS will no doubt increase the interpretability of commercial satellite imagery. Several states are now entering into the satellite imagery market. Thus, the availability of data from different sources could remove some of the political sensitivity attached to satellite monitoring. There is no doubt that when satellites with hyperspectral sensors on board become operational, this technology will be applied for extensive monitoring of the earth's surface for all kinds of applications.

If images from satellites are geocoded using accurate maps, the precision with which each target on the ground can be located is improved – for example, to only a few tens of metres. Thus, even a civil remote sensing satellite with relatively coarse resolution could be used rather effectively for acquiring targeting information. All of these developments raise a number of issues. For example, states are particularly sensitive about the availability for sale of data on their economic resources and on activities related to their national security. This has provoked much discussion on whether and what kind of technical parameters exist to permit classification of 'sensitive' and 'non-sensitive' data. One parameter, the spatial resolution, has been discussed at length but no clear decision seems to have emerged. Yet, the resolution of sensors on board the civil remote sensing satellites is continuously improving.

It has been argued that a civil remote sensing satellite could detect a potential invasion of a nation before it takes place, thus affecting the would-be aggressor's foreign policy through military actions (US Congress, 1987). Furthermore, the military potential of nations could become transparent to the world at large. While such concerns are understandable, it is difficult to see how the dissemination of high-resolution images could be prevented now.

4.7 REFERENCES

Bishop, C. M. (1995) *Neural Networks for Pattern Recognition*, Oxford University Press, Oxford.

Jasani, B., 'Commercial observation satellites and verification', in Commercial Observation Satellites and International Security, Krepon, M., *et al.* (ed.), (MacMillan, Basingstoke, 1990), pp. 142–150.

United Nations (1986) Document 41/65, 3 December.

US Congress, Office of Technology Assessment (1987) *Commercial Newsgathering from Space: A Technical Memorandum*, OTA-TM-ISC-40, US Government Printing Office, Washington, DC, May.

5

Analysis of digital satellite imagery

M. J. Canty and I. Niemeyer

It is of course not possible to do justice to the topic of digital image analysis in one short chapter. We restrict ourselves here to a brief description of some basic operations especially associated with statistical change detection. These include the digital representation of multispectral satellite imagery, simple image enhancement methods, Fourier transforms and filtering, image fusion techniques for panchromatic sharpening, and multivariate alteration detection.

5.1 IMAGE REPRESENTATION

A satellite image can be represented as an array of pixel intensities or gray-scale values

$$g_k(i,j), \qquad 0 \le i \le r-1, 0 \le j \le c-1 \tag{1}$$

where r is the number of pixel rows and c is the number of pixel columns. If we are dealing with an N-channel multispectral image, such as Landsat-TM or IKONOS-2, then the index k, $1 \le k \le N$, denotes the channel. Usually a pixel intensity is stored in a single byte, so that $0 \le g_k \le 255$, often referred to simply as *digital number*. It is more economical to use a vector representation for multispectral imagery, i.e.

$$\mathbf{g}(i,j) = (g_1(i,j), g_2(i,j) \dots g_N(i,j))^{\mathrm{T}}$$

Here T means transpose and $\mathbf{g}$ is a column vector of multispectral pixel intensities. The *average intensity* of the image is given by

$$\bar{\mathbf{g}} = \frac{1}{rc} \sum_{i,j=0}^{r-1,c-1} \mathbf{g}(i,j)$$

and the *dispersion matrix* by

$$\mathbf{S} = \frac{1}{rc}\sum_{i,j=0}^{r-1,c-1}(\mathbf{g}-\bar{\mathbf{g}})(\mathbf{g}-\bar{\mathbf{g}})^{\mathrm{T}}$$

It is often useful to think of an image approximately as a realisation of a multivariate normally distributed random vector **G**. We say that $\bar{\mathbf{g}}$ and S are estimates of the *expected value* and *covariance matrix* of **G**,

$$\boldsymbol{\mu} = \langle \mathbf{G} \rangle \qquad \text{and} \qquad \Sigma = \langle (\mathbf{G}-\boldsymbol{\mu})(\mathbf{G}-\boldsymbol{\mu})^{\mathrm{T}} \rangle$$

respectively, where $\langle\cdot\rangle$ is an ensemble average.

Various conventions can be used for storing the image array $g(i,j)$ in computer memory or on storage media. In *pixel-interleaved* format, for example, a two channel, 3×3 pixel image would be stored as

$$\begin{array}{cccccc} g_1(0,0) & g_2(0,0) & g_1(0,1) & g_2(0,1) & g_1(0,2) & g_2(0,2) \\ g_1(1,0) & g_2(1,0) & g_1(1,1) & g_2(1,1) & g_1(1,2) & g_2(1,2) \\ g_1(2,0) & g_2(2,0) & g_1(2,1) & g_2(2,1) & g_1(2,2) & g_2(2,2) \end{array}$$

whereas in *channel* or *band interleaved* format it would be stored as

$$\begin{array}{ccc} g_1(0,0) & g_1(0,1) & g_1(0,2) \\ g_2(0,0) & g_2(0,1) & g_2(0,2) \\ g_1(1,0) & g_1(1,1) & g_1(1,2) \\ g_2(1,0) & g_2(1,1) & g_2(1,2) \\ g_1(2,0) & g_1(2,1) & g_1(2,2) \\ g_2(2,0) & g_2(2,1) & g_2(2,2) \end{array}$$

Auxiliary information, such as image acquisition parameters and georeferencing, is normally included with the image data on the same file, and the format may or may not make use of compression algorithms. Examples are the geoTIFF[1] file format used by Space Imaging Inc. for distributing Carterra(c) imagery and which includes lossless compression and the cross-platform PCDSK format employed by PCI Geomatics with its image processing software, which is in plain ASCII code and not compressed.

In computer graphics memory, any three channels of a multispectral image can be stored (e.g. in 24 bits, 8 bits for each of the primary colours red green and blue). The monitor displays this as a so-called *color composite* image which, depending on the choice of image channels and their relative intensities, may or may not appear to be natural.

Intensity enhancement of an image is best accomplished by means of *lookup tables*. The pixel intensities g_k are used to index an array $LUT[i]$, $i = 0 \ldots 255$, the

1 geoTIFF refers to TIFF files which have geographic (or cartographic) data embedded as tags within the TIFF file. The geographic data can then be used to position the image in the correct location and geometry on the screen of a geographic information display.

entries of which also lie between 0 and 255. These entries can be chosen to implement linear stretching, saturation, histogram equalisation, etc. according to

$$\hat{g}_k(i,j) = LUT[g_k(i,j)], \qquad 0 \leq i \leq r-1, 0 \leq j \leq c-1$$

5.2 PRINCIPAL COMPONENTS TRANSFORMATION

The principal components transformation forms linear combinations of multispectral pixel intensities which are mutually uncorrelated and which have maximum variance. We can assume without loss of generality that $\langle \mathbf{G} \rangle = 0$, so that the covariance matrix is $\Sigma = \langle \mathbf{G}\mathbf{G}^{\mathrm{T}} \rangle$, and look for a linear combination $Y = \mathbf{a}^{\mathrm{T}}\mathbf{G}$ with maximum variance, subject to the normalisation condition $\mathbf{a}^{\mathrm{T}}\mathbf{a} = 1$. Since the covariance of Y is a $\mathbf{a}^{\mathrm{T}}\Sigma\mathbf{a}$, this is equivalent to maximising an unconstrained Lagrange function

$$L = \mathbf{a}^{\mathrm{T}}\Sigma\mathbf{a} - 2\lambda(\mathbf{a}^{\mathrm{T}}\mathbf{a} - 1)$$

Differentiating,

$$\frac{\partial L}{\partial \mathbf{a}} = 2\Sigma\mathbf{a} - 2\lambda\mathbf{a} = 0$$

which is the eigenvalue problem

$$\Sigma\mathbf{a} = \lambda\mathbf{a}$$

Since $\mathbf{S}$ is real and symmetric, the eigenvectors are orthonormal. Denote them $\mathbf{a}_1 \ldots \mathbf{a}_N$ for eigenvalues $\lambda_1 \geq \ldots \geq \lambda_N$. Define the matrix

$$\mathbf{A} = (\mathbf{a}_1 \ldots \mathbf{a}_N), \qquad \mathbf{A}\mathbf{A}^{\mathrm{T}} = \mathbf{I}$$

and the transformed *principal component* vector $\mathbf{Y} = \mathbf{A}^{\mathrm{T}}\mathbf{G}$ with covariance matrix Σ'. Then we have

$$\Sigma' = \langle \mathbf{Y}\mathbf{Y}^{\mathrm{T}} \rangle = \langle \mathbf{A}^{\mathrm{T}}\mathbf{G}\mathbf{G}^{\mathrm{T}}\mathbf{A} \rangle = \mathbf{A}^{\mathrm{T}}\Sigma\mathbf{A} = \begin{pmatrix} \lambda_1 & 0 & \cdots & 0 \\ 0 & \lambda_2 & \cdots & 0 \\ \vdots & \vdots & \ddots & \vdots \\ 0 & 0 & \cdots & \lambda_N \end{pmatrix}$$

The fraction of the total variance in the original multispectral image which is described by the first i principal components is

$$\frac{\lambda_1 + \cdots + \lambda_i}{\lambda_1 + \cdots + \lambda_i + \cdots + \lambda_N}$$

If the original multispectral channels are highly corelated, as is usually the case, the first few principal components will account for >90% of the variance the image. For example, a color composite of the first three principal components of a Landsat TM scene displays essentially all of the information contained in the six spectral components in one single image. Nevertheless, because of the assumption of a normal distribution, higher order principal components may also contain significant information (Richards, 1995).

5.3 FOURIER ANALYSIS AND FILTERS

Let $g(j)$ be a discrete sample of the real function $g(x)$, sampled c times at the sampling interval Δ, i.e.

$$g(j) = g(x = j\Delta), \qquad j = 0 \ldots c-1$$

We might think of g as the radiance at the satellite sensor and Δ as the spatial resolution. Then, ignoring sensor calibration, $g(j)$ is a row of pixel intensities or digital numbers. It can be represented approximately by a *discrete Fourier series*

$$g(j) = \sum_{k=-c/2}^{c/2-1} \hat{g}(k) e^{i2\pi kj/c}, \qquad j = 0 \ldots c-1$$

a set of c equations in the c unknown complex frequency components $\hat{g}(k)$. The solution for the frequency components is called the *discrete Fourier transform* of g and is given by

$$\hat{g}(k) = \frac{1}{c} \sum_{j=0}^{c-1} g(j) e^{-i2\pi kj/c}, \qquad k = -c/2 \ldots c/2-1 \tag{2}$$

Determining the frequency components in (2) would appear to involve, in all, c^2 floating point multiplications. The *fast Fourier transform* (FFT) exploits the structure of the complex e-functions to reduce this to order $c \log c$ (see for example, Press, *et al.*, 1986).

Fourier analysis is easily generalised to two dimensions for the purpose of image analysis. Let $g(i,j)$, $i, j = 0 \ldots c-1$, represent a (quadratic) gray scale image. Its discrete Fourier transform is

$$\hat{g}(k,l) = \frac{1}{c^2} \sum_{i=0}^{c-1} \sum_{j=0}^{c-1} g(i,j) e^{-i2\pi(ik+jl)/c} \tag{3}$$

The most common application of Fourier analysis in digital image processing is for spatial frequency filtering. Let $t(i,j)$ represent a filter or template for modifying an image $g(i,j)$. A simple example is the average over a 3×3 window:

$$t(i,j) = \begin{cases} 1/9 & \text{for } i,j = -1, 0, 1 \\ 0 & \text{otherwise} \end{cases}$$

which is a low-pass filter. Operating with this template to give a new image $f(i,j)$ can be represented as the *convolution* $g * t$

$$f(i,j) = \sum_{k,l} g(k,l) t(i-k, j-l), \qquad 0 \le i \le r-1,\ 0 \le j \le c-1$$

According to the *convolution theorem*, this calculation can also be accomplished by taking the Fourier transforms $\hat{g}$ and $\hat{t}$ of image and filter, respectively, performing a simple multiplication of the transforms

$$\hat{f}(k,l) = \hat{g}(k,l)\hat{t}(k,l), \qquad -r/2 \leq k \leq r/2 - 1, \ -c/2 \leq l \leq c/2 - 1$$

and then taking the inverse transform of $\hat{f}$. Depending on the size of the template, this may be more efficient than performing the convolution directly.

A typical example of a high-pass filter used for edge detection is the *Laplacian* template

$$\begin{matrix} 0 & 1 & 0 \\ 1 & -4 & 1 \\ 0 & 1 & 0 \end{matrix} \qquad \text{which approximates} \qquad \frac{\partial^2 g}{\partial x^2} + \frac{\partial^2 g}{\partial y^2}$$

Zero values in the filtered image signify to edges or contours. Because this filter approximates a second derivative measurement on the image, it is sensitive to noise. To counter this, the image is often Gaussian smoothed beforehand in order to reduce the high frequency noise prior to the differentiation step. In fact both operations, Gaussian smoothing and differentiation, can be combined into a single step, the so-called *Laplacian of Gaussian* or LoG filter.

5.4 IMAGE FUSION

The statistical change detection algorithm which we shall discuss shortly exploits the multispectral information of satellite imagery. The IKONOS-2 platform, for example, offers panchromatic images with 1 m resolution, but the multispectral channels have only 4 m spatial resolution. Similarly, the Landsat 7 satellite provides panchromatic images with a ground resolution of 15 m, whereas the multispectral and thermal channels are at 30 m and 60 m, respectively. Multispectral change detection would therefore seem to be restricted to the lower resolutions. However, one can resort to fusion techniques to include the higher resolution panchromatic information in the analysis, as we shall now discuss.

5.4.1 IHS transformation

The *HIS transformation* is the most well-known technique for fusion of high-resolution with lower resolution images. In the case of IKONOS imagery, for example, we can choose any combination of three of the four available multispectral channels, re-sample them to 1 m resolution and represent each pixel as a point in Cartesian RGB (red-green-blue) space. We then perform a transformation to IHS (intensity-hue-saturation) space, replacing the intensity component of each pixel with the intensity of the corresponding pixel in the panchromatic image. Then we transform back to RGB coordinates to display the image on a graphics screen. The disadvantage of this method, if one wishes to use the spectral information in the RGB components of the fused image, is that the components are 'contaminated' with the panchromatic image. Another disadvantage is that we are restricted to using three of the available multispectral channels.

5.4.2 Brovey transformation

In the Bovey transformation (Prinz, *et al.*, 1997), we multiply each re-sampled multispectral pixel by the ratio of the corresponding panchromatic pixel intensity to the sum of all the multispectral intensities. The pixel intensities $\hat{g}_k$ in the kth fused multispectral channel are thus given by

$$\hat{g} = g_k \frac{g_p}{\Sigma g_{k'}}$$

where g_k is the (re-sampled) pixel intensity in the kth channel and g_p is the corresponding pixel intensity in the panchromatic image. This technique assumes that the spectral range spanned by the panchromatic image is essentially the same as that covered by the multispectral channels. This is seldom the case. Moreover, to avoid bias because of differences in sensor calibration, the intensities used should be the radiances at the satellite sensors, not the digital numbers supplied by the provider.

5.4.3 Wavelet transformation

A very elegant method for fusion of multispectral with higher resolution panchromatic images involves the *wavelet transformation* (Ranchin and Wald, 2000). The wavelet transform of a two-dimensional image can be shown (Mallat, 1989) to be equivalent to an iterative application of the high-low-pass filter bank illustrated in Figure 5.1. The high-frequency information is stored in the arrays C^H, C^V and C^D, while the original image with its resolution degraded by a factor two is the array $\hat{g}_p$. Applying the filter bank iteratively yields an image compressed by a factor of 4, and so on.

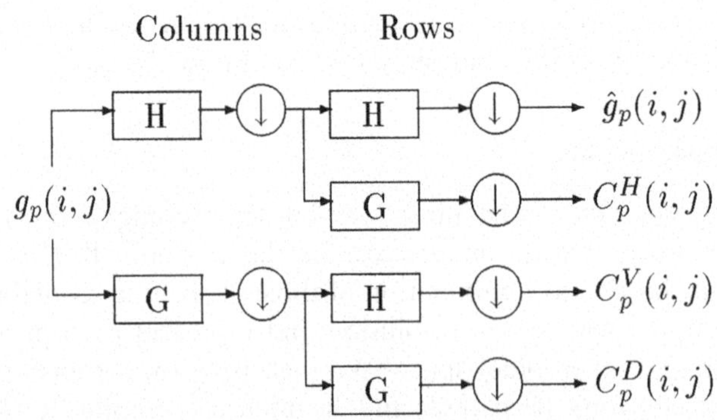

Figure 5.1. A wavelet filter bank applied to a panchromatic image g_p. H is a low-pass and G a high-pass filter derived from the coefficients of the Daubechies D_4 wavelet (Ranchin and Wald, 2000). The symbol ↓ signifies downsampling. The image can be exactly reconstructed by inverting the filter.

As an example, the fusion procedure for IKONOS imagery is as follows: both the twofold compressed panchromatic image and the four multispectral images are compressed once again (to 8 m spatial resolution) and the high-frequency components C^2 are sampled to obtain the correction coefficients

$$a_k^z = \sigma_k^z / \sigma_p^z, \qquad b_k^z = m_k^z - a^z m_p^z, \qquad z = H, V, D, k = 1 \dots 4$$

where m and σ denote mean and standard deviation, respectively. These coefficients are then used to normalise the wavelet coefficients for the panchromatic image to those of the multispectral image:

$$[\hat{C}_p^z]_k = a_k^z C_p^z + b_k^z, \qquad z = H, V, D, k = 1 \dots 4$$

The 4 m degraded panchromatic image is then replaced by the each of the four 4 m multispectral images and the normalised wavelet coefficients are used to reconstruct the original 1 m resolution. We thus obtain what would be seen if the multispectral sensors had the resolution of the panchromatic sensor. The spectral content of the fused multispectral channels is very similar to that of the original channels (see Ranchin and Wald).

5.5 CHANGE DETECTION

The principal steps involved in extracting change detection information are image-to-image registration of the bitemporal scene to sub-pixel accuracy, extraction of change signals and determination of significance thresholds.

5.5.1 Image registration

The most tedious of these tasks is the registration of the images involved, in particular the setting of ground control points (GCPs). It is essential that registration be carried through at sub-pixel accuracy to avoid as much as possible false change signals. In general, it is necessary to resort to be manual entry of GCPs; however, various techniques for their automatic determination have been suggested in the literature. For example (Li *et al.*, 1995) determine significant contours obtained from the Laplacian-of-Gaussian filter (see 5.5.3), which are then matched to a second, later image to determine GCPs for registration. A typical RMS error the nearest neighbor polynomial re-sampling based on GCPs determined in this way is about ±0.5 pixel. We have found the technique to function very reliably in bitemporal scenes for which seasonal vegetation changes do not dominate, most notably nuclear test sites, which are often located in remote, arid regions.

5.5.2 Multivariate alteration detection

In order to see changes in the two multispectral images represented by N-dimensional random vectors **F** and **G**, a simple procedure is to subtract them from each other component-by-component, examining the N differenced images characterised by

$$\mathbf{F} - \mathbf{G} = (F_1 - G_1, F_2 - G_2 \ldots F_N - G_N)^{\mathrm{T}}$$

for significant changes. Pixel intensities near zero indicate no change, large positive or negative values indicate change, and decision thresholds can be set to define significant changes. There is a better way to proceed, however. Suppose we make a linear combination of the intensities for all N channels in the first image, that is we create a *single* image whose pixel intensities are $U = \mathbf{a}^{\mathrm{T}}\mathbf{F}$, where the vector of coefficients $\mathbf{a}$ is as yet unspecified. We do the same for t_2, i.e. we make the linear combination $V = \mathbf{b}^{\mathrm{T}}\mathbf{G}$, and then look at the single difference image $U - V$. The advantage of this procedure is that it combines all the information into a single image, and one is free to choose the coefficients $\mathbf{a}$ and $\mathbf{b}$ in any suitable way. Nielsen, *et al.* (1998) suggest determining the coefficients so that the positive correlation between U and V is minimised. This means that the resulting difference image $U - V$ will show maximum spread in its pixel intensities. If we assume that the spread is primarily due to actual changes that have taken place in the scene over the interval $t_2 - t_1$, then this procedure will enhance those changes as much as possible.

Specifically we seek linear combinations such that

$$\mathrm{var}(U - V) = \mathrm{var}(U) + \mathrm{var}(V) - 2\mathrm{cov}(U, V) \rightarrow \text{maximum}$$

subject to the constraints

$$\mathrm{var}(U) = \mathrm{var}(V) = 1 \tag{4}$$

Note that under those constraints

$$\mathrm{var}(U - V) = 2(1 - \mathrm{corr}(UmV))$$

Since we are dealing with change detection, we require that the random variables U and V be positively correlated, that is $\mathrm{cov}(U, V) = \mathrm{cov}(a^{\mathrm{T}}F, b^{\mathrm{T}}G) > 0$. We now seek vectors $\mathbf{a}$ and $\mathbf{b}$ which minimise the positive covariance $\mathrm{cov}(U, V)$ under the constraints (2).

For multivariate normally distributed data the combined random vector is distributed as

$$\begin{pmatrix} \mathbf{F} \\ \hline \mathbf{G} \end{pmatrix} \sim N\left[\begin{pmatrix} 0 \\ 0 \end{pmatrix}, \begin{pmatrix} \Sigma_{ff} & \Sigma_{fg} \\ \Sigma_{gf} & \Sigma_{gg} \end{pmatrix}\right]$$

With the definition of variance we have

$$\mathrm{var}(U) = \mathbf{a}^{\mathrm{T}}\Sigma_{ff}\mathbf{a}, \qquad \mathrm{var}(V) = \mathbf{b}^{\mathrm{T}}\Sigma_{gg}\mathbf{b}$$

Introducing the Lagrange multipliers $\nu/2$ and $\mu/2$, extremalising the covariance $\mathrm{cov}(U, V)$ under the constraints (2) is equivalent to extremalising the unconstrained Lagrange function

$$L = \mathbf{a}^{\mathrm{T}}\Sigma_{fg}\mathbf{b} - \frac{\nu}{2}(\mathbf{a}^{\mathrm{T}}\Sigma_{ff}\mathbf{a} - 1) - \frac{\mu}{2}(\mathbf{b}^{\mathrm{T}}\Sigma_{gg}\mathbf{b} - 1)$$

Differentiating, we obtain

$$\frac{\partial L}{\partial \mathbf{a}} = \Sigma_{fg}\mathbf{b} - \frac{\nu}{2}(\mathbf{a}^{\mathrm{T}}\Sigma_{ff}\mathbf{a} - 1) - \frac{\mu}{2}(\mathbf{b}^{\mathrm{T}}\Sigma_{gg}\mathbf{b} = 0$$

or

$$\mathbf{a} = \frac{1}{\nu}\Sigma_{ff}^{-1}\Sigma_{fg}\mathbf{b}, \qquad \mathbf{b} = \frac{1}{\mu}\Sigma_{gg}^{-1}\Sigma_{fg}\mathbf{a}$$

The correlation between the random variable U and V is

$$\rho = \frac{\mathrm{cov}(U, V)}{\sqrt{\mathrm{var}(U)\mathrm{var}(V)}} = \frac{\mathbf{a}^{\mathrm{T}}\Sigma_{fg}\mathbf{b}}{\sqrt{\mathbf{a}^{\mathrm{T}}\Sigma_{ff}\mathbf{a}\,\mathbf{b}^{\mathrm{T}}\Sigma_{gg}\mathbf{b}}}$$

Substituting for **a** and **b** in this equation gives (with $\Sigma_{fg}^{\mathrm{T}} = \Sigma_{gf}$)

$$\rho^2 = \frac{\mathbf{a}^{\mathrm{T}}\Sigma_{fg}\Sigma_{gg}^{-1}\Sigma_{gf}\mathbf{a}}{\mathbf{a}^{\mathrm{T}}\Sigma_{ff}\mathbf{a}}, \qquad \rho^2 = \frac{\mathbf{b}^{\mathrm{T}}\Sigma_{gf}\Sigma_{ff}^{-1}\Sigma_{fg}\mathbf{b}}{\mathbf{b}^{\mathrm{T}}\Sigma_{gg}\mathbf{b}}$$

which are equivalent to the two *generalized eigenvalue problems*

$$\Sigma_{fg}\Sigma_{gg}^{-1}\Sigma_{gf}\mathbf{a} = \rho^2\Sigma_{ff}\mathbf{a}$$

$$\Sigma_{gf}\Sigma_{ff}^{-1}\Sigma_{fg}\mathbf{b} = \rho^2\Sigma_{gg}\mathbf{b} \tag{5}$$

Thus the desired projections $U = \mathbf{a}^{\mathrm{T}}\mathbf{F}$ are given by the eigenvectors $\mathbf{a}^1 \ldots \mathbf{a}^N$ corresponding to the generalised eigenvalues $\rho^2 \sim \lambda^1 \geq \ldots \geq \lambda^N$ of $\Sigma_{fg}\Sigma_{gg}^{-1}\Sigma_{gf}$ with respect to Σ_{ff}. Similarly the desired projections $V = \mathbf{b}^{\mathrm{T}}\mathbf{G}$ are given by the eigenvectors $\mathbf{b}^1 \ldots \mathbf{b}^N$ of $\Sigma_{gf}\Sigma_{ff}^{-1}\Sigma_{fg}$ with respect to Σ_{gg} corresponding to the *same* eigenvalues. Under these transformations, the variance of $U - V$ is given by

$$\mathrm{var}(U - V) = 2(1 - \rho) \sim 2(1 - \sqrt{\lambda^p}), \qquad p = 1 \ldots N$$

The transformation corresponding to the smallest eigenvalue, namely $(\mathbf{a}^N, \mathbf{b}^N)$ will thus give maximal variance for the difference $U - V$. Nielsen *et al.* refer to the N different components

$$U^p - V^p = \mathbf{a}^{p\mathrm{T}}\mathbf{F} - \mathbf{b}^{p\mathrm{T}}\mathbf{G}, \qquad p = 1 \ldots N \tag{6}$$

as the *multivariate alteration detection* (MAD) components of the combined bitemporal image.

The last multivariate alteration detection (MAD) component has maximum spread in its pixel intensities and, ideally, maximum change information. However, depending on the type of change one is looking for, the other components may also be extremely useful. The second-to-last image has maximum spread subject to the condition that the pixel intensities are statistically uncorrelated with those in the first image, and so on. Since interesting anthropomorphic changes will generally be uncorrelated with dominating seasonal vegetation changes or stochastic image noise, it is quite common that such changes will be concentrated in higher order components. This in fact is one of the nicest aspects of the method – it sorts different categories of change into different image components.

An additional advantage of the MAD procedure stems from the fact that the calculations involved are invariant under linear transformations of the original image intensities. This implies that the method is insensitive to differences in atmospheric conditions or sensor calibrations at the two acquisition times.

5.5.3 Significance thresholds

Having determined the MAD difference components, it is necessary to distinguish between the change and no-change pixels. Consider the following *mixture model* for a random variable X representing one of the MAD components:

$$p(x) = p(x|NC)p(NC) + p(x|C-)p(C-) + p(x|C+)p(C+) \tag{7}$$

where $C+$, $C-$ and NC denote positive change, negative change and no change, respectively. The set of pixels with component value x_i, $S = \{x_i\}$, may be grouped into four disjoint sets:

$$S_{NC}, S_{C-}, S_{C+} S_U = S \setminus S_{NC} \cup S_{C-} \cup S_{C+}$$

with S_U denoting the set of ambiguous pixels.[2] From the sample mean and sample variance, we estimate initially the moments for the distribution of no-change pixels:

$$\mu_{NC} = \frac{1}{|S_{NC}|} \cdot \sum_{i \in S_{NC}} x_i$$

$$(\sigma_{NC})^2 = \frac{1}{S_{NC}} \cdot \sum_{i \in S_{NC}} (x_i - \mu_{NC})^2$$

($|S|$ denotes set cardinality) and similarly for $C-$ and $C+$. Bruzzone and Prieto (1999) suggest improving these estimates by using the pixels in S_U and applying the so-called *EM algorithm* (see Bishop for a good explanation:

$$\mu'_{NC} = \sum_{i \in S} p(NC|x_i)x_i / \sum_{i \in S} p(NC|x_i)$$

$$(\sigma'_{NC})^2 = \sum_{i \in S} p(NC|x_i)(x_i - \mu'_{NC})^2 / \sum_{i \in S} p(NC|x_i)$$

$$p'(NC) = \frac{1}{|S|} \cdot \sum_{i \in S} p(NC|x_i) \tag{8}$$

2 The symbols $\cup$ and $\setminus$ denote set union and set difference, respectively. These sets can be determined in practice by setting conservative, scene-independent thresholds for change and no-change pixel intensities, see Bruzzone and Prieto.

where $P(NC|x_i)$ is the *a posteriori* probability for a no-change pixel conditional on measurement x_i. We can use the following rules for determining the *a posteriori* probabilities:

1. $i \in S_{NC}$: $p(NC|x_i) = 1$
2. $i \in S_{C\pm}$: $p(NC|x_i) = 0$
3. $i \in S_U$: $p(NC|x_i) = p(x_i|NC)p(NC)/p(x_i)$ (Bayes' rule)

where

$$p(x_i|NC) = \frac{1}{2\pi\sigma_{NC}^2} \cdot \exp\left(\frac{-(x_i - \mu_{NC})^2}{2\sigma_{NC}^2}\right)$$

and

$$p(x_i) = p(x_i|NC)p(NC) + p(x_i|C-)p(C-) + p(x_i|C+)p(C+)$$

Similar equations apply for the positive and negative change distributions. We thus obtain a set of equations which can be iterated numerically to improve the initial estimates of the distributions. Finally, we determine the change thresholds as the appropriate solutions of

$$p(x|NC)p(NC) = p(x|C\pm)p(C\pm)$$

5.6 SOFTWARE FOR CHANGE DETECTION

CDSat (Change **D**etection with **Sat**ellite imagery) is a graphical environment for examining changes in multispectral satellite images. It implements the MAD method (described in 5.5.3) as proposed by Nielson *et al.* In addition, it implements automatic significance threshold determination, contour and corner matching algorithms for the determination of GCPs for fast image-to-image registration as well as the ARSIS wavelet fusion model for resolution enhancement of multispectral channels (see Ranchin and Wald). It adheres to the PCIDSK image file format used by PCI Geomatics Inc. and therefore can run as an extension module for their EASI/PACE(c) image processing software.

CDSat is written in Object Pascal with the rapid application development environment DELPHI(c) of Borland Inprise Corp. Currently it is available on CD-ROM or over the internet for the Windows 95/98/NT/2000 operating systems. For more information, contact m.canty@fz-juelich.de.

5.7 REFERENCES

Bishop, C. M. (1995) *Neural Networks for Pattern Recognition*, Oxford University Press, Oxford.

Bruzzone, L., Prieto, D. F. (1999) A Bayesian Approach to Automatic Change Detection, *Proceedings of the IEEE 1999 Int. Geoscience and Remote Sensing Symposium (IGARSS99)*, Hamburg, Germany, 28 June–2 July, Vol. III, 1816–1818.

Li, H., Manjunath, B. S., Mitra, S. K. (1995) A contour-based approach to multisensor image registration, *IEEE Transactions on Image Processing*, **4**(3), 320–334.

Mallat, S. G. (1989) A theory for multiresolution signal decomposition: the wavelet representation, *IEEE Transactions on Pattern Analysis and Machine Intelligence*, **11**(7), 674–693.

Nielsen, A. A., Conradsen, K. and Simpson, J. J. (1998) Multivariate alteration detection (MAD) and MAF processing in multispectral, bitemporal image data: New approaches to change detection studies, *Remote Sens. Environ.*, **64**(1), 1–19.

Press, W. H., Flannery, B. P., Teukolsky, S. A. and Vetterling, W. W. (1986) *Numerical Recipes*, Cambridge.

Prinz, B., Wiemker, R., Spitzer, H. (1997) Simulation of high resolution satellite imagery from multispectral airborne scanner imagery for accuracy assessment of fusion algorithms, *Proceedings of the Joint Workshop of ISPRS WG I/1, I/3 and IV/4, Hannover*, Institute for Photogrammetry and Engineering Surveys, University of Hannover, Vol. 17, 223–231.

Ranchin, T., Wald, L. (2000) Fusion of high spatial and spectral resolution images: the ARSIS concept and its implementation, *Photogrammetric Engineering and Remote Sensing*, **66**(1), 49–61.

Richards, J. A. (1995) *Remote Sensing Digital Image Analysis*, Springer, Berlin.

6

Commercial remote satellite programmes

Q. S. Bob Truong, R. Keeffe, A. Baudoin, R. Krishnan, G. M. Steinberg, T. Hamazaki, M. Onoda, O. Zhdanovich and R. Pearson

6.1 CANADA (*Q. S. Bob Truong and R. Keeffe*)

6.1.1 Introduction

The Canadian Nuclear Safety Commission (CNSC), through the Canadian Safeguards Support programme (CSSP), has actively undertaken a research and development effort, in support of the IAEA, during the past 25 years. This effort (Truong and Keeffe, 2001) covers several major areas, ranging from the provision of cost-free experts, training of inspectors, equipment development, and system studies, which deal with several themes including remote sensing and GIS. The Canadian programme on remote sensing, with particular emphasis on potential safeguards applications, is presented in this chapter.

6.1.2 Canadian remote sensing programme – overview

The Canadian remote sensing programme consists of two main areas:

- Commercial satellite; and
- Airborne survey.

Figure 6.1.1 shows the locations of nuclear facilities in Canada for which imagery has been acquired for studies carried out under the Canadian remote sensing program for safeguards. Results of the work can be found in several papers and presentations by Truong *et al.*, 1998, 1999, 2000 and 2001.

6.1.2.1 Commercial satellite

The Canadian remote sensing programme began in 1997, initially with a small project to investigate some potential applications of commercial satellite imagery in international safeguards using commercial satellite imagery. At that time, only panchromatic imagery of up to 5 m resolution was considered for the study. The

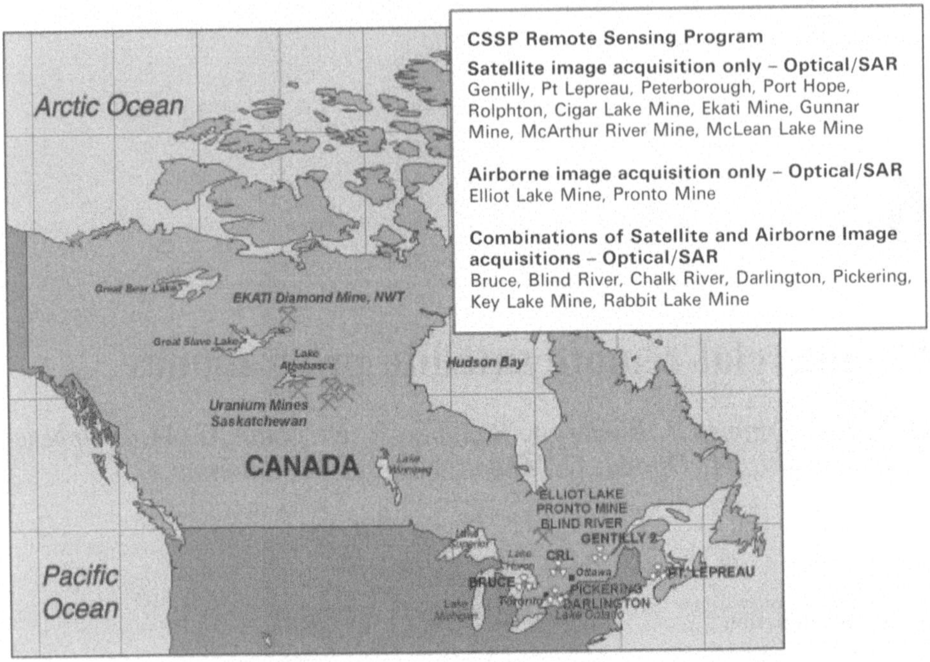

Figure 6.1.1. Canadian study sites map.

initial emphasis was on the details of objects and structures that could be observed with high-resolution commercial satellite imagery. The programme has since expanded to include:

- Other sensors such as Synthetic Aperture Radar (SAR), multispectral (including thermal infrared), and hyperspectral.
- Better than 5 m resolution imagery using high-resolution airphotos and KVR-1000 satellite imagery, in order to simulate the quality of the 1 m commercial satellite imagery, followed up with the IKONOS 1 m
- Various processing and analysis techniques that could be used to extract useful information from imagery acquired from a multi-sensor program.
- Integration tools such as GIS, which could be used to warehouse, organise, analyse and display information, utilising a user-friendly map-based system
- In addition to the R&D work, the CSSP has given awareness training to IAEA inspectors, provided to the IAEA cost-free experts for imagery and GIS needs analyses, and collaborated with the UK and the German governments' support programmes.

6.1.2.2 Airborne survey

The airborne programme is a collaborative effort between the CSSP and various Canadian government departments. The objective of the airborne component is to gain experience with the analysis methods and the SAR and hyperspectral sensors themselves, using airborne platforms, while waiting for the availability of RADARSAT-2 (multi-polarised SAR) and ORBVIEW-4 (hyperspectral sensor). In contrast with other remote sensing programmes in Canada, which have wider scopes and contain other objectives, the airborne survey under the CSSP in support of the IAEA focuses specifically on investigations that could lead to safeguards applications. In essence, the CSSP and its partners pool available resources, and exploit 'dual use' situations where airborne survey data could be of mutual benefit to both parties. For example, hyperspectral survey data over uranium deposits or uranium mine tailings could provide useful information for both safeguards verification and commercial applications such as exploration of mineral deposits, or monitoring of vegetation coverage on mine tailings in land reclamation projects carried out by mining companies.

6.1.2.3 Budget

The budget of the Canadian remote sensing programme is quite modest, and represents only a small percentage of the total CSSP budget. However, significant results have been achieved through pooling resources with other support programmes and other Canadian government departments.

6.1.2.4 Deliverables

The deliverables to the IAEA, under the Canadian remote sensing programme, during the past few years, include:

- reports written with other support programmes, a GIS-needs analysis report, and a comprehensive CSSP report for the IAEA, 1998;
- international conference and symposium papers;
- satellite imagery awareness workshops; and
- service of an imagery analyst on a part-time basis.

More details of the Canadian remote sensing programme for safeguards follow, beginning with the commercial satellite work in Section 6.1.3. The airborne component is discussed in Section 6.1.4. Future direction and some concluding remarks are presented respectively in Section 6.1.5. Additional details can be found in the References, Section 6.1.7.

6.1.3 Commercial satellite programme

The results of the work under the Canadian commercial satellite programme have been presented at several conferences and workshops, and documented fully in a comprehensive CSSP report for the IAEA. The work can be grouped under the following activities:

- case studies;
- imagery analysis and GIS support to the IAEA;

- site diagram review;
- baseline report;
- advanced processing and analysis techniques; and
- collaboration with other support programmes.

6.1.3.1 Case studies

Several case studies were carried out to illustrate possible applications of commercial imagery for evaluation of member states' declaration, change detection, detection of undeclared activities, monitoring of changes within a declared facility, monitoring activity at a decommissioned facility, and assisting in inspection planning and reporting. The case studies are discussed in considerable detail in Chapter 10 of this book.

6.1.3.2 Imagery analysis and GIS support to the IAEA

The CSSP has provided the service of an analyst to the IAEA to assist the IAEA in the production of imagery analysis and to assist in defining the requirements for the IAEA Geospatial Laboratory. In addition, a GIS-needs analysis has been carried out for the IAEA. The details of the Canadian experience with satellite imagery and GIS are presented in Chapter 9.

6.1.3.3 Site diagram review

Comparison between IKONOS imagery and site diagrams for several major nuclear facilities was carried out to demonstrate the usefulness of satellite imagery in the preparation of site diagrams for submission under the Additional Protocol. These exercises indicate that satellite imagery provides accurate and up-to-date information, which could be very useful in the review of the site diagrams.

6.1.3.4 Baseline Report

Acquisition of satellite imagery of various resolutions has continued with the aim of producing a baseline report for major nuclear facilities in Canada. In addition, satellite imagery is being investigated for its potential role in developing a baseline for an underground nuclear fuel repository. By using change detection techniques, undeclared excavations or evidence of illegal diversion should become apparent in future images.

6.1.3.5 Advanced image processing and analysis techniques using limnological and optical knowledge for discharge detection

Previous work carried out under the CSSP, has shown that thermal imagery from American Landsat satellites could be used to detect cooling water discharges, and could therefore be used to verify the operational status of nuclear facilities. In some images, thermal plumes could be easily detected in single band imagery with no mathematical manipulation and little image enhancement because there was a very strong thermal contrast between the effluent and the receiving water. However, for certain situations such as discharges into well-mixed conditions (cold water and violent tides) the thermal plume may be subtler. We have shown that the visible bands of Landsat and IKONOS images often contain additional information, and that the thermal signature of a discharge from a nuclear facility is not the only signal available to describe its operation.

This work introduces some important hydrological phenomena that govern the biological and physical organisation of water bodies, and discusses some basic concepts of marine and aquatic optics that are relevant to the safeguards problem. Using image analysis techniques that have been used widely in ocean optics work and in applications in the mapping and monitoring of water quality, we have re-analysed data that were obtained under a joint project with other Canadian government departments. We present a preliminary examination of imagery from both satellite multispectral and aircraft hyperspectral sensors, and discuss methods to extract information that could be useful in the detection and verification of declared, or undeclared, nuclear activities. In one example of an IKONOS image of the Canadian Bruce Nuclear Generating Facility, simple enhancement techniques failed to find any plume other than a small jet visible in the surface wave field. With knowledge of limnology, oceanography and aquatic optics, we have been able to separate and remove the surface reflection, and detect a large plume of slightly less turbid water emanating from the facility. Figure 6.1.2 displays these observations (Borstad *et al.*, 2001).

It should be noted that similar analysis of the water surface has been performed on multispectral satellite imagery (Hough, 1999) to confirm South Korean and Japanese reports of the existence of an underground plutonium production facility on an island in a man-made reservoir in North Korea. By analysing the red spectrum, a plume of pollution coming out of the island was detected. High-resolution imagery was also used to confirm the existence of a helicopter landing pad, a large pier, a road leading to an underground facility and a ventilation shaft (as reported by Mitchell, 1999).

6.1.3.6 Collaboration with other support programmes

Collaborations with the UK Support programmes on SAR imagery work and with the German support programme on change detection using Landsat and IKONOS multispectral imagery have continued. Results of these collaborations have been presented in joint papers presented by Canty *et al.*, at the ESARDA annual meetings in Dresden (May 2000) and Bruges (May 2001), and at the ISSSR meeting in Quebec City (June 2001). A comprehensive report by Canty M., Jasani B. and Truong Q. S. Bob, 2001 on this collaboration was published in Germany in a Juelich Institute report. Four areas have been investigated: detection of cooling water discharges from a nuclear facility, change detection using MAD technique, image fusion and thermal detection with SAR.

Details of the work on detection of cooling water discharges are discussed in the Juelich report and Chapter 10 of this book. Examples of MAD change detection and image fusion are shown in Figures 6.1.3 to 6.1.5.

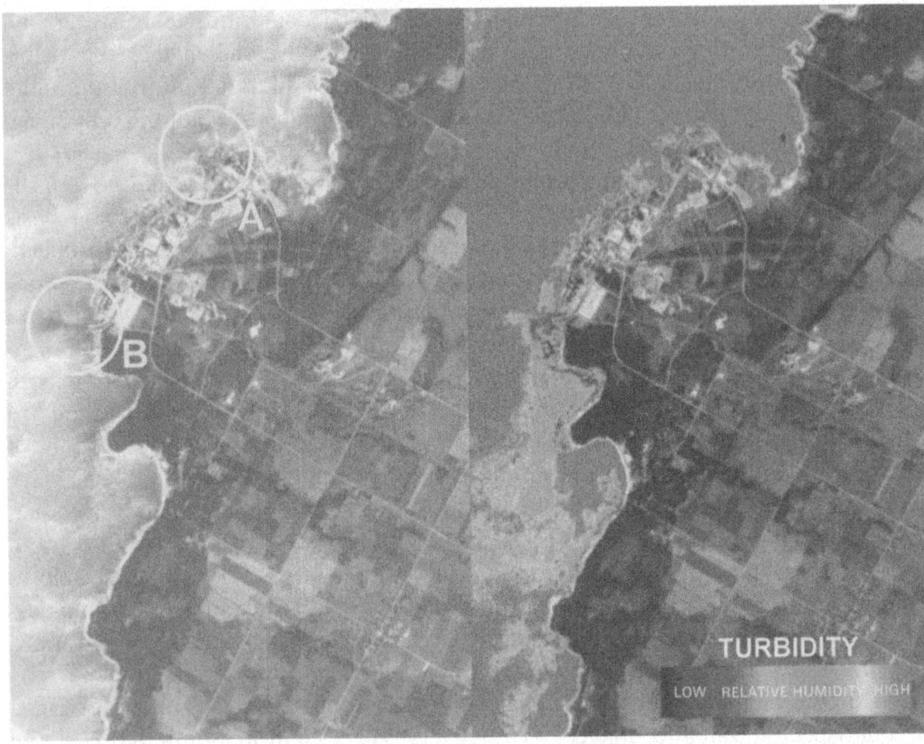

Figure 6.1.2. Image processing comparison for hydrological phenomena. *Left side*: A natural colour image of the Bruce Nuclear Generating Facility, acquired by IKONOS on 11 November 1999. The circles indicate the position of discharge points. There is a small surface effect at Bruce B and hints of a plume but it is difficult to see. The pattern of the waves visible in the original data shows clearly that the wind is blowing from the north. *Right side*: The same image after removing the sun glint of the surface waves, and enhancement. A large plume is seen flowing south of Bruce B and curling into the small bays. The slightly less turbid water along the coast between A and B may be a smaller plume originating at Bruce A (see Plate 1).

6.1.4 Airborne programme in support of satellite programme

In collaboration with several Canadian government departments, an airborne survey programme over several Canadian nuclear facilities, (see Figure 6.1.6) has been carried out using various SAR and hyperspectral techniques. The main aim of the airborne programme is to gain experience with sensor capabilities and analysis methods, while waiting for the availability of RADARSAT-2 (multi-polarised SAR) and ORBVIEW-4 (hyperspectral sensors).

A multipolarised SAR sensor has been flown on a Canadian government Convair 580 aircraft to verify our findings using SAR satellite imagery. For example, the effects of the radar energy beam polarisation in detecting the

Figure 6.1.3. MAD change detection for Space Imaging Landsat-5 August 1998–August 1996, fused to the negative of an IRS-1C panchromatic (5 m) image taken in September 1997. MAD component 1 and 2 exceeding the significance thresholds are in red and green [courtesy of M. Canty] (see Plate 3).

signature of cooling water discharges and surface waves caused by underwater features.

Airborne data have also been collected using hyperspectral sensors over several Canadian uranium mines and refinery. The hyperspectral data is being analysed for possible identification of substances that could be associated with a nuclear operation. Further details of the airborne work will be discussed in the sections below.

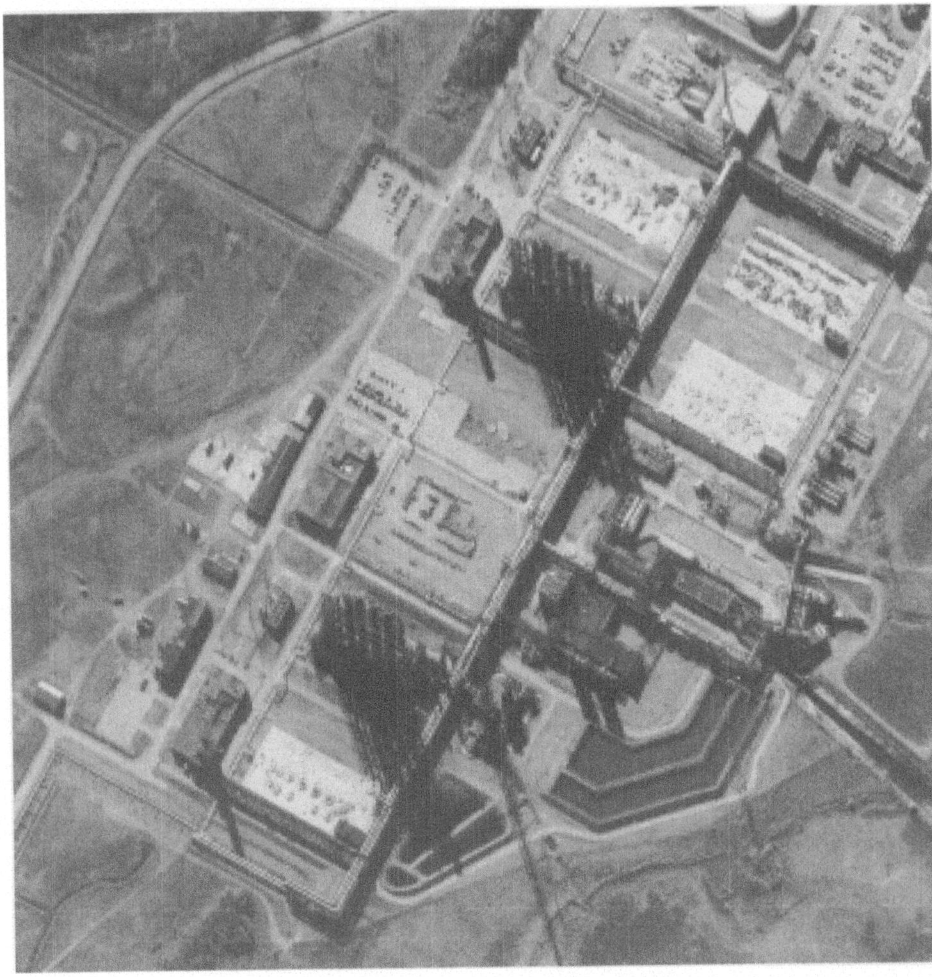

Figure 6.1.4. Space Imaging IKONOS image from 11 November 1999. False colour composite of part of the BNPD site using the first, second and third multispectral channels (4 m) fused to the panchromatic image (1 m) [courtesy of M. Canty].

6.1.4.1 Airborne survey with SAR sensors

Background

The main objective of the airborne SAR programme is to carry out a systematic investigation and to gain some experience with SAR data processing techniques and analysis methods, while waiting for the availability of the multi-polarised SAR sensor on RADARSAT-2.

It was found that SAR sensors were able to detect surface effects caused by underwater features such as sandbars or disturbances created by the cooling

Figure 6.1.5. A USGS Landsat 7 Image from 26 August 1999. False colour composite of the BNPD site using the first, second and third principal components (30 m) fused with panchromatic image (15 m) [courtesy of M. Canty] (see Plate 12).

water discharge such as that from the open channel at the Bruce nuclear generating station. The surface effect of the cooling water discharge at Bruce was clearly visible on one ERS image (see Figure 6.1.7). This same effect was not observed on another SAR image (RADARSAT), which seems to capture the surface effects caused by underwater sandbars. It was speculated that the different observations could probably be attributed to several factors, such as the radar energy beam polarisation (V-V for ERS and H-H for RADARSAT-1) or the look angle of the energy beam. The various RADAR beam polarisations are shown in Figure 6.1.8.

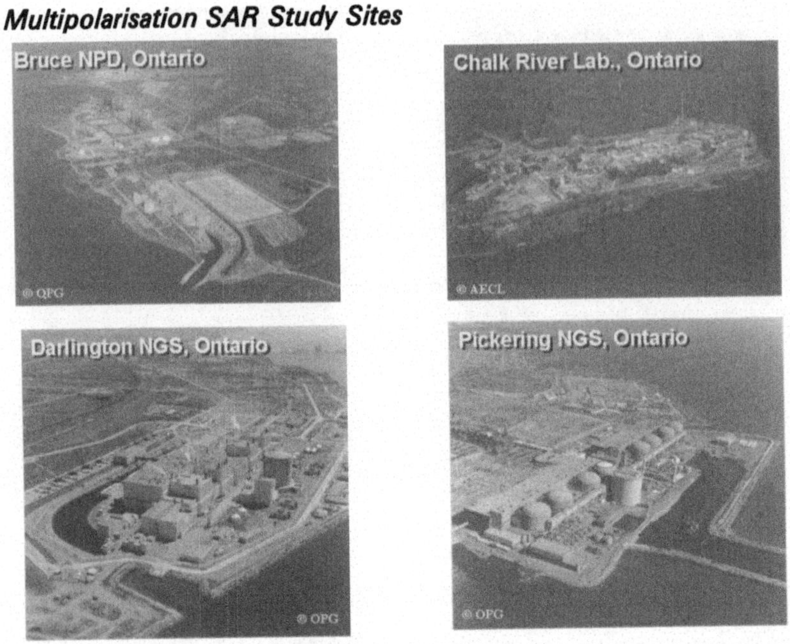

Figure 6.1.6. Four multipolarisation study sites.

Figure 6.1.7. ERS-2 versus RADARSAT-1 detection comparison.

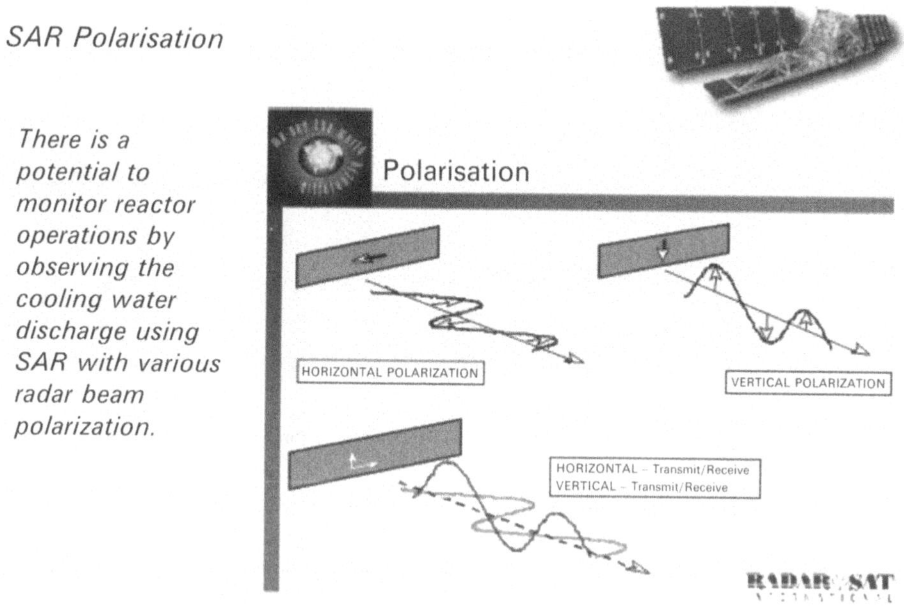

Figure 6.1.8. SAR beam polarisations.

Data Acquisition

In collaboration with several Canadian government departments, the SAR airborne survey has been carried out with a multipolarised SAR sensor on board a Canadian government Convair 580 airplane. This aircraft is normally flown at an altitude of about 6,000 m during this type of survey. Over-flights were carried out over four nuclear facilities: Bruce, Pickering, Darlington nuclear power generating stations, and Chalk River Laboratories. Collateral information such as weather data, water temperature, ground truth photographs, satellite imagery, weather pattern simulation, etc. has also been collected.

Initial results

The SAR airborne data have not yet been fully analysed, but initial results based on available processed data appear to indicate that detection of cooling water discharge surface effects are dependent on several factors, not only the look angle, the beam polarisation, surface turbulence caused by wind or wave effect, but also on the design and the strength of the discharge. The cooling water discharge at Bruce seems to be detected by all beam modes (V-V, H-H, or a combination of the two – V: Vertical; H: Horizontal). Since some difficulties were previously encountered in the detection of submerged discharges (e.g. at Point Lepreau), it was a surprise that the submerged discharge at Darlington was also detected by the SAR airborne sensor. This detection at Darlington was verified by IKONOS and Landsat-7 imagery (Figure 6.1.9).

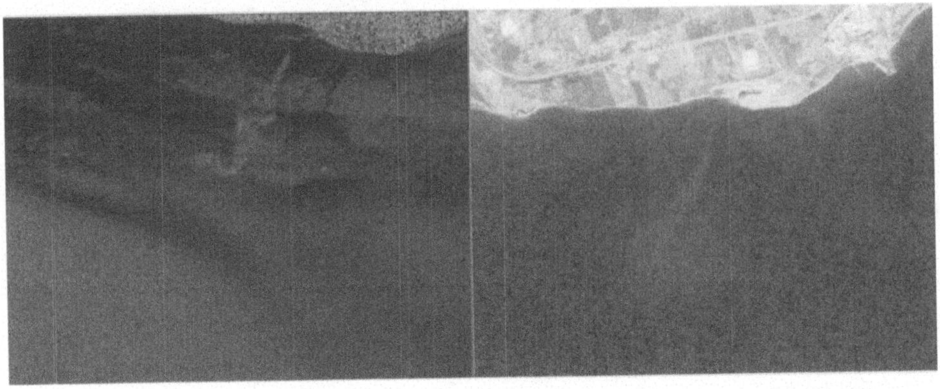

Convair-580 SAR VH Polarisation image of 14 March 2000.

Landsat-7 thermal image of 6 March 2000.

Figure 6.1.9. Convair SAR to Landsat 7 detection comparison.

For Pickering and Chalk River the surface effects caused by discharge are not so pronounced. The faint signature for Pickering could be caused by a weaker discharge, when compared to Bruce and Darlington. For Chalk River an opposite effect was observed. Instead of a bright return, a dark area is seen on the SAR image over the area where the submerged discharge meets the water surface. This calm zone at the reactor cooling water discharge location on the Ottawa River was also observed, and photographed, from a boat.

It was found that the flow direction of the cooling water discharge is strongly dependent on the local surface wind conditions, which could be affected by regional/global weather systems. For example, the unusual flow direction at Bruce was traced back to Hurricane Floyd (September 1999) with the help of a three-dimensional simulation, which was carried out by Environment Canada using supercomputers designed for handling complex hydrodynamic calculations, of the Eastern Canada/US regional weather system at the time of the airborne survey. The situation is illustrated in Figures 6.1.10 and 6.1.11. The first figure shows the discharge at Bruce heading in the 'wrong' direction. However, the wind vector (circle on the right of Figure 6.1.11) 'during Floyd' is pointing south, confirming that direction captured on the SAR image is in fact correct. The prevailing direction for the Bruce cooling water discharge is generally in the north-west direction as indicated by the vector in the circle on the left.

6.1.4.2 Airborne survey with hyperspectral sensors

Background

The hyperspectral project is a collaborative effort between the CSSP and Canada Centre for Remote Sensing (CCRS). Various locations around a uranium mine

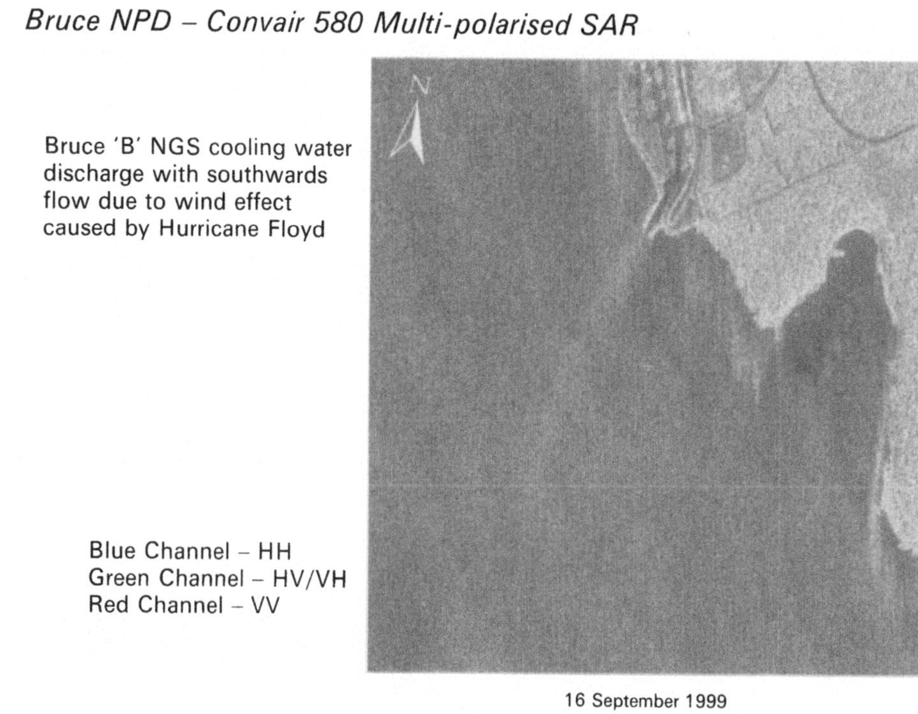

Figure 6.1.10. SAR airborne image showing the discharge at Bruce (see Plate 17).

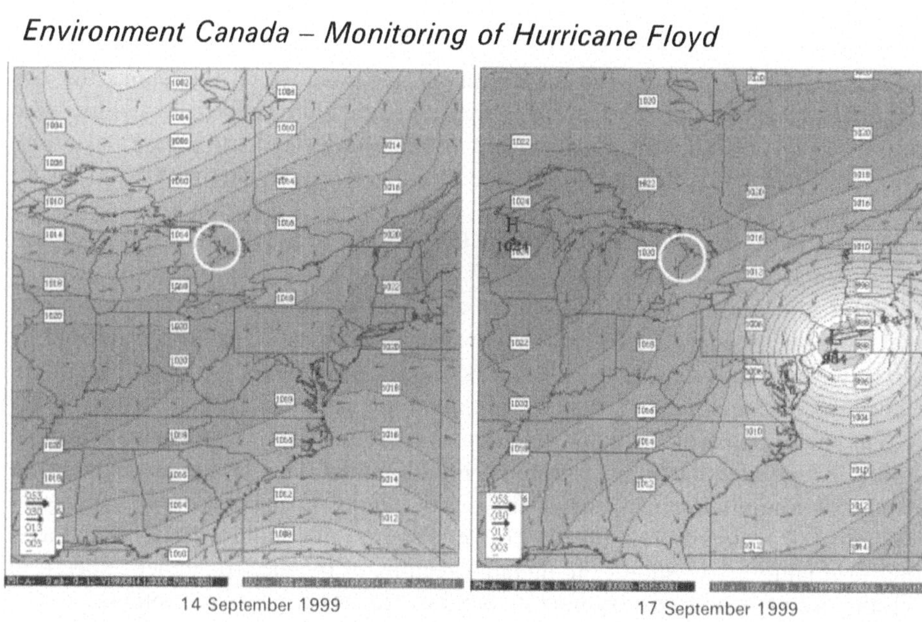

Figure 6.1.11. Surface wind velocity vectors before and during Hurricane Floyd.

Figure 6.1.12. Four hyperspectral study sites.

such as the settling ponds, the mine tailings, the waste rock piles, and the ore stockpiles are of interest. The airborne hyperspectral surveys over mine sites were taken at about 2,700 m and a team on-site obtaining ground truth information by taking photographs, collecting samples and taking measurements with a hand-held spectrometer.

The initial surveys were done with the Probe-1 sensor over Pronto Mine tailings, and the Blind River refinery. These two facilities are located on the north shore of Lake Huron, Ontario. Subsequent flights were carried out with Compact Airborne Spectrographic Imager (CASI) and Short wavelength infrared Full Spectrum Imager (SFSI) hyperspectral sensors over Key Lake and Rabbit Lake, two operating mines in northern Saskatchewan. The sites are shown in Figure 6.1.12.

The initial phase of the work investigates the potential of hyperspectral remote sensing for the identification of uranium mine tailings. This project endeavours to distinguish uranium mining from other types of mining. Probe-1 airborne hyperspectral data was acquired on 8 July 1999 over Pronto mine tailings near Elliot Lake, Ontario. Figure 6.1.13 shows the imaging regions of the electromagnetic spectrum with the visible band in a very narrow band of frequency relative to others.

Analysis of the Probe-1 data includes radiometric correction, surface reflectance retrieval, end-member selection and identification, constrained linear spectral

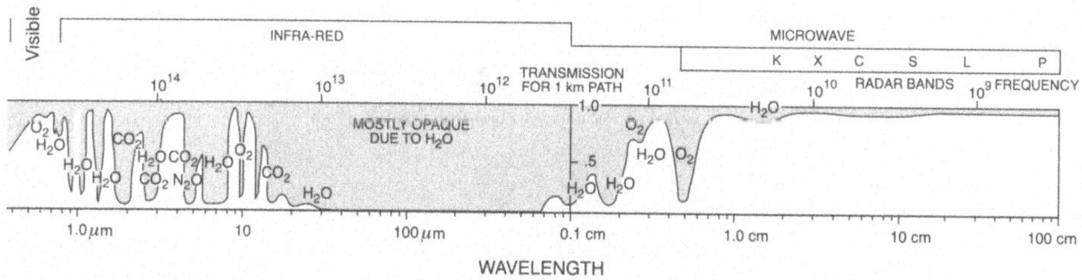

Figure 6.1.13. Electromagnetic spectrum.

unmixing analysis, and the production of an abundance map of each end-member. Reflectance spectra of laboratory standards of uranium ore, uranium/thorium ore, uranium tailings, and yellow cake were measured using a portable spectroradiometer to help in the identification of the end-member spectra.

Spectral unmixing of Pronto data and initial results

The constrained linear spectral unmixing was performed on the Probe-1 data using an algorithm implemented in the Imaging Spectrometer Data Analysis System (ISDAS). In linear unmixing, the spectrum from each pixel in the data cube (see Figure 6.1.14) is decomposed into a linear combination of end-member spectra, which represents, in this case the spectra of the purest pixels in the scene.

An example of spectral unmixing for two end-member spectra is shown in Figure 6.1.15. The selection of end-member spectra was performed using an automatic extraction method based on an iterative error analysis recently implemented in ISDAS. This method performs a series of linear constrained unmixing with, as end-members, the pixel spectra that minimise the unmixing error. The algorithm reaches an end when a predetermined condition is reached, such as the number of end-members or an error tolerance. In this case, 25 end-members were requested.

Both the uranium and copper tailings came out as end-members, Figure 6.1.16 Analysis has also been performed on the hyperspectral data acquired over the water surfaces.

6.1.5 Concluding remarks

From the results available to date the following statements could be made.

- Although high-resolution imagery can provide excellent details of facilities, the entire remote sensing range of the electromagnetic spectrum is of benefit to international safeguards applications. For example, it would more efficient and cost effective to use Landsat or SAR imagery to search for a facility. Once the facility has been located then high-resolution IKNONOS imagery

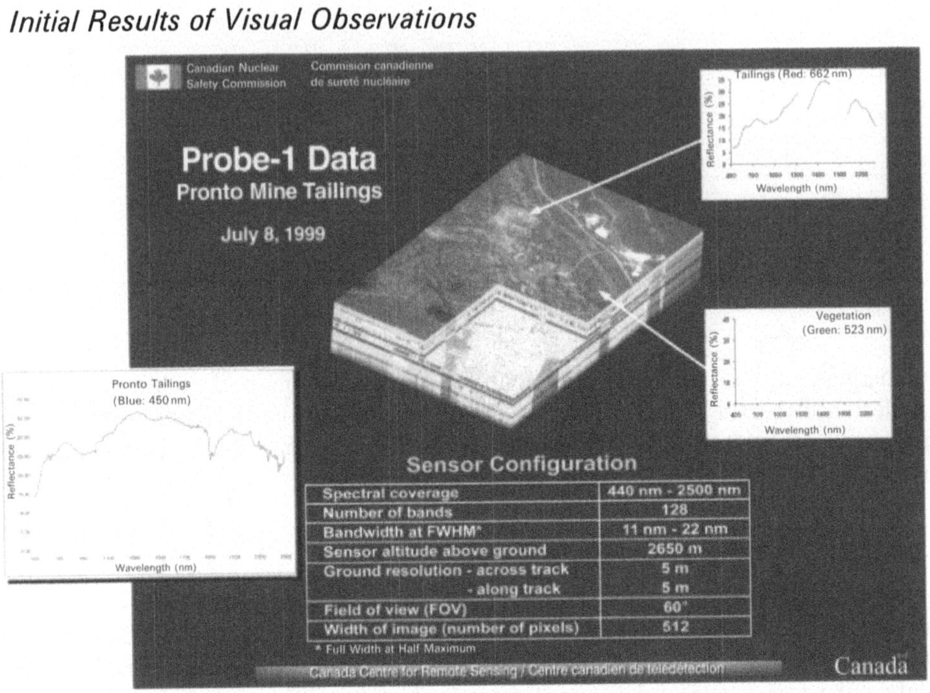

Figure 6.1.14. Probe-1 Data, Pronto Mine Tailings.

would be used to identify the details. It is anticipated that resolution better than 1 m will soon become available, and this will undoubtedly assist in the identification of small objects.

- Unlike the initial indication from ERS-2 and RADARSAT satellite imagery, the airborne SAR results indicate that the sensor's ability to detect open channel cooling water discharge (e.g. Bruce design) does not seem to be strongly dependent on the radar energy beam polarisation. However, further work is needed to identify if the look angle, as well as other factors (e.g. the design of the discharge, the local current and surface wind conditions) would play an important role in the detection of cooling water discharge from the facilities.
- The flow direction of the cooling water discharge at Bruce nuclear generating station is strongly dependent on the local wind conditions, which could be affected by regional/global weather systems. For example, the unusual flow direction at Bruce was traced back to Hurricane Floyd with the help of a three-dimensional simulation of the regional weather system (Eastern Canada/US) at the time of the airborne survey. The dependence of the flow direction may not have any significant bearing on safeguards. However, this exercise shows that apparently anomalous results, which would have been discarded, were successfully resolved with relevant collateral information.

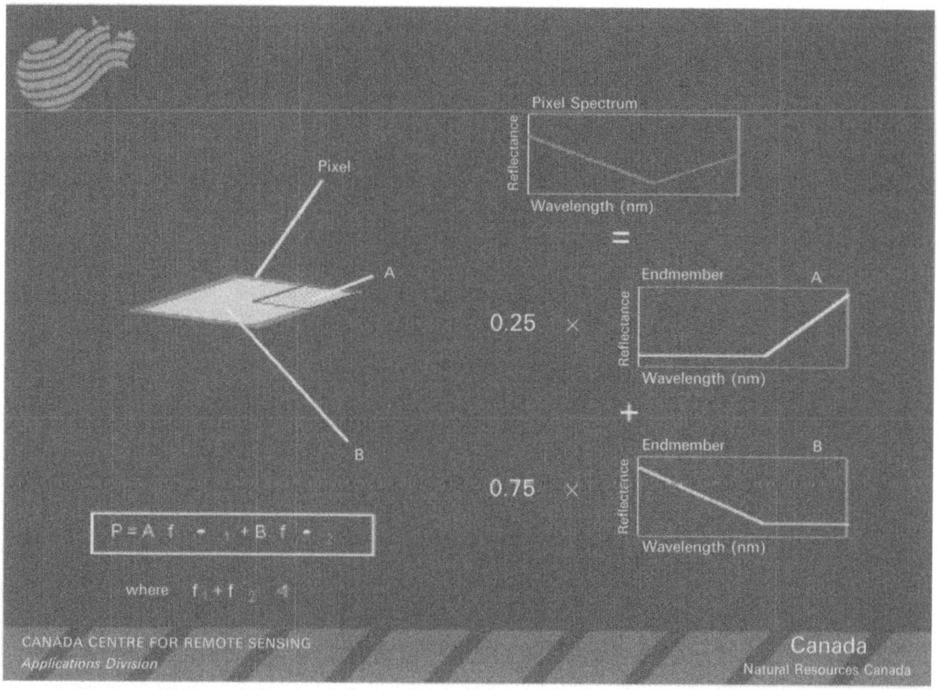

Figure 6.1.15. Spectral unmixing.

- The airborne SAR sensor detected the surface effects caused by the cooling water discharge from a large submerged pipe at the Darlington nuclear generating station. With tone enhancement, this 'signature' of the cooling water discharge was also visible from IKONOS 1 m panchromatic imagery. This success in the detection of the Darlington plume is in contrast with the Point Lepreau situation where the plume has not yet been successfully detected. The reason for this could be due to the design and location of the Point Lepreau discharge pipe, which is located much deeper in well-mixed and much colder water.
- For 'difficult cases' (e.g. Point Lepreau) advanced techniques using ocean optics in the analysis of multispectral and hyperspectral data (discussed in Section 6.1.3.5) may be helpful in the detection of the plume from the facility.
- Initial hyperspectral results from the Pronto Mine tailings and the Key Lake Mine are promising. It would appear that this analysis method could be used to identify material from remotely acquired hyperspectral data. The ability to 'do chemistry from space' would add another dimension to the knowledge base. In other words, in addition to the location, shape, size, and colour, the material could also be identified through remote sensing.

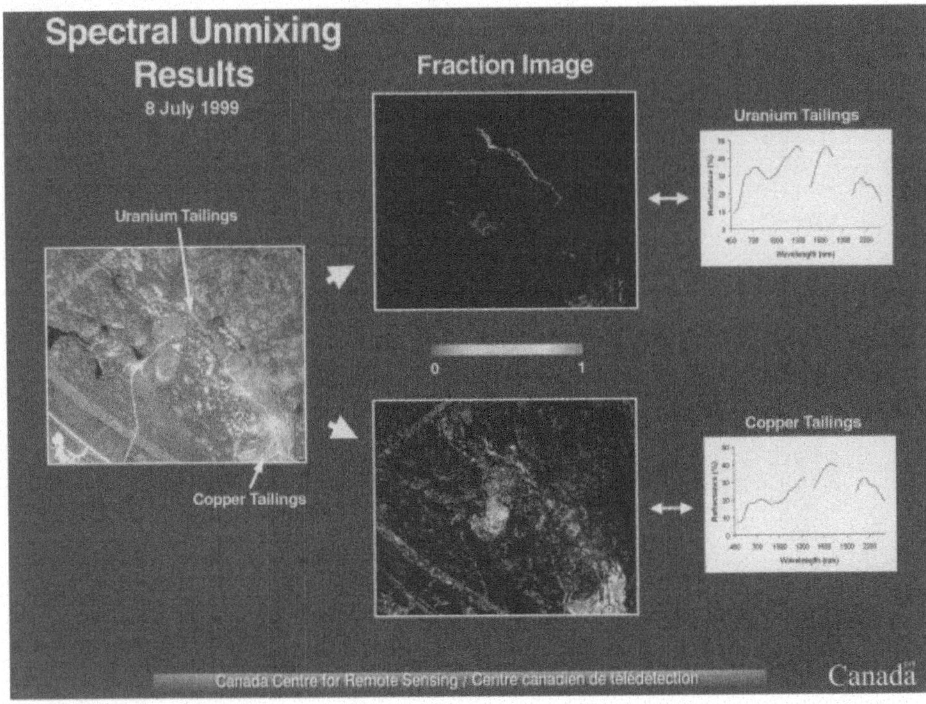

Figure 6.1.16. Spectral unmixing results – uranium and copper tailings (see Plate 19).

- Similarly, height extraction from high-resolution stereo pair imagery and terrain models would also contribute to this knowledge base. Further discussion on this and GIS-related topics is discussed in Chapter 9.

6.1.6 Acknowledgements

The work presented in this chapter is a compilation of the collaborative effort between the CSSP of the Canadian Nuclear Safety Commission and the following organisations: International Security Research and Outreach programme of the Department of Foreign Affairs and International Trade (DFAIT), CCRS, Environment Canada, the UK and German support programmes.

Technical contribution, information and advice provided by Messrs. Phillip Baines of DFAIT, Jean-Pierre Paquette of ImStrat Corporation, Michel Jean of EC, Gary Borstad of G.A. Borstad Associates Ltd., Mort Canty of Juelich Institute, Bhupendra Jasani of King's College, Robert Neville and Karl Staenz of CCRS, Vincent Tao of University of Calgary, and Ms. Josée Lévesque of MIR Télédétection Inc. are much appreciated.

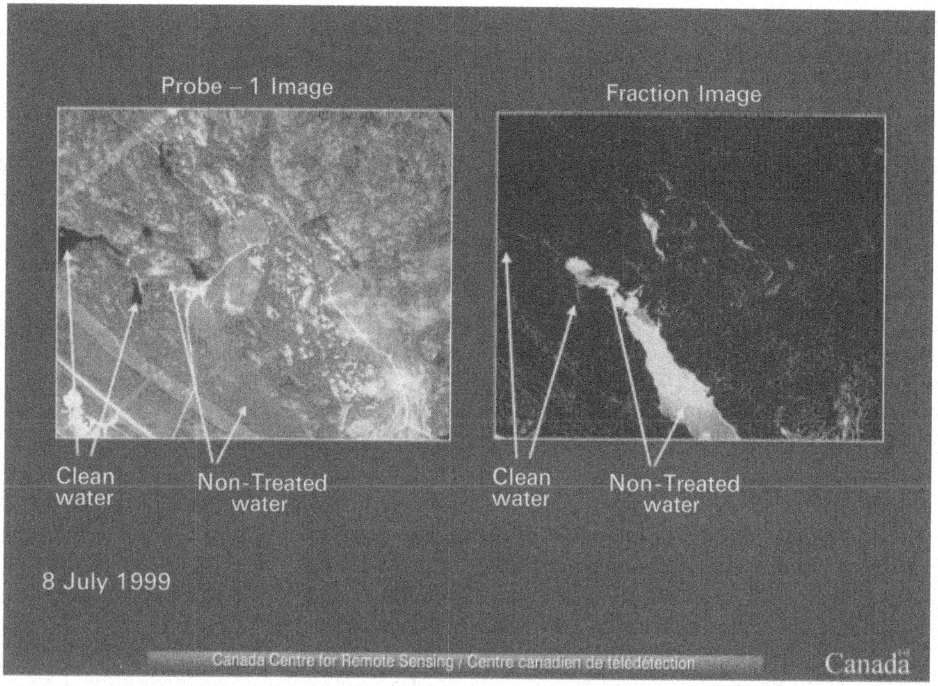

Figure 6.1.17. Spectral unmixing results – environmental effects in water (see Plate 20).

In addition, the authors would like to acknowledge the excellent co-operation provided by Ontario Power Generation, Atomic Energy of Canada Limited and Cameco Corporation.

6.1.7 References

Borstad, G. A., Truong, Q. S. Bob, Keeffe, R., Staenz, K., and Neville, R. A. (2001) Using Limnological and Optical Knowledge to Detect Discharges from Nuclear Facilities – Potential Application for International Safeguards. Presentation at the IAEA Symposium on International Safeguards, Verification and Nuclear Material Security, Vienna.

Canty, M., Jasani, B., and Truong, Q. S. Bob (2000) Satellite Imagery for Nuclear Safeguards: The Bruce Nuclear Power Development Site. Proceedings of the ESARDA 22nd Annual Meeting, Dresden.

Canty, M., Jasani, B., and Truong, Q. S. Bob (2001) Remote Sensing and Change Detection for Nuclear Safeguards: The Bruce Nuclear Power Development Site. Juelich Institute Report # 3843, Juelich.

Canty, M., and Truong, Q. S. Bob (2001) High Resolution Detection for Remote Verification. Presentation at the ESARDA 23rd Annual Meeting and Symposium on Safeguards and Nuclear Material Management, Bruges.

Canty, M., Niemeyer, I., and Truong, Q.S. Bob (2001). Change Detection with Multispectral Satellite Imagery: Application to Nuclear Verification. Presentation at the ISSSR Conference, Quebec City.

Hough, H. (1999) North Korean site raises questions. *Jane's Intelligence Review*, 12 November.

Mitchell, C. (1999) North Korea Underground Nuclear Facility Poses Threat. CBN News, 13 September.

Keeffe, R., and Truong, Q. S. Bob (2001) Evolution of the Canadian Safeguards Support programme. Paper presented at the INMM 42nd Annual Meeting, Indian Wells, California, USA.

Neville, R.A., Staenz, K., Lévesque, J., Truong, Q. S. Bob, and Borstad, G. A. (2001) Hyperspectral Analysis of Imagery of a Uranium Mine Site. Presentation at the ISSSR Conference, Quebec City, Canada.

Truong, Q. S. Bob, Keeffe, R., Baines, P., and Paquette, J. P. (1998) Potential Applications of Commercial Imagery in International Safeguards. Proceedings of the INMM 39th Annual Meeting, Naples, Florida, USA.

Truong, Q. S. Bob, Keeffe, R., Baines, P., and Paquette, J. P. (1999) Potential Application of Commercial Imagery in International Safeguards. Journal of Nuclear Material Management. Volume XXVII, Number 2, pp. 13–18, USA.

Truong, Q. S. Bob, Keeffe, R., Baines, P., and Paquette, J. P. (1999) Potential Applications of Commercial Satellite Imagery in International Safeguards – A Progress Report. Proceedings of the INMM 40th Annual Meeting, Phoenix, Arizona, USA.

Truong, Q. S. Bob (2000) Potential Applications of Satellite Imagery in International Safeguards. Proceedings of the ESARDA 22nd Annual Meeting, Dresden, Germany.

Truong, Q. S. Bob (2000) Canadian Remote Sensing programme. Presentation at the INMM/ESARDA 3rd Workshop on Science and Modern Technology for Safeguards, Tokyo, Japan.

Truong, Q. S. Bob, *et al.* (2001) Canadian Remote Sensing programme – Hyperspectral Survey. Presentation at the INMM 42nd Annual Meeting, Indian Wells, California, USA.

Truong, Q. S. Bob and Keeffe, R. (2001) Canadian Safeguards Support programme – A Future Outlook. Presentation at the IAEA Symposium on International Safeguards, Verification and Nuclear Material Security, Vienna, Austria.

6.2 THE FRENCH EARTH OBSERVATION PROGRAMME (*A. Baudoin*)

6.2.1 Introduction

After the Ariane programme for launcher development, earth observation is the second largest French space activity in terms of programme and budget. The French space programme for earth observation aims altogether at satisfying knowledge and research needs, and at developing applications of remote sensing in many sectors of socio-economic activity.

It includes important scientific missions and is largely based on the use of families of satellites and instruments suited to various user needs. The policy for earth observation of the French Space Agency, CNES (Centre National d'Etudes Spatiales), is based on international cooperation, with particular attention to help develop a European coherence and synergy in this domain. A major research challenge for earth sciences is to provide decision makers with a sound base about environmental problems and natural hazards. Weather forecast, climate evolution, protection of air, water quality and the food chain in oceans and soils are some of the applications where space observations can be used to monitor global sustainable development over the next decades.

Beyond these activities related to scientific and environmental applications, the importance of which should increase in the future, a continuing effort to develop the use of remote sensing in cartography, geology, agriculture and forest as well as in many other applications is maintained.

In this chapter, an overview on earth observation activities in France, which are conducted in several programme frameworks, mainly in cooperation, is given. Examples of these are meteorology through ESA and EUMETSAT programmes, environment and ocean missions mainly conducted in ESA, earth sciences programmes through bilateral cooperation (Fellous *et al.*, 2000) and defence programmes. Emphasis is given on the latter and civil observation satellites such as the current SPOT and the future Pléiades programmes, whose principal interest is for IAEA and treaty verification.

6.2.2 French activities in earth observation

The French contribution to ESA programmes

France has been a participant in the European Space Agency's ERS 1 and 2 satellites, which carry synthetic aperture radar, together with active and passive sensors devoted to surface observation of marine parameters (temperature, wind, waves, topography). ESA is also developing the ENVISAT satellite, planned for launch in 2002. ENVISAT is carrying a C-Band SAR, an advanced version of the ERS SAR and several other instruments aimed at measuring atmospheric, oceanic, or terrestrial parameters.

Defence programmes

Defence has been always an important application domain for space technology, since the beginning of the space race between the USA and the former USSR during the 1960s. While developing civilian satellites, which became the SPOT project, France also undertook feasibility studies for a defence observation system called SAMRO (Satellite Militaire de Reconnaissance Optique) during the 1970s. This project was postponed until 1986 when a new programme called Hélios emerged.

A first set of optical surveillance satellites, Hélios 1A and Hélios 1B, took advantage of the developments of Spot 4 to look for synergies as far as possible between the two programmes. Common developments were possible on the platform, the image telemetry and the onboard recorder and in the Control Centre (Brachet, 2000).

The first satellite, Hélios 1A, was launched on 7 July 1995, and Hélios 1B was placed in orbit on 3 December 1999. These two satellites provide real service continuity not only to their governmental and military users in France but also to Italy and Spain. These three countries are partners in the programme.

To prepare for the future, an improved set of satellites, Hélios II, has been developed since 1994 and this programme was given the go-ahead in April 1998. These satellites have been developed in synergy with Spot 5, and Hélios IIA is due to be launched in 2004.

6.2.3 Brief history of the SPOT programme

As early as the 1970s, several activities and programmes were conducted to develop new sensors and to evaluate their value for scientific or commercial applications. When the first operational satellite for earth observation, ERTS (Landsat 1), was launched in July 1972, NASA had already selected several French teams as principal investigators.

In parallel, many experiments with cameras, multispectral scanners and thermal radiometers were performed, using aircraft and balloons operated by GDTA (Groupement pour le Développement de la Télédétection Aérospatiale), founded by CNES and IGN (Institut Géographique National) in 1971. French scientists and engineers then had the opportunity to study these new types of image and to develop new applications.

Feasibility studies of an earth observation satellite began in 1974 and, in December 1976, the first SPOT (Satellite Probatoire d'Observation de la Terre) project was proposed to ESA with two main sensors, a MRVIR (Medium Resolution Visible and Infrared,) and a HRV (High Resolution Visible). The MRVIR was a mechanical scanner, like the one on Landsat, with 100-m ground resolution, seven spectral bands and 300-km swathe width. The HRV was a push-broom scanner with 20-m ground resolution, three spectral bands and 60-km swathe width. Within this HRV instrument, a pointing mirror was proposed to observe any area in the field of view of the MRVIR instrument (Brachet, 1985).

Unfortunately, this project was not given the go-ahead by the member states of ESA, and a new project, without the MRVIR instrument but with two modified HRVs, was proposed and accepted by the French Government in February 1978. The modifications proposed on the HRV were related to ground resolution, improved to 10 m with an additional panchromatic band, and to side looking angle, raised up to 27° in order to improve stereoscopic capabilities requested by the cartographic and geologic communities.

In spite of the ESA refusal, some European countries were still interested in the SPOT project, especially Sweden and Belgium. They confirmed this interest a few months after the French decision by signing cooperation agreements to become partners in the programme, at the level of 4% each. These partnerships, since November 1978 for Sweden and June 1979 for Belgium, have continued throughout the successive decision stages up to SPOT 5.

A few years after the SPOT 1 decision, two main decisions were taken, having a major impact on the success of the programme.

In October 1981, SPOT 2 was chosen, even before the launch of SPOT 1, to secure the programme and ensure data continuity. This decision transformed an experimental programme into a fully operational one, with the same name SPOT, but where the letter P now changed to *pour* (for). SPOT became the earth observation satellite (Satellite pour l'Observation de la Terre). In July 1982, a commercial company, Spot Image, was created to distribute SPOT data on the world market.

These two decisions were taken in accordance with the two main principles of the SPOT programme:

- continuity of the service to users;
- steadily improving quality of SPOT data and services in response to changing customer needs.

The first principle implies a series of satellites operating continuously. Possible launch failure or early satellite breakdown should be anticipated by having a satellite ready as soon as possible after the previous one, even if this satellite should then wait several years before being launched.

The second principle implies improving the performances of instruments, mainly in terms of ground resolution, spectral bands and stereoscopic capabilities but also in terms of delays and services, thanks to ground segment improvements.

Five SPOT satellites were given the go-ahead, four have been launched and three are still operational. They were given the go-ahead, and launched according to the two above principles:

- SPOT 1 was launched on 22 February 1986, and withdrawn from active service on 31 December 1990. It was then reactivated in March 1994 to meet the increasing demand for SPOT imagery, notably during the northern hemisphere growing season.
- SPOT 2 was launched on 22 January 1990 and is still operational.
- SPOT 3 was launched on 26 September 1993. A failure of the attitude control system occurred on 14 November 1997 and after 3 years in orbit the satellite stopped functioning.
- SPOT 4, with additional capabilities, was launched on 24 March 1998 and has been fully operational since May 1998, after its commissioning phase.
- SPOT 5, with major improvements, is now (in December 2001) ready to be launched in April 2002.

6.2.4 Characteristics of the SPOT system

The SPOT System (Goudy, 1999) is not only a space system, with a constellation of satellites on specific orbits and with appropriate sensors, but also a ground system with all the facilities necessary to operate the satellites and to acquire, process and deliver SPOT products on a worldwide basis.

SPOT constellation

All SPOT satellites are located in the same sun synchronous orbit. There are 26 possible positions (at a given time), corresponding to each day of the orbital cycle,

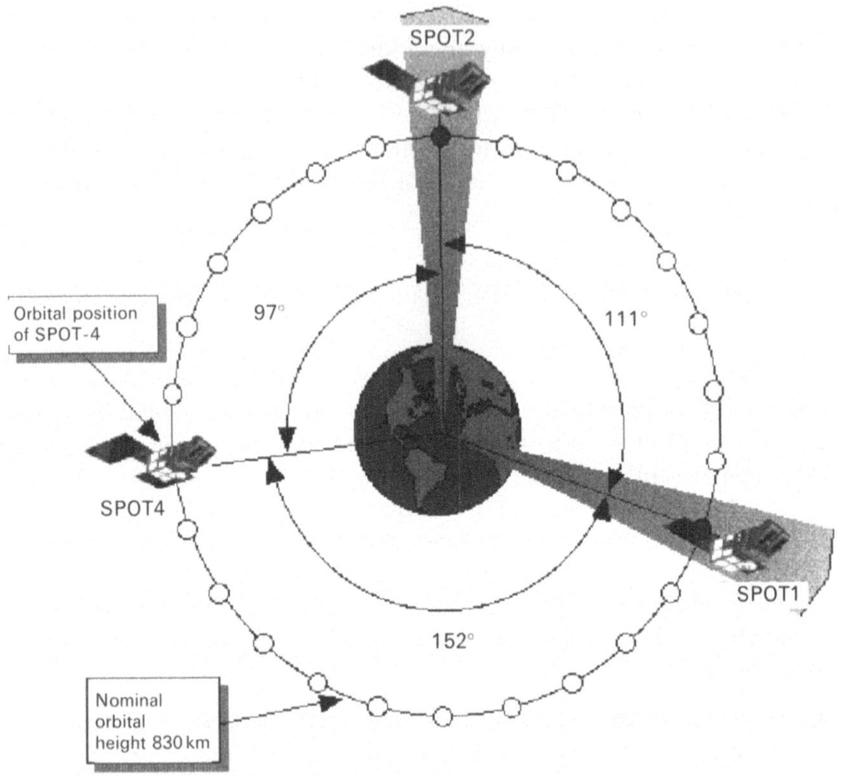

Figure 6.2.1. SPOT 1, 2 and 4 in their orbits.

Table 6.2.1. The SPOT orbit is polar, circular, sun synchronous and phased with the characteristics shown below.

Parameter	Nominal value
Altitude	818–833 km
Inclination	98, 72°
Period	101 min
Descending node local time	10 h 15
Revolutions per day	14+ 5/26
Cycle duration	26 days

and SPOT 1, SPOT 2 and SPOT 4 are positioned (see Figure 6.2.1 and Table 6.2.1) in order to optimise revisit time and stereo acquisition.

Above 40° latitude, any point is reachable every day. At the equator, on a given day, only a small strip of about, 250 km width (out of the 2,800 km separating two consecutive SPOT tracks) is not accessible.

Table 6.2.2. Stereo access per cycle (26 days) with SPOT 1, 2, 4 as a function of latitude.

Latitude	0°	30°	45°	60°
Stereo access opportunities	10	14	26	44
Number of tandem stereo acquisitions	1	3	10	15

Stereo acquisition can be performed in 'tandem mode' the same day using two of the three satellites within a 26-day cycle (the number of opportunities for stereo acquisitions are given in Table 6.2.2).

The SPOT payload

SPOT 1, 2 and 3 payload: The SPOT payload comprises two identical HRV imaging instruments, two tape recorders for image data, and a payload telemetry package for image transmission to ground receiving stations. The position of each HRV front-end mirror can be steered by ground control to observe any region of interest not necessarily vertically beneath the satellite. So, each HRV offers an oblique viewing capability, the viewing angle being adjustable through ±27° relative to the vertical. Two spectral modes of acquisition are employed, panchromatic (P) and multispectral (XS). Both HRVs can operate in either mode, either simultaneously or individually.

SPOT-4 payload: SPOT 4 is a new generation SPOT satellite with some improvements to the platform (5-year lifetime instead of 3) and to the payload including:

- Addition of a new band in the short-wave infrared (SWIR: 1.58–1.75 μm). The name of the two instruments has changed from HRV to HRVIR.
- On-board registration of all spectral bands. This has been achieved by replacing the panchromatic band (0.51–0.73 μm) by band B2 (0.61–0.68 μm) operating in both 10-m and 20-m resolution mode.
- Improved knowledge of ground reflectance, acquired by SPOT 1 and 2, is used to introduce electronic sensor gains matching according to landscape type and season, thus ensuring greater dynamic range. In addition, the imaging instruments are no longer susceptible to glare or are affected by polarisation of incident light.
- The two HRVIR imaging instruments are programmable for independent image acquisition, increasing significantly the total number of imaging opportunities. In particular, it is possible to change the viewing direction of one instrument without affecting the quality of the images acquired at the same time by the other instrument.
- The recording capacity of each of the two on-board recorders has been increased from 22 to 40 min. In addition, a solid-state memory of about 10 Gbits has been added to increase the overall reliability of on-board recording and extend design life, while ensuring greater storage capacity.

- Payload telemetry transmission to direct receiving stations is made secure in order to protect the commercial confidentiality of transmitted data.

VEGETATION: In addition to the two HRVIRs, a new instrument, VEGETATION, developed as a European cooperative project including the European Union, Belgium, Sweden, and Italy, gives a very wide view of the earth surface (2,000 km wide swathe) with a spatial resolution of about 1 km and high radiometric resolution. It uses the same spectral bands as the HRVIR instruments (B2, B3 and SWIR) plus an additional band known as B0 (0.43–0.47 μm) for oceanographic applications and for atmospheric correction. The VEGETATION programme is devoted to operational monitoring of vegetation and the study of the biosphere. The VEGETATION system acquires daily a complete picture of the continental masses in four spectral bands (blue, red, NIR, and SWIR) with ground resolution of 1 km at nadir, a radiometric resolution of 0.003 and absolute calibration accuracy better than 5%. The first model of the VEGETATION instrument was launched on board the SPOT 4 satellite. A second flight model will be flown on board SPOT 5. The data acquired are stored on board and dumped to the system's main ground station located in Kiruna (Sweden). The data are then forwarded to a dedicated centre located in Mol (Belgium) for processing, distribution, and archiving. The main system products are individual passes or 10-day global synthesis. Data distribution and the user interface are managed by Spot Image.

6.2.5 The SPOT ground segment

Ground infrastructures: The responsibilities for provision and upkeep of SPOT ground infrastructures are shared between CNES and Spot Image. CNES is in charge of:

- a Control Centre (CMP, Centre de Mission et de Maintien à Poste) for checking and controlling the satellite constellation and for maintenance of the payloads;
- an Image Quality entity (QIS, Qualité des Images SPOT) for ensuring high data quality and determining system parameters needed for image processing.

Spot Image responsibilities are about:

- image tasking;
- data acquisition;
- image archiving and processing;
- product generation and distribution.

SPOT programming: The use of multi-viewing instruments (with pointing mirrors) is greatly advantageous for optimising data acquisition of the three satellites of the constellation, each satellite being able to take two images (60–80 km wide) within a 900 km wide strip centred on the satellite track. Each user can specify his request in terms of priority (three levels are possible), geographical location, size and shape of areas to be covered, specific viewing conditions (especially for stereoscopy) and time conditions. Taking into account all these requests, from users around the world,

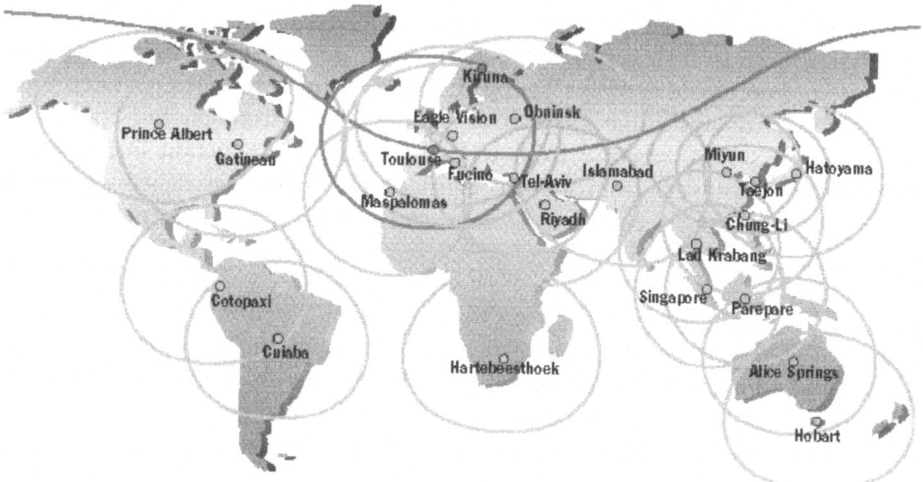

Figure 6.2.2. SPOT receiving stations.

Spot Image generates for each satellite an acquisition plan containing for each sensor a sequence of image acquisition parameters. These acquisition plans are then transmitted to the corresponding satellites through the CMP on a daily basis.

Image acquisition: According to the previous acquisition plan, each sensor of the satellites is tuned to collect images requested by users. These images can be either recorded on board or (sometimes and) transmitted in real time to the ground. A network of receiving stations (see Figure 6.2.2), covering almost all parts of the world, is used to acquire these images from the SPOT satellites. Each station can receive data from a satellite that is within its visibility circle. Two kinds of receiving station are used:

- The two main receiving stations are located at Spot Image headquarters (Toulouse, France) and at Esrange (near Kiruna in Sweden). They can receive either recorded data (acquired anywhere in the world) or real-time images.
- More than 20 other stations (SRD: Stations de Réception Directe) can only receive real-time data (corresponding to images acquired within a circle of about 2,500 km around the station).

Image archiving and processing: All data received through the main receiving stations are transformed into images using systematic and automatic processing for calibration and correction. Each image is then archived after quality assessment, location checking and cloud cover estimation. All parameters describing the image are published within the SIRIUS system, a digital catalogue of all SPOT acquisitions, which has replaced the former DALI system. This catalogue can be browsed on the Internet (http://www.spotimage.fr). It contained at the end of 2001 more than seven million SPOT images acquired since 1986.

From the archive, several types of product can be derived at several geometrical quality levels up to high cartographic accuracy, in several digital or photographic formats, and on media such as CD-ROM, film or paper printouts.

Taking advantage of the commercial and operational experience of Spot Image and of the technical expertise from CNES, the SPOT production line has been continuously improved: from archive consultation to product delivery. In 1986 delivery delay was between 2 and 3 weeks, and nowadays it has dropped down to a few days or even hours.

User services: Spot Image was set up to develop the use of SPOT data through a market-driven approach. This means that operational products and services should be provided to users, taking account of their specific needs, which are different for cartographers, farmers or telecom operator and also different in Europe, Asia, or Africa.

Dedicated products and services are provided by Spot Image (Munier, 2000) to users, who buy them directly from Spot Image in Toulouse or through its commercial network. Four subsidiaries (Spot Image Corporation in the USA, Spot Imaging Services in Australia, Spot Asia in Singapore and Beijing Spot Image in China), 23 receiving stations and 90 distributors in 80 countries are serving users from any country to satisfy their requests about any place of the world.

For mapping and the defence community, which represents about 40% of the SPOT market, images can be used to produce spacemaps at various scales from 1:100,000 to 1:50,000. Some cartographic work has been done to produce maps at larger scales for updating old maps, such as cadastres in agricultural areas like the one produced in Argentina. More and more digital images are integrated in GIS (Geographical Information Systems) with other geographic databases (road network, land cover, etc.). As an example, in France the SPOTView is compatible with BD CARTO (the cartographic database of IGN).

In agriculture, two main applications are already using SPOT data: crop statistics, especially for the European Commission or for the US Ministry of Agriculture, and control of landowners declarations, for surveillance by the PAC (Common European Agriculture Policy). Precision farming is increasingly looking at the use of satellite data on a regular basis to optimise the use of seeds and fertilisers. For this application, SPOT is already used, associated with other data sources.

Cellular phone operators are new customers of DEM (Digital Elevation Model) and Land Cover maps in order to optimise their base network. Specific SPOT Telecom packages are produced with industrial partners to satisfy this quickly growing market.

For land planning, SPOT Thema is also available, which gives a land cover map. Two levels of details are possible: the map contains either eight land cover classes or 29 classes.

Note that Spot Image is proposing not only SPOT data and products but also radar images from ERS (and soon from ENVISAT) and from RADARSAT as well as other optical images.

It has been acknowledged that the market-driven approach, well suited to providing services that go beyond data distribution, should be complemented by plans to help scientists to use space-based observational data for their research. Therefore, a dedicated program, ISIS, was implemented in 2000. It permits French scientists who need SPOT data to get images from archives or new acquisitions at small costs. This program was extended in February 2001 to European scientists and also to any demonstration work within the GMES (Global Monitoring for Environment and Security) initiative.

Another initiative is the Charter on Major Risks and Disasters, introduced by CNES and ESA in 1999 during the UNISPACE-III conference in Vienna and signed in June 2000. Other space agencies such as CSA for Canada, ISRO for India and NOAA for the USA then joined the two founders. This Charter is aimed at providing satellite images as quickly as possible when a disaster occurs, at the request of rescue organisations.

6.2.6 The near future: SPOT 5

SPOT 5 programme

The SPOT 5 satellite is undergoing a final development stage so as to be ready for launch in April 2002. It will bring new improvements, especially in terms of ground resolution.

The SPOT-5 programme was given the go-ahead by the French government in 1994 and was redefined in 1996 with authorisation to improve ground resolution below 5 m (Baudoin, 1999). An innovative image acquisition and processing method was developed by CNES to reach 3 m resolution from two 5 m images. More recently, an agreement between CNES, Spot Image and Matra-Marconi-Space (now Astrium), signed in January 1999, gives SPOT 5 full stereoscopic capacity with the HRS instrument. It was also decided to continue with the use of the VEGETATION instrument on board SPOT 5.

SPOT 5 mission

The main mission requirements, set up by the user community and Spot Image who together considered their experience from exploitation of the previous satellite, are the following:

- Image acquisition and service continuity with previous SPOT satellites, in order to satisfy current users who have already invested in ground equipment and software for their applications. This is facilitated using the same sun synchronous orbit with a 26-day cycle, the same off-track viewing capacity of $\pm 27°$, the same spectral bands and the same 60 km swathe width for a vertical view.
- Improvement of ground resolution: at 3 m in the panchromatic band and at 10 m in the multispectral mode (except for the SWIR which remains at 20 m).
- Simultaneous stereoscopic acquisition (fore and aft instead of side looking) for generating DEM (Digital Elevation Model) at an accuracy of 10 m.

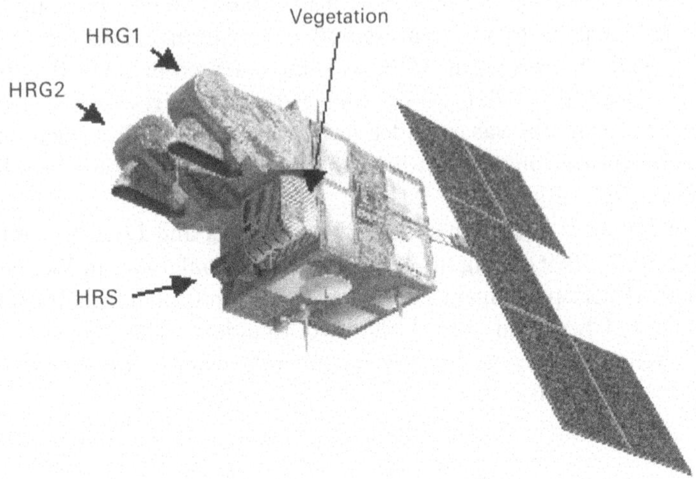

Figure 6.2.3. SPOT 5 with its payload.

- Lifetime of the mission more than 5 years with a useful viewing capacity of about 42,000 km^2 per year, including 6,000 km^2 covered by stereoscopic data.

SPOT 5 technical characteristics

System components: The SPOT 5 system is composed of:

- A space segment: the satellite and its payloads (see Figure 6.2.3).
- Two main ground segments: SSOS (segment sol de l'opérateur satellite: ground segment of the satellite operator) under the responsibility of CNES, which is in charge of keeping the satellite on its nominal orbit and all the instruments with their nominal functioning parameters; SSOC (segment sol de l'opérateur commercial: ground segment of the commercial operator) under the responsibility of Spot Image, which is in charge of acquisition, programming, data reception, image processing and product distribution.

High-Resolution Geometric instruments (HRG): Two High-Resolution Geometric instruments (HRG) will give SPOT 5 the capacity to pursue and improve the HRVIR SPOT 4 service. Each instrument has the same field of view (4.13°) and the same off-track pointing capabilities (±27°) as the HRVIR instrument on SPOT 4, giving the opportunity to cover 60 km (vertical view) and up to 80 km for an oblique view. Improvements from SPOT 1, 2, 3 and SPOT 4 to SPOT 5 are the following:

- Panchromatic images are acquired using two lines of 12,000 detectors at 6.5 μm giving 5-m resolution at nadir. These two lines can be combined to achieve better resolution and a pixel size of 2.5 m (see Table 6.2.3).
- Multispectral images are acquired with 6,000 pixels per line in three bands B1, B2, B3 and 3,000 pixels in the SWIR band.
- Compression rate and transmission (up to a factor of 2).

Table 6.2.3. Spectral band continuity and resolution improvement from SPOT 1 to SPOT 5.

Detector line	Spectral range (μm)		Ground resolution			
			Spot 1, 2, 3 (m)	SPOT 4 (m)	SPOT 5	
					On board (m)	Supermode
PA1	Panchromatic	0.49–0.69	10		5	2.5 m
PA2	Panchromatic	0.49–0.69			5	2.5 m
B1	Green	0.49–0.61	20	20	10	From 10 m down to 2.5 m
B2	Red	0.61–0.68	20	10	10	From 10 m down to 2.5 m
B3	Near-infrared	0.78–0.89	20	20	10	From 10 m down to 2.5 m
MIR	Short wave infrared	1.58–1.75	–	20	20	From 10 m down to 2.5 m

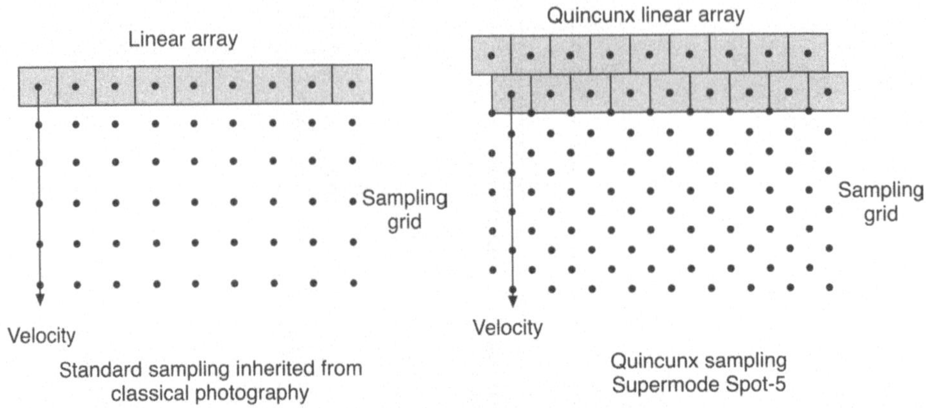

Figure 6.2.4. Supermode principle.

Improved location accuracy (10 m after ground processing) is reached using a star tracker, not present on previous SPOT satellites.

Supermode – a new high-resolution acquisition mode: Each of the two HRG instruments has two new panchromatic linear arrays of detectors (instead of one in previous SPOT sensors). Each of the detector lines can produce 5-m pixel size images, and it is possible to combine the two 5-m images of the same area into a new 2.5-m pixel size image (see Figure 6.2.4). Specific image processing software, developed by CNES, is used to reconstruct the final image after three steps: interleaving, interpolation and restoration. This new type of image, simulated with all its geometric and radiometric characteristics, has been compared by several users with other types of digital image at different ground resolutions. These users (cartographers, urban planners, environmental experts, foresters, agronomists) have concluded that images derived from Supermode technology have the same identification capability as classical images with 2.5–3.5-m resolution, depending on application and analysed features.

The High-Resolution Stereoscopic instrument (HRS): This new instrument (Rouziès *et al.*, 1999) will be used to generate DEM over very large areas with good altimetric accuracy (5–10 m for height measurements, 10–15 m for absolute elevation). The main characteristics of this instrument are given in Table 6.2.4.

SPOT 5 product and service evolution: Products derived from SPOT data are defined by Spot Image and continuously upgraded according to user needs and technical evolution. Products from new SPOT 5 instruments, HRV and HRS will offer SPOT continuity with improved accuracy due to better instrument performance (especially in terms of resolution and location accuracy). Among possible products, still to be defined in detail, are basic products such as:

- X1: multispectral (10-m B1-B2-B3 and 20-m SWIR);
- TH1: Supermode, 2.5-m sampling for two 5-m panchromatic images;

or derived products such as:

Table 6.2.4. HRS characteristics.

Parameter	Nominal value
Spectral band	Panchromatic: 0.5–0.7 μm
Swathe width	120 km
Maximum continuous acquisition length	600 km
Pointing angles	±20°
Ground resolution	10 m
Line sampling	10 m
Column sampling (parallax direction)	5 m

- H4: 5-m panchromatic combined with 10-m B1-B2-B3 and 20-m SWIR;
- TH2: 2.5-m panchromatic combined with 10-m B1-B2-B3.

Several geometric quality levels are envisaged, using system corrections on their own (level 1) or combined with Ground Control Points (level 2) and even Digital Elevation Models (orthoimage).

Services will be improved using new computer and communication technologies, especially for helping clients:

- to formulate their requests (from archives, through a catalogue or for a new acquisition by programming the satellite to look at a requested target);
- to follow the production status of their request;
- to receive their products as quickly as possible.

6.2.7 A new programme: Pléiades

Pléiades (Baudoin *et al.*, 2001) is a new earth observation satellite concept proposed by CNES, based on the 15-year-old SPOT experience and taking into account the evolution of user requirements, technological capabilities and worldwide scope. This concept was defined between 1998 and 2000. The main objectives and orientations of the programme are to fulfil user needs and to look for a larger European cooperation and for dual exploitation by civilian and defence partners while encouraging technical innovation.

Overall user needs and market demand

Analysis of the Pléiades mission has been performed by CNES with the help of a User Advisory Committee. The main French actors in earth observation have provided their support to this analysis. Results and recommendations from several meetings with different user communities organised by the Ministry of Education, Research and Technology (MENRT) within the framework of its 'Espace et Société' initiative were taken into account, as well as those from two dedicated workshops on Pléiades. The GMES initiative has also been taken into account to assess the

Table 6.2.5. Pléiades component user requirements.

Sensor	Resolution (m)	Swathe width (km)	Number of bands*	Revisit time (days)	Main applications
Wide field	2–5	40–100	3–4	3–7	Cartography, geology, agriculture, forestry, hydrology
Optical HR	≤1	10–30	3–4	1–2	Cartography, risk, forestry, geology
Superspectral	3–10	100–300	6–20	1–2	Agriculture, forestry, geology
Hyperspectral	5–20	50–300	30–200	2–7	Geology
Thermal	1–40	100	to be determined	<1	Forest fires, geology, oceans
SAR C	2–4	50–300	1–2	1–5	Geology, oceans, hydrology, geology
SAR X	1–5	10–300	1–4	<1	Hydrology, risk, forestry, oceans
SAR L	2–10	50–100	1–4	1–7	Geology, forestry, hydrology, oceans
SAR P	5–10	70–100	1–4	1–7	Hydrology, geology, forestry
Interferometric cartwheel	1–5	70–100	NA	NA	DEM for cartography, risk, hydrology

evolution of institutional demand beyond that of the European and worldwide market.

Systems to be envisaged

Ten types of system have been identified to fulfil most user needs, as described in Table 6.2.5.

SAR and optical sensors are both required for most applications, the first mainly for its all-weather capacity and the second for its better visual interpretation. It is also acknowledged that external data from ground or airborne surveys are needed to get reliable and accurate information from space imagery in most cases.

Pléiades implementation

The implementation of these ten possible missions has been envisaged through several programme frameworks. Systems not yet considered as operational in the near future should be studied within ESA research programmes: P band SAR, thermal and hyperspectral imagery. When systems are already available or under development in Europe, they should be considered as a primary source of data and their continuity should be envisaged in the mid-term future: C band SAR (with ERS and ENVISAT) and Wide Field (with SPOT).

Development of dual systems is not considered the responsibility of ESA, due to legal constraints in its Convention. This applies to high-resolution systems (optical and X band radar), which will be provided through a French–Italian agreement (see below). Nevertheless, minor participation of ESA member states to development of these systems is envisaged and European users could benefit from them.

New operational systems for which there is a great economic interest, such as the superspectral system (and the Wide Field system to be built after SPOT 5), could be brought about by a Public Private Partnership (PPP). This PPP could be implemented within the ESA earthWatch programme or through bilateral or multilateral cooperation schemes.

Cooperation between France and Italy on a dual system

Since the beginning of 1999, the possibility of joint cooperation between nations was looked at and the most fruitful cooperation scheme was found with Italy, as its approach to its own COSMO-SkyMed programme was very similar to Pléiades, resulting in responsibilities and work being shared between the two countries. Both countries were looking at dual systems and the French and Italian ministries of Research and Ministries of Defence agreed in March 2000 to study possible joint cooperation on earth observation, taking as reference points the Pléiades and COSMO-SkyMed projects of the French and Italian space agencies.

This possibility was confirmed by the Turin Agreement, signed by Lionel Jospin and Giuliano Amato, the heads of government of the two countries, on 29 January 2001. This intergovernmental agreement is an Umbrella Memorandum of Understanding, which outlines the overall objectives and implementation rules of the cooperation. Its main purpose is the development and exploitation of a Dual System, with two high-resolution components: one with *two optical satellites under the leadership of France*, the other with *four SAR-X satellites under the leadership of Italy*. Extension of the cooperation to other countries has been looked into, and both France and Italy are proposing such a possibility to other ESA member states.

Overall mission specifications of the Dual System: Pléiades and COSMO-SkyMed mission studies have identified the need for high-resolution data, both from optical and from radar systems. The mission requirements and technical performances for the Dual System are now ready and should be used as a reference for the definition phase of the Dual System. Beside specific defence needs and constraints, the Dual System is specified to fulfil a broad spectrum of applications, in the field of cartography, agriculture and forestry, geology and hydrology, marine applications, earth science, resource management, land use and law enforcement, according to scientific, institutional and commercial users. Risk management, monitoring of coastal zones and control of sea pollution are considered important applications to be satisfied. The system will have night/daylight and all-weather observation capability all over the world. It will provide multi-sensor observation capacities that:

- ensure, at lower cost, service continuity of existing optical observation capacities (SPOT) and improve theie performance (resolution, stereoscopic capacity, image quality, acquisition times, amount of data,...);
- give access to a radar observation capacity with interferometric capabilities.

Figure 6.2.5. Pléiades HR satellite.

In order to complement these multi-mission capabilities, the system will be able to integrate data/products coming from external optical/radar systems (access to catalogue, programme requests, product processing and delivery, . . .).

The high-resolution optical component

Dual System optical high-resolution specifications: Taking user requirements into account (both civilian and defence) as well as technical feasibility and cost trade-off, the following proposed mission specifications have been agreed between France and Italy, for a constellation of two satellites (see Figure 6.2.5) to be launched in 2005 and 2006:

- Bands and ground resolution: 0.7 m at nadir for panchromatic band and 2.8 m at nadir for blue, green, red and near-infrared (four bands).
- Swathe: 21 km at nadir.
- Revisit time: equal or better than 24 hours.
- Coverage capacity: better than 250 images per day and per satellite (20,000 km^2 per orbit).

Technological performance and innovation: Phase A study of the Pléiades optical HR system took place in 2000 and ended at the beginning of 2001. The satellite was designed and built under contract, with Astrium-F acting as the prime contractor for

Table 6.2.6. Comparison of SPOT 1, SPOT 5 and Pléiades satellite characteristics and performances.

Parameter	SPOT 1	SPOT 5	Pléiades HR
Mass	1.8 t	3.0 t	0.9 t
Ground resolution	10 m	2.5 m	0.7 m
Data transmission rate	50 Mbit/s	300 Mbit/s	2 Gbit/s
Image compression rate	1.3	2.8	5
Processing rate	8 Mops/s	750 Mops/s	10 Gops/s

the satellite and Alcatel Space Industries responsible for the instrument. The technical feasibility of the satellite, despite its very demanding specifications, was demonstrated with new technological developments among which were high-performance fibre-optic gyroscopes, control moment gyros, TDIs and new generation solid-state recorders.

For recurring missions, the availability of smaller and more energy-economic equipment or instruments has allowed the size (and cost) of new generation satellites to decrease dramatically compared with those of previous generations. More compact telescopes and smaller equipment have allowed rigid yet agile imagery satellites to be designed where off-nadir pointing is achieved by steering the entire spacecraft body. The absence of a tilt mirror, in turn, enables the telescope to be more compact.

Table 6.2.6 shows the evolution of some characteristics and performances from SPOT 1 to Pléiades HR satellites.

The SAR-X component

The SAR-X component of the Dual System will be developed under the leadership of Italy. The four satellites (see Figure 6.2.6) are to be launched between the end of 2003 and the end of 2005. With these four satellites, the revisit time requirement should be achieved (i.e. better than 12 hours).

The fundamental characteristics of SAR payload operational modes can be broken down into those of modes with one polarisation selectable from HH, VV, HV or VH:

- SPOTLIGHT (resolution: order of 1 m and less; spot area: 10×10^2).
- HIMAGE (Stripmap) (resolution: few meters; swathe width: several tens of km).
- WIDEREGION (ScanSAR) (resolution: few tens of meters; swathe width: hundreds of km).
- HUGEREGION (ScanSAR) (resolution: several tens of meters; swathe width: few hundreds of km).

Figure 6.2.6. SAR-X Cosmo-SkyMed satellite.

Modes with two polarisations selectable from HH, VV, HV or VH:

- PING PONG (Stripmap) (resolution: few meters; swathe width: several tens of km).

The interferometric cartwheel

Concept: The interferometric cartwheel is an innovative concept, proposed by CNES (Massonnet, 1999), using several (three or four) micro-satellites as SAR passive receivers 'cooperating' with a main satellite acting as an active SAR transmitter. The satellites are placed on an orbit very similar to that of the main satellite, within the same orbital plane and with the same orbital period but with slightly different elliptical shapes. This gives the micro-satellites a relative motion (evoking a wheel with a diameter of a few kilometres rotating typically 150 km ahead of the main satellite). Its main interest is its capability to generate low-cost high-accuracy DEM (typically 1 m accuracy), but it can also be used for many applications, such as ocean current studies.

Development: Studies on this new concept have been undertaken by CNES in co-operation with NASDA (for using ALOS L band PALSAR), as well as with ESA and DLR (for using ENVISAT C band SAR). The definition phase of the project is in progress, while the final decision about system configuration (using either ALOS or ENVISAT) is expected in 2002 with the objective of launching the satellite constellation around 2005.

Other components

The wide field: Besides the need of high (metric) resolution satellite there is still a need for sensors with wider coverage (typically 50 km) with even lower ground resolution (2–3 m). This need has been evaluated by CNES and Spot Image and confirmed by the ERSIS study performed by Alcatel, Alénia, and Astrium for ESA (ERSIS Final Report, 2000).

SPOT 5, to be launched in 2002, should fulfil this need right up to 2007. To

Table 6.2.7. User requirements for a superspectral component.

Application	Agriculture	Forest	Land, littoral, urban	Geology and risks
Ground resolution (m)	3–20	5–20	2–5	10
Image coverage (km)	50–300	100	50	50
Revisit time (days)	1–2	6–15	–	15–30

ensure mission continuity, a new small satellite system, named 3S (successor to the SPOT satellites) has been defined and may be given the go-ahead next year.

Superspectral components: For many years, CNES in partnership with industry and research centres has conducted intensive studies into superspectral remote sensing, with particular emphasis on agricultural and ecological applications. Theoretical studies have been supported by experimental campaigns with ground, airborne and satellite data. Requirements for such a superspectral system are given in Table 6.2.7.

Technical studies have shown that this superspectral mission could be achieved either with small satellite (synergies with the previous Wide Field mission could be envisaged) or with micro-satellites.

6.2.8 Conclusion

The French space programme of earth observation satellites, vigorously developed in the 1980s, has reached the position of being the first 'user' programme within CNES, which takes up about 30% of its budget. Deeply based on international cooperation, mainly European, this programme covers a wide range of interests: from advanced research in all the earth science disciplines to the development of applications of all kinds and from cartography to natural resource survey and risk management. This breadth of interest has played an important role in the development of the value-added service industry, using satellite data to obtain information of use to increasingly many users.

Earth observation is a domain of excellence of both the French space industry and the scientific community, and will certainly stay at the forefront of the national space programme in the coming decades.

In the very near future, SPOT 5 will transmit better quality images of the Earth. This will be followed by Pléiades symbolising the French contribution, within a European cooperation framework, to the development of space applications.

ACKNOWLEDGEMENTS

The author expresses his thanks to Gérard Brachet, CNES Director General, and to Michel Arnaud, Deputy Director at CNES for reviewing this paper and to all authors of the references for their contribution to this work.

6.2.9 References

Baudoin, A. (1999) The current and future SPOT program, ISPRS Workshop, Hanover, Germany, September.

Baudoin, A., Boussarie, E., Damilano, P., Rum, G. and Caltagirone, F. (2001) Pléiades: a Multi Mission and Multi Cooperative Program, 52nd International Astronautical Congress, Toulouse, France, October.

Brachet, G. (1985) Le programme SPOT, historique, objectifs, description du système et organisation générale. *Bulletin de la SFPT*, **100**, 4.

Brachet, G. (2000) 'SPOT/HELIOS systems: An established know-how. *Air & Space Europe*, **2**(4), August.

ERSIS final report (2000) ESA Contract 13592/99/NL/SF (April).

Fellous, J. L., Massonnet, D., Baudoin, A., Buthaud, H., Papineau, N. and Podaire, A. (2000) The French space program for earth science and applications, ISTS 2000, Morioka, Japan, May.

Goudy, P. (1999) De Spot 1 à Spot 4. *Bulletin de la SFPT*, **154**, 2.

Munier, P. (2000) L'aventure SPOT, du temps des pionniers à la haute résolution. *Géomètre*, **10** (October).

Massonnet, D. (1999) Capabilities and limitations of the interferometric cartwheel, CEOS SAR Workshop, Toulouse, France, October.

Rouziès, C., Fratter, C., Bernard, M. and Maggiori, L. (1999) HRS: A massive along track stereoscopic capability for SPOT 5, ISPRS Workshop, Hanover, Germany, September.

6.3 THE COMMERCIAL REMOTE SENSING SATELLITE PROGRAMME IN INDIA *(by R. Krishnan)*

6.3.1 Introduction

The Indian Remote Sensing Satellite programme is conceptualised, developed and operated by the Indian Space Research Organisation (ISRO), which is under the Department of Space, Government of India. India has developed capabilities to design, build and launch remote sensing satellites during the last 25 years. Agricultural Resources Inventory and Survey Experiment (ARISE) was the first step taken by ISRO in remote sensing applications. A comprehensive inventory of major agricultural crops and crop acreage was the motivation for this airborne experiment, which used colour infrared film and a Dakota aircraft. Balloons were also used from the height of about 25 km to gather multispectral imagery. The launch of Aryabhata on 19 April 1975 from a Soviet rocket signalled the beginning of India's satellite endeavour (Kasturirangan 1999).

In this section, a brief history of the Indian remote sensing Satellite programme as well as the future plans is given.

6.3.2 Early missions

The early remote sensing missions launched by ISRO comprised of the two Bhaskara satellites (1979 and 1981). The imaging sensor used in the Bhaskara satellite was the

Frame Imager, which had a two-channel videocon. Its ground resolution was 1000 m and FOV 341 × 341 m. The data rate was <100 kilobits/second. The Bhaskara missions gave ISRO the necessary hands-on experience in the area of remote sensing satellites. ISRO also conducted a Joint Experiments Project to understand the user requirements for a remote sensing satellite. This enabled ISRO to plan the next generation satellites, which has put ISRO in the forefront of nations having the remote sensing capabilities.

6.3.3 The IRS missions: the first generation

The Indian Remote Sensing Satellite (IRS) programme began with the launch of the IRS-1A satellite on 17 March 1988. This was a sun synchronous satellite with an altitude of 980 km. It carried two sensors namely Linear Imaging Self-scanned Sensor (LISS-1) and LISS-2 with 72.5 and 36.25 ground resolution. This was a 3-axis stabilised satellite. An image acquired by this sensor over Goa, India, is shown in Figure 6.3.1. The second satellite, IRS-1B, was launched in 1991. Table 6.3.1 shows the characteristics of IRS-1A and IRS-1B. These satellites essentially served the purpose of gathering information regarding agricultural, geological and hydrological resources in association with the user agencies. This enabled the design of a future set of operational remote sensing missions.

6.3.4 IRS: second generation

The launch of IRS-1C in 1995 gave India the unique distinction of having, at that time, the best spatial resolution of 5.8 m among all available commercial satellites (see Figure 6.3.2). This high resolution with a swath width of 70 km called for several design innovations in optics like the fabrication of aspheric mirror surfaces, camera tilt mechanism in preference to a scan mirror, etc. It also carried a LISS-3 sensor (see Figure 6.3.3) with a 23.6 m spatial resolution and a WIFS sensor with 188 m. IRS-1D, launched in 1997 was a satellite identical to IRS-1C. The characteristics of these are summarised in Table 6.3.2.

Figure 6.3.4 shows an image acquired by the LISS-III sensor. This is a mosaic of the North Easter Days. Mosaics are prepared by joining adjacent scenes and performing the required radiometric adjustments; by using different sensors; and they can be prepared in different scales.

A panchromatic image (5.8 m resolution) over Mumbai, India, was merged with that acquired by the LISS-3 sensor (23.5 m resolution). This was to take the advantage of the good spatial and spectral resolutions of the panchromatic sensor and the LISS-3 respectively. The merged image is shown in Figure 6.3.5. Such products are available at different scales (1:50,000; 1:25,000; 1:12,000).

6.3.5 IRS: P series

ISRO has developed an indigenous Polar Satellite Launch Vehicle called PSLV. The developmental flights of PSLV were used to launch a number of experimental/

Figure 6.3.1. This shows an image acquired over Goa, India, by the IRS-1A LISS-1 sensor (source: CPG/NDC/RDQE).

Table 6.3.1. The characteristics of IRA 1A and 1B satellites.

IRS 1A / 1B	
Sun Synchronous, 980 km altitude	
10.30 AM Descending node	
LISS-1: 4 bands	
	0.45–0.52 μ
	0.52–0.59 μ
	0.62–0.68 μ
	0.77–0.86 μ
	Spatial resolution 72.5 m; Swath: 148 km
LISS-2: 4 bands same as LISS-1	
	Spatial resolution 36.25 m; Swath: 145 km

quasi-operational missions. These were called the IRS-P series consisting of IRS-P2, -P3 and -P4. The IRS-P2 carried only a LISS-2 camera with 32.74 m spatial resolution. The IRS-P3 had the German Modular Opto-electronic system, which spanned a large number of narrow spectral channels for ocean observations. IRS-P4 carried

Figure 6.3.2. This is an extract from a full IRS Pan scene over the city of Houston, USA (resolution 5.8 m).

the Ocean Colour Monitor (OCM) which is superior to those available from other missions. Tables 6.3.3, 6.3.4 and 6.3.5 show the characteristics of the P series of sensors.

Examples of images acquired by some of these sensors are shown in Figures 6.3.6, 6.3.7, 6.3.8 and 6.3.9.

6.3.6 Future missions

IRS-PS and IRS-P6 are scheduled for launch in early February and June 2002, respectively. The latter will carry a LISS-4 sensor with three bands and have a spatial resolution of 5.8 m. IRS-P5, slated for 2002 will have a 2.5 m, IFOV and fore and aft cameras and is specially suited for digital elevation generation. The CARTOSAT-2 will have a 1 m ground resolution with a PAN sensor. ISRO also plans to launch a satellite called MEGHA-TROPIQUES with a view to allow frequent simultaneous observations of the atmospheric water cycle.

6.3.7 Use of IRS data

The IRS data is available to Indian users through the National Remote Sensing Agency (NRSA). Besides the Indian stations, there are also international ground

Figure 6.3.3. This shows a scene over Goa, India acquired by the LISS-III sensor on board IRS-1C satellite.

Table 6.3.2. Summary of the characteristics of IRS-1C and -1D remote sensing satellites.

IRS-1C/1D	
PAN:	1 band
	0.5–0.75 μ
	Spatial resolution 5.8 m; Swath: 70.5 km
LISS-3:	4 bands
	0.45–0.52 μ
	0.52–0.59 μ with spatial resolution 23.6 m; Swath: 141 km
	0.62–0.68 μ
	1.55–1.70 μ
	with spatial resolution 70.8 m; Swath: 148 km
WIFS:	2 bands
	0.62–0.68 μ
	0.77–0.86 μ
	with spatial resolution 188 m: Swath: 774 km

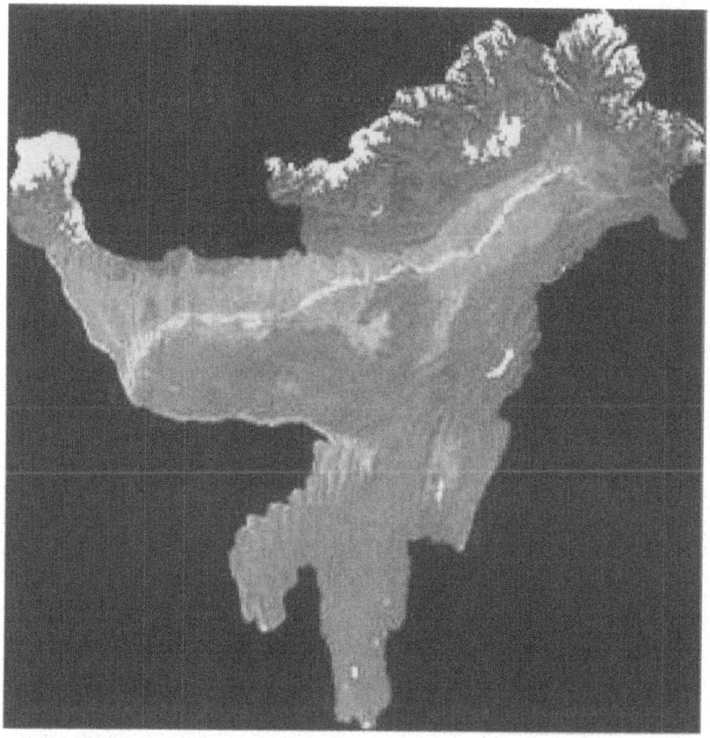

Figure 6.3.4. Mosaic of the North Easter Days by LISS III sensor.

stations in Southern Korea, Japan, Germany, Abu Dhabi, Thailand, Ecuador and the USA, which receive IRS data.

The use of IRS data in India spans a wide spectrum that includes agriculture, urban studies, cartography, forestry, land use, hydrology, and ocean applications. Several operational packages for remote sensing data application in different fields have been developed. The sales figures available from the NRSA data centre show that:

1. The usage of IRS PAN data has overtaken that of LISS-III. From a 60:30 ratio for LISS-III: PAN, a couple of years ago, it is now a 40:50.
2. The photographic product sale compared to digital products has gone from 53:47 in 1996 to 18:82 in 2000. This shows a dramatic increase in the users' preference for digital products.
3. The latter satellites, like 1C and 1D, are preferred to the earlier 1A and 1B for obvious reasons.
4. There has been an eight-fold increase in the revenue earned from satellite data over the period 1987–2000.
5. The private sector's usage of satellite data has increased by a factor of 3 in the last six years, indicating the maturity of the remote sensing technology. In fact,

Figure 6.3.5. A panchromatic image (5.8 m resolution) over the city of Mumbai was merged with that acquired by the LISS-3 sensor.

in revenue terms, the increase has been more than 15 times indicating the readiness of the sector to pay for customised high-quality products.

6 The average turn around time for products has decreased by a factor of 5 in the last decade.

The IRS data is available without any restrictions. Data acquired within the footprint of the Indian station (essentially India and neighbouring countries) can be got from the NRSA, Hyderabad, India. All other data could be obtained from the Space Imaging Corporation, USA. The data is essentially available within the reach of the ground stations. Pricing information, can be obtained from either the NRSA or Space Imaging.

6.3.8 Conclusion

This section has given brief outline of the indian remote sensing programme. Studies and technology development efforts are underway on future satellites with higher spatial and spectral resolutions. As regards to the future, Vikram Sarabhai, the father of the Indian space programme, has said: '...if we are convinced that we are to play a meaningful role nationally, and in the community of nations, we must

Table 6.3.3. The characteristics of the IRS-P2 and -3 satellites.

IRS-P2	
LISS-2	with spatial resolution 32.74 m; Swath 145 km
IRS-P3	
WIFS:	3 bands
	0.62–0.68 μ
	0.77–0.86 μ
	1.50–1.75 μ
	with spatial resolution 188 m; Swath: 810 km
MOS-A:	4 bands
	757 nano-meters
	760.6 nano-meters
	763.3 nano-meters
	766.6 nano-meters
	with spatial resolution 1570 m; Swath: 195 km
MOS-B:	13 bands
	408 nano-meters
	443 nano-meters
	485 nano-meters
	520 nano-meters
	570 nano-meters
	615 nano-meters
	650 nano-meters
	685 nano-meters
	750 nano-meters
	815 nano-meters
	870 nano-meters
	945 nano-meters
	1010 nano-meters
	with spatial resolution 525 m; Swath: 200 km
MOS-C:	1600 nano-meters
	with spatial resolution 645 m: Swath: 192 km

be second to none in the application of advanced technologies to the real problems of man and society . . .'. To this end, the future Indian remote sensing satellites may well play an important role in the verification of treaties.

6.3.9 Acknowledgements

The author wishes to acknowledge the inputs provided by Sri Mukund Rao, ISRO and Sri V. Raghu, NRSA.

Table 6.3.4. The characteristics of the IRS-P4 satellites.

IRS-P4
OCM: 8 narrow bands in (VNIR)
(Ocean Colour Monitor)
402–422 nano-meters
433–453 nano-meters
480–500 nano-meters
500–520 nano-meters
545–565 nano-meters
660–680 nano-meters
745–785 nano-meters
845–885 nano-meters
with spatial resolution 360 m; Swath: 1420 km

MSMR: 4 frequencies
6.6 GHz range
10.65 GHz range
18.00 GHz range
21.00 GHz range
with spatial resolution 120 km, 80 km and 40 km; Swath: 1360 km

Table 6.3.5. The characteristics of the IRS-P5 and -P6 satellites.

IRS-P6
LISS-4: with spectral bands
0.52–0.59 μ
0.62–0.68 μ
0.77–0.86 μ
with spatial resolution 5.8 m; Swath: 23.9 km
\+ LISS-3
IRS-P5
2 PAN cameras with Fore and Aft (26° and −5°) viewing
with spatial resolution 2.5 m; Swath: 30 km

CARTOSAT-2
PAN with one band 0.5–0.8 μ
With spatial resolution 1 m nominal; Swath: 12 km

Megha-Tropiques
6 channels
S1: 183.31+/− 0.20 GHz
S2: 183.31+/− 1.1 GHz
S3: 183.31+/− 2.5 GHz
S4: 183.31+/− 4 GHz
S5: 183.31+/− 6.4 GHz
S6: 183.31+/− 9 GHz
With spatial resolution 10 m.

Figure 6.3.6. This is an image acquired by the Wide Field Sensor (WiFS) on board the IRS-P3 satellite.

6.3.10 References

Kasturirangan, K., Remote Sensing Satellites – Challenges to Applied Physics (1999) Shri Murali M. Chugani Memorial Award Lecture, Mumbai.

6.4 ISRAEL'S COMMERCIAL REMOTE SENSING SATELLITE PROGRAM *(by G. M. Steinberg)*

6.4.1 Introduction

Israel has become a significant actor in the area of space imaging, both in terms of its own dedicated military reconnaissance satellites, and also with the development of an advanced ultra-high resolution commercial space imaging system. These capabilities were developed of necessity, based on the threat environment in the Middle East and Persian Gulf, and the need for early warning and the ability to track military

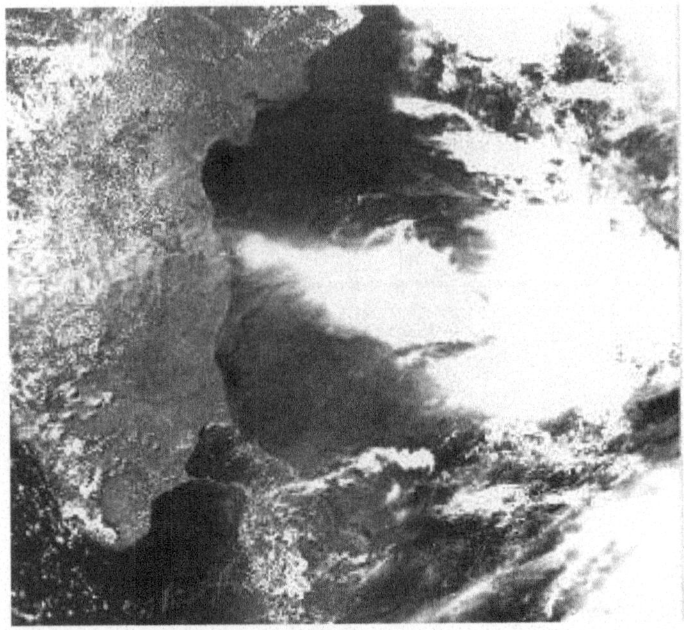

Figure 6.3.7. This is an image over south-east India and Sri Lanka acquired by the Ocean Colour Monitor (OCM) on board the IRS-P4 satellite (see Plate 18).

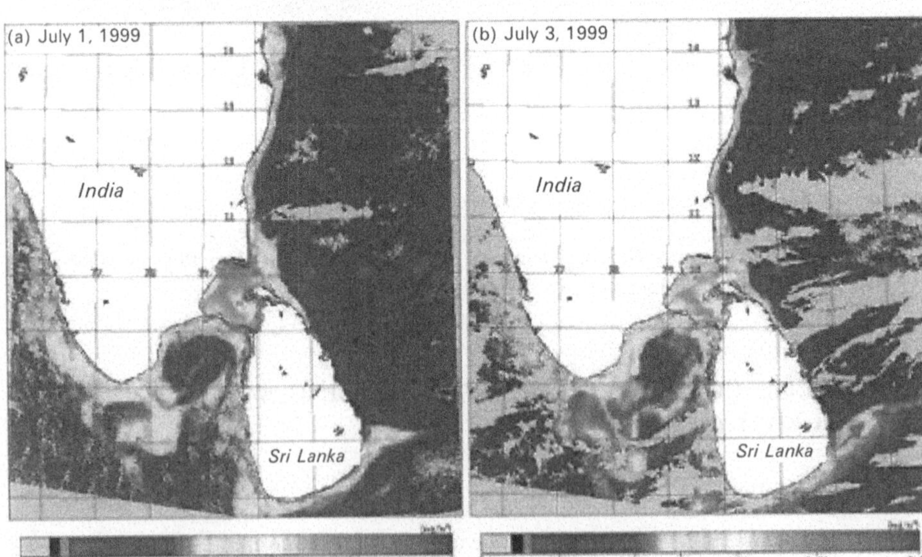

Figure 6.3.8. Two images over the southern Indian peninsular acquired over two days to determine chlorophyll-a using the OCM sensor (source: SAC/ISRO).

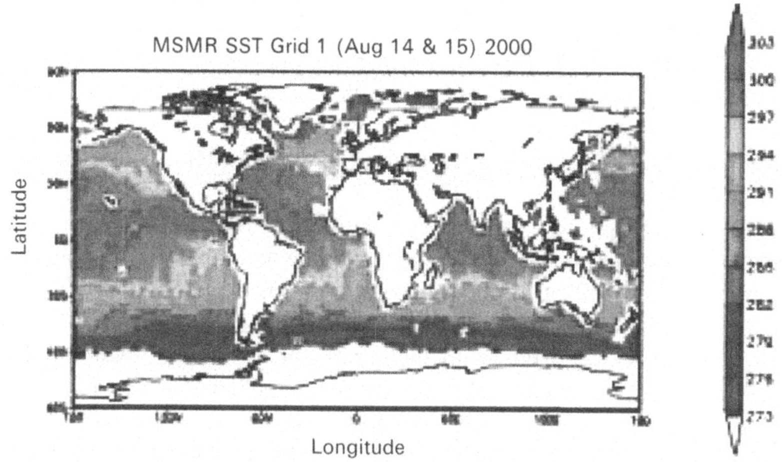

Figure 6.3.9. Sea surface temperature observed by the Multi-frequency Scanning Microwave Radiometer (MSMR) on board the IRS-P4 satellite.

developments, particularly with respect to the acquisition of missiles and weapons of mass destruction.

In this chapter, the technical nature of Israel's remote sensing capability will be described and analysed. A particular emphasis is given to the Earth Remote Observation System (EROS), Israel's first commercial imaging satellite system. Israeli views and policies with respect to the dual-use nature of such commercial imaging satellites are also considered, including the concerns regarding the destabilising impacts of the proliferation of this technology in high-conflict areas such as the Middle East. In the third section, we will examine the potential application of commercial imaging from space to verification of arms limitation and regional security frameworks in the Middle East, with the understanding that under conditions of intense conflict, the prospects for agreements on co-operative monitoring and verification are extremely remote.

6.4.2 The origins of the Israeli space imaging efforts

Israel has had an active and growing space programme for two decades. As in the case of the USA and the former USSR during the Cold War, the environment of conflict led Israel to develop imaging satellites capable of real-time intelligence for reconnaissance purposes. The centrality of intelligence and early warning was emphasised as the countries in the region began to acquire ballistic missiles and weapons of mass destruction (Steinberg, 2000). On occasion, the USA shared strategic and space-based intelligence information with Israel. However, despite the close defence co-operation with the American government, Israel did not have routine access to real-time satellite intelligence data (Rotem, 1994).

Defence systems

As a result of these factors, Israel developed an indigenous high-resolution space imaging satellite, beginning with dedicated Ofeq (Horizon) military satellite series, which served as the technical foundation for the EROS commercial system. For more details on satellite imaging programmes of various Middle Eastern states see Steinberg (2001a).

In 1988, the Ofeq-1 test satellite was launched using the three-stage Israeli-designed and manufactured Shavit launcher. The launch site is located on the Mediterranean Coast and, in order to avoid flying over hostile countries, a highly unusual flight path was used (north-west over the Mediterranean) placing the satellite into a retrograde orbit at an inclination of 143° (Congressional Research Report, 1994). Since additional thrust was needed to achieve the extra 1,950 km/h of velocity required to escape into orbit from a westward launch, this restricted the size of the payload (Simpson *et al.*, 1989). Ofeq-1 was reported to be a test vehicle designed to lead to the development of an orbital reconnaissance capability, and it re-entered the earth's atmosphere in January 1989. Ofeq-2 was launched in April 1990, similar in weight and technical characteristics to Ofeq-1, and had an orbital lifetime of three months (Congressional Research Report).

Ofeq-3, launched on 5 April 1995, was apparently the first operational reconnaissance system, with a payload containing ultraviolet and high-resolution imaging sensors. Its higher perigee (369 km) and orbital maneuvering capability allows for a longer lifetime. According to reports in the Israel press, this version of the Shavit launcher included a small new Israel Aircraft Industry (IAI) third-stage rocket engine with 674 lb. of thrust (*Aviation Week and Space Technology*, 1994). Its orbit takes it over sites in the Middle East, including Iraq. On 22 January 1998, the attempted launch of Ofeq-4 (reportedly equipped with an advanced imaging system) ended in failure when the booster malfunctioned (Reuters 1998; Rodan, 1998). Had the last launch been successful, it would have provided Israel with two simultaneously operating imaging systems, significantly enhancing capability. The future of the Ofeq series is in doubt, as many of its capabilities have been incorporated in the EROS satellites (Schiff, 1998).

Commercial systems

In December 2000, Israel placed the first of its EROS series into orbit. Because Israeli launch capacity is currently insufficient for the orbital altitude and weight of the EROS payload, EROS A1 was launched by a Russian START –1 LV from Siberia, and began transmitting images a few weeks later (Barzilai, 2000). Images released for public distribution included Seoul, in South Korea, Brest in France and Izmir in Turkey (ImageSat International, online at www.imagesatintl.com).

The EROS series is based on technology developed by IAI on the foundation of the Ofeq military reconnaissance systems. Work on this project began with the establishment of the Space Technology Division at IAI in 1984. IAI's first proposal for the EROS programme was submitted in 1993. In 1996, IAI signed a contract with Core Software Technology (CST) for the development of the EROS system.[1] The satellite and related components are provided by Israeli-based firms, and CST provides assistance in marketing and processing the real-time imaging to ground stations around the world (Cole and Marcus, 1990; Rodan, 1995; Lardier, 1997).[2] In 1998, a programme for a constellation of eight commercial light Low Earth Orbit (LEO) high-resolution imaging satellites was completed. This system is designed to provide potential customers with very frequent coverage of any point on the globe, compared to the more sporadic coverage that would be available from a system of one or two satellites.

The consortium involved in the EROS project, including IAI, CST, and Electro Optical Industries, Ltd. (Israel) (ELOP) combined to form Imagesat International to manage development, launch and operations for the EROS project. EROS is a low-weight 'light-sat' incorporating a breakthrough in systems concepts that allow a 200 kg satellite to match the performance characteristics of satellites weighing several tonnes. (Similarly, IAI's Amos Communications satellite was the lightest communications satellite built at the time of launch, in 1996.) Customers can purchase 'turn-key' earth observation systems, based on 'vertical integration of IAI's satellite technology and CST's archiving, distribution, image management and exploitation technology.' Imagesat International is offering to retrofit existing ground stations, used for accessing SPOT and Landsat images, at a cost of $10 million (de Selding, 1999).

Initially, the first stage of the programme was to have consisted of two enhanced EROS A+ satellites, to be launched in 2000 and 2001, followed by the EROS B satellites. However, delays in launching EROS A1 as well as accelerated progress in the development of EROS B led to the decision to remove the second EROS A satellite from the programme, and to move directly to the EROS B satellite, to be launched in 2001 or 2002. In July 2001, MBT, a division of Israel Aircraft Industries Ltd. (IAI) Electronics Group, signed a $110 million agreement with ImageSat International for the first EROS B satellite (Barzilai, 2001).

By 2004, the full system of eight satellites is scheduled to be fully operational. This constellation will provide data to customers and Satellite Operating Partners

1 Based in Pasadena, California and founded in 1991, Core Software Technology is headed by Steve Wilson. Core developed a global on-line visual indexing and distribution infrastructure, known as ImageNet, for users in geospatial information products (i.e. satellite and aerial imagery, computer cartography and deomographic products). ImageNet claims to be 'the single largest on-line source of electronic indexed satellite imagery archive' with 'over 20,000 registered subscribers through franchised Primary Service and Data Providers in North America, Europe, Russia, the Middle East, Australia and the Southern Pacific Rim'. See also Core Software Technology Homepage http://www.coresw.com/

2 See also press release, 'Israel Aircraft Industries and Core Software Technology Announce Formation of a Joint Venture Company to Enter High-Resolution Satellite Imagery Market', IAI Electronics Group and Core Software Technologies, Tel Aviv, 28 February 28, Imagesat International Homepage: http://westindianspace.com/home.html

(SOPs) in GIS-ready form. The satellites will be launched into LEOs, at altitudes of between 480–600 km.

The EROS A delivers imagery at a resolution of 1.8 m, using the 'pushbroom sweep method' of imaging, and covers a 12.5 km swath with panchromatic imagery, provided to customer at near-real-time. EROS B1 will deliver 0.82 m resolution from an altitude of 600 km, covering a swath of 16.5 km, followed by EROS B2 through B6 (five satellites), with 0.82 m panchromatic imagery and 3.68 m multispectral imagery. The EROS A+ sensors will consist of 10,000 detector elements in arrays that provide day-only or day/night performance, with time delay integration. The EROS-B series will incorporate 15,000-element modules on a single focal plane, to provide for different sources such as an IR layer over conventional (visible light) images (Esche, 1999). The SOPs in the EROS network will receive worldwide access to high-resolution imaging, archiving and distribution, based on existing remote sensing ground infrastructure, as well as additional installations, as necessary. The SOPs will receive 'a dedicated regional satellite with local customer tasking to meet national and civilian satellite program requirements.' In 1999, WIS and the Israeli government reportedly signed an agreement providing exclusive access to images of the Middle East region obtained by the first three EROS satellites for eight years (de Selding).

Customers can also purchase Priority Acquisition Service (PAS) programmes to provide priority tasking of EROS satellites worldwide. Daily coverage (at least one revisit per day) will be provided for North and South America, Europe, Asia, the Middle East, Africa and Australia. West Indian Space (WIS) is offering imaging services for use in civilian mapping, infrastructure management, and military intelligence via long-term contracts at $10–20 million annually.[3]

In addition to selling a range of services to investors and customers, WIS is also planning to issue bonds via a private placement to help finance the $250 million needed to launch and operate their first three satellites. The full cost of the eight satellite system is estimated at $750 million, and WIS is seeking to raise most of this sum through private investors (Rosenberg, 1999). States that are subject to export restrictions, such as Libya, Iraq, and Iran, are barred from purchasing EROS imaging (de Selding, 1999).

[3] ELOP is also participating in the DAVID programme, funded by the European Community's R&D Fifth Program. The group comprises Israeli and German participants including German satellite maker ONB and GAF, an image processing company who will also partner in a joint venture for the commercialisation of the technology. The service will be based on an array of small satellites, each weighing only 180 kg. It will provide 5 m resolution earth resources imagery, viewing at specific narrow-band wavelengths. The service will also be able to conduct automatic monitoring of land-use, issue alerts when detectable hazards are spotted, and assist national or municipal authorities in the enforcement of licensing and regulation. The group is negotiating with investors for private investment that will enable full-scale development and a commitment for a launch window.

6.4.3 Dual-use technology concerns

While often advertised as commercial high-resolution imaging systems, the dual-use dimensions of these systems are prominent. As a result of the development of these capabilities, formerly secret military facilities are becoming widely visible, not only to highly-classified American and Russian reconnaissance satellites, as in the past, but to a much wider audience (Steinberg, 1998).

While some analysts view this development as positive, in terms of transparency and its applications to arms limitation agreements and prevention of surprise attack (Williamson, 2000), there are also potentially very significant negative impacts on security and stability, particularly in regions characterised by a high level of conflict, such as the Middle East. Instead of contributing to confidence building, high-resolution commercial imaging also has the potential for aggravating conflicts and international instability, changing the balance of power, sharpening existing asymmetries in military capabilities, and making regional and international conflicts harder to manage.

However, the impetus for high-resolution commercial satellites was derived in the past by very optimistic market forecasts that included annual sales estimates of billions of dollars in imagery, technology, and software (O'Connell and Lachman, 2000). The enthusiasm was tempered, to a limited degree, by warnings regarding the military applications that were likely to follow, and the implications, particularly for US security interests. For the first time, many countries, as well as non-governmental actors in areas such as the Middle East, North Asia, Central Europe, and South America, are gaining access to very detailed and almost real-time images of neighbouring states. Former US government officials, such as Henry Sokolski, included satellites in a list of 'non-apocalyptic weapons' and warned of the consequences had satellite imaging been available to Saddam Hussein in the Gulf War (Sokoloski, 1994; Florini, 1988). Vipin Gupta noted that unlimited sales of high-resolution imaging could disrupt 'delicate balances of power', complicate the containment of international crises, and fuel developments in offensive weapons capability (Gupta, 1995). Former CIA director James Wollsey counseled caution, noting that, 'This very comfortable world people have been living in where fixed target installations on land are safe' will vanish with the proliferation of high-resolution commercial imaging (Wollsey, 1996). Other analysts and policy makers warned of the dangers linked to terrorist access to this data and capability (Cohen, 1996). 'Islamic Jihad could get its hands on a 1 m resolution picture of...a US Air Force General's headquarters in Turkey, convert the shot to a precise three-dimensional image, combine it with data from a GPS device...and transmit it to Baghdad, where a primitive cruise missile, purchased secretly from China could await its targeting coordinates' (Lane, 1996).

The impacts, both stabilising and destabilising, will depend on factors such as resolution, the form in which the data is sold (original digital data, or derivatives), the nature of the distribution system (direct real-time ground links to receivers, or delayed transmission via filtering stations), available software, shutter control, and similar factors (Hays and Houchin, 1999). In addition, as ultra-high resolution imaging becomes more widespread, countermeasures are likely to be developed in order to camouflage critical military installations and activities, and to limit the ability of overhead satellite imaging to degrade central military capabilities deemed vital for national security and survival.

Such concerns led to considerable negotiations between the Israeli and US governments and, in 1995, they agreed to coordinate policies. The US accepted the Israeli position on the need to prevent Arab countries (such as Saudi Arabia) from obtaining the ability to control the tracks of high-resolution imaging satellites from the ground, and also placed limits on the sale of state-of-the-art software for image enhancement. In addition, the details of shutter control were considered in greater detail for Middle East war scenarios. On 4 March 1996, the Israeli Ministry of Defense issued a formal public statement of Israeli policy, which incorporated limitations in a manner that was similar to the policies adopted by the US.[4]

In June 1996, the US Senate passed an amendment to the 1997 Defense Authorization Act entitled, 'Prohibition on Collection and Release of Detailed Satellite Imagery Relating to Israel and Other Countries and Areas' (Bingman, 1996). The final version of the amendment prohibited the sale of imaging data over Israel with a resolution below that provided by other 'commercially available' sources. The precise definition and implementation of this language was unclear, and the Israelis agreed to a 2 m limit, based on the availability of Russian KVR-1000 images (although there are questions regarding the degree to which these can be called 'commercially available'). In 1998, the US placed the limit at 1 m. Israel protested and, in July 1998, an agreement was reached blacking-out Israel at resolutions below 2 m. The images of Israel released by IKONOS have complied with this limitation, but a re-evaluation of US policy is expected to reduce the limit to 1 m or less, and 0.5 m resolution systems are being planned for launching in 2004 (Williamson, 2000 cited in Loeb, 2000). While the impacts of these changes remain unclear, there are implications and these can be discerned by the activities and policies of the key states in the region, including Israel, Iran, Iraq, Egypt, and Saudi Arabia with respect to high-resolution satellite imaging.

6.4.4 Application to Middle East arms options

Ten years ago, during the optimistic period of Middle East peace efforts following the 1991 Madrid conference and the meetings of the multinational working group on Arms Control and Regional Security (ACRS) that ended in 1995, the potential for using satellite imaging to verify limitation agreements was given considerable attention (Steinberg, 2001b). During this period, the potential for Confidence and Security Building Measures (CSBMs), including mechanisms to prevent accidental war, surprise attack, and crisis instability was considered to be quite high, and the Sinai I and II disengagement agreements (1974 and

4 Lane argues that it is in America's interest to continue to lead the regulation in remote sensing, and claims that the loss of US international leadership in launches has contributed to lower disincentives for missile proliferation (see Lane, 1996).

1975) and the 1979 Israel-Egypt Peace Treaty established useful precedents.[5] The models provided by the CSCE and CFE systems were also discussed, including elements such as pre-notification, conventional arms limitations, and limits on military exercises. CSBMs and measures to increase stability and mutual security are also incorporated in the 1994 Israel-Jordan peace treaty, and are important elements in negotiations between Israel and Syria.

In the context of the ACRS discussions, the participants discussed limited force zones, including agreed ceilings on the types and numbers of large platforms (i.e. tanks, self-propelled artillery, towed guns, combat aircraft, mobile air defence systems, attack helicopters, naval craft, ballistic missiles launchers, etc). To monitor compliance, the states involved would agree on a system of electronic tags and tag scanners, video, weight sensors, motion sensors, etc., to monitor the movement and location of treaty-limited items. Overhead monitoring, including the use of space-based imaging is important in these contexts, in order to insure compliance and provide assurance against the possibilities of breakout and sudden abrogation of these limitations. In addition, the space monitoring technologies have the additional benefit of providing data for environmental monitoring and regional economic coordination.

However, the conditions in the Middle East, with the inherent instabilities and continuing conflicts, are very different from Europe after the Cold War and the demise of the former USSR. As a result, the nature of CSBMs and monitoring regimes must be designed specifically to meet conditions in the region. With continuing threats of conflict and the need for security, consideration of arms control verification applications based on co-operative monitoring or national technical means and *openness* were always tempered by the need for *military secrecy* that remains the ultimate guarantor of national security and survival in the often violent circumstances of the Middle East (Pregenzer, 1994). All forms of information on military deployments and the location of targets can be exploited to provide data for offensive military operations.

It is therefore necessary to ensure that a regional satellite verification centre cannot become a source of instability or a basis for changing the balance of power in the region. In part, this can be accomplished through careful *definition of the field of view and targets of satellite images*. The coordinates and resolutions selected for transmission would have to be defined clearly and systematically in advance, to assure that the data that is relevant for treaty verification is provided, while unrelated information outside the agreed parameters is excluded. (This would not prevent the individual states from ordering images or purchasing data from satellite imaging services outside the framework of the regional centre.)

At the same time, the potential impact of a Middle East co-operative satellite monitoring centre should not be exaggerated. Like other technologies, satellites can provide a useful tool in the implementation of regional agreements, but these agreements themselves are dependent on political factors. At most, the technology can

[5] These issues were discussed during the Co-operative Monitoring Workshop, Scandia National Laboratories, July 1994.

support the political process, and enhance regional co-operation and confidence building, but it cannot substitute for political agreements. Under current conditions of intense conflict in the Middle East, such political agreements seem highly unrealistic.

6.4.5 References

Aviation Week and Space Technology (1994) October 17.

Barzilai, A. (2000) Israel to launch civilian satellite today, Israeli Defense Ministry expected to be customer. *Haaretz*, 5 December.

Barzilai, A. (2001) MBT to build new EROS satellite', *Haaretz*, July 29.

Bingman (1996) Statement of Senator Bingman, *Congressional Record*, p. S6924-S6925.

Cohen, E. A. (1996) A Revolution in warfare. *Foreign Affairs*, **75**(2) March/April, pp. 37–54.

Cole, J. and Marcus, A. D. (1990) Israeli-led venture in satellite imaging poses challenge to US aerospace firms. *Wall Street Journal*, 28 February.

Congressional Research Report (1994) Space Activities of the United States, Soviet Union and Other Launching Countries/Organizations: 1957–1993. Report presented to the Committee on Science, Space Technology, House of Representatives, 103rd Congress, US Government Printing Office, Washington, D.C., p. 161.

deSelding, P. B. (1999) US-Israeli venture plans bold offer, *Space News*, March 15, 1999.

Esche, T. (1999) El-Op aims high for space business: US-Israeli venture plans to launch eight high-resolution satellites. *Imaging Notes*, May/June, Vol. 14, No. 3, p. 6.

Florini, A. M. (1988) The opening skies: third party imaging and US security. *International Security*, Vol. 13, No. 2 (Fall), p. 10.

Gupta, V. (1995) New satellite images for sale: the opportunities and risks ahead. *International Security* 20 (Summer).

Hayes, Lt.-Col., P. L. and Houchin, II, Lt.-Col., R. F. (1999) Commercial spysats shutter control: the military implications of US policy on selling and restricting commercial remote sensing data. School of Advanced Airpower Studies, Maxwell AFB, Alabama. Prepared for the Institute for National Security Studies, USAF Academy, Colorado (draft version).

Lane, C. (1996) The satellite revolution. *The New Republic*, 12 August, p. 24.

Lardier, C. (1997) Proliferation of remote sensing satellite as market expands. *Air & Cosmos/Aviation International*, 4 April, pp. 34–35.

O'Connell, K. and Lachman, B. E. (2000) From space imagery to information: commercial remote sensing market factors and trends. In: *Commercial Observation Satellites: At the Leading Edge of Global Transparency*, edited by Baker, J. C., O'Connell, K. M. and Williamson, R. A. (eds) ASPRS – the American Society for Photogrammetry and Remote Sensing (ASPRS) and RAND, Bethesda, Maryland.

Pregenzer, A. L. (1994) Crisis prevention centers as confidence building measures: suggestions for northeast Asia. Verification and Monitoring Analysis Department, Sandia National Laboratories, New Mexico.

Reuters (1998) Israel to launch new satellite before old one expires. 7 July.

Rodan, S. (1995) Space Wars. *Jerusalem Post Magazine*, 10 March.

Rodan, S. (1998) Ofeq satellite budget approved. *Jerusalem Post*, 26 June.

Rosenberg, D. (1999) US-Israel satellite-imaging firm plans bond issue. *Reuters*, 28 February.

Colour plates

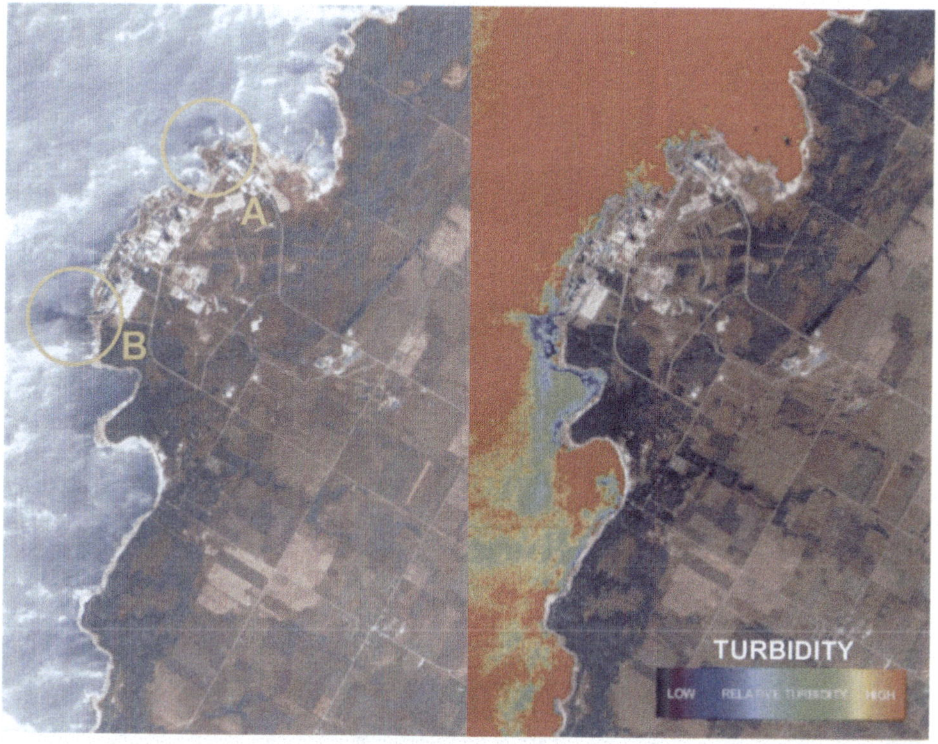

Plate 1 Image processing comparison for hydrological phenomena. *Left side*: A natural colour image of the Bruce Nuclear Generating Facility, acquired by IKONOS on 11 November 1999. The circles indicate the position of discharge points. There is a small surface effect at Bruce B and hints of a plume but it is difficult to see. The pattern of the waves visible in the original data shows clearly that the wind is blowing from the north. *Right side*: The same image after removing the sun glint of the surface waves, and enhancement. A large plume is seen flowing south of Bruce B and curling into the small bays. The slightly less turbid water along the coast between A and B may be a smaller plume originating at Bruce A.

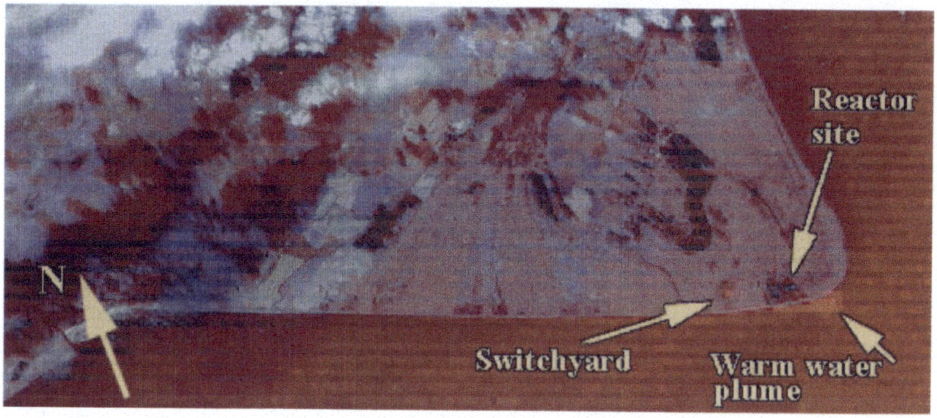

Plate 2 An extract from a quarter scene acquired on 3 November 1997 by the US Landsat 5 satellite (resolution of 28.5 m) over the UK Dungeness Nuclear Power Reactor station. The warm water discharged into the sea can be seen in this image which is a combination of the thermal band 6 (red) and bands 5 (green) and 4 (blue). Some heat escaping the switchyard building can also be detected while clearly the reactor buildings are reasonably well insulated. Some of the buildings in the reactor complex can be distinguished as separate buildings (source: ESA/Eurimage).

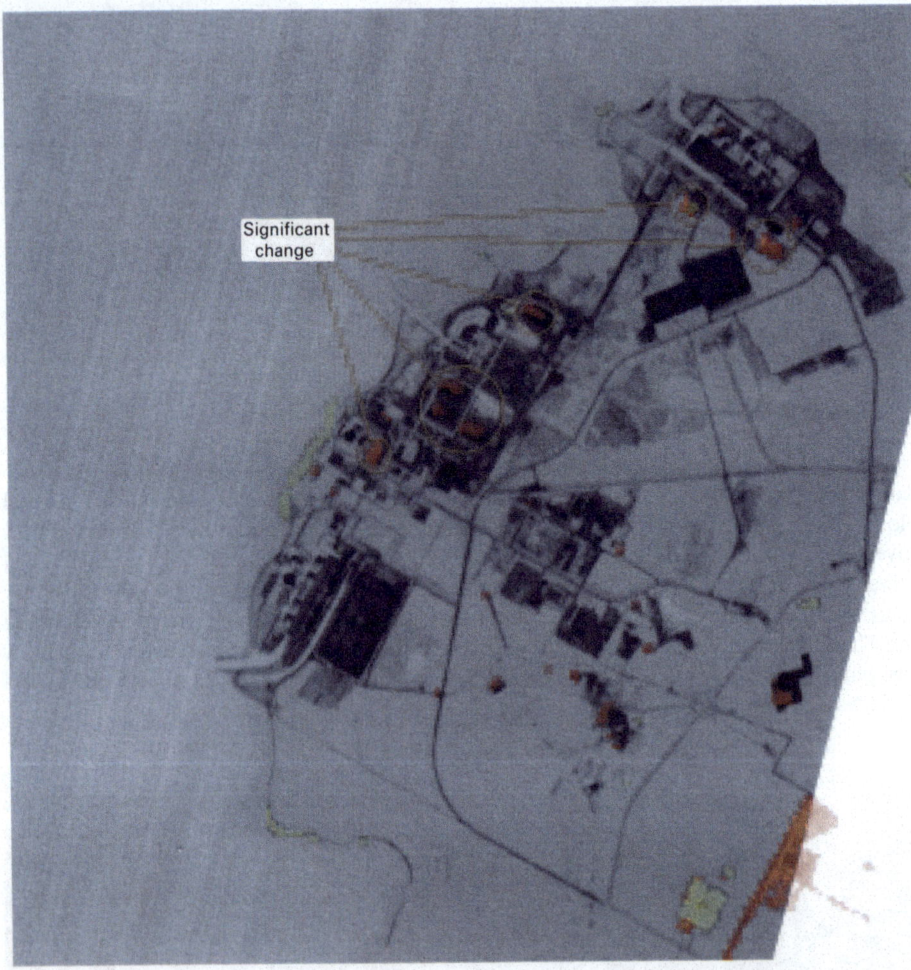

Plate 3 MAD change detection for Space Imaging Landsat-5 August 1998–August 1996, fused to the negative of an IRS-1C panchromatic (5 m) image taken in September 1997. MAD component 1 and 2 exceeding the significance thresholds are in red and green [courtesy of M. Canty].

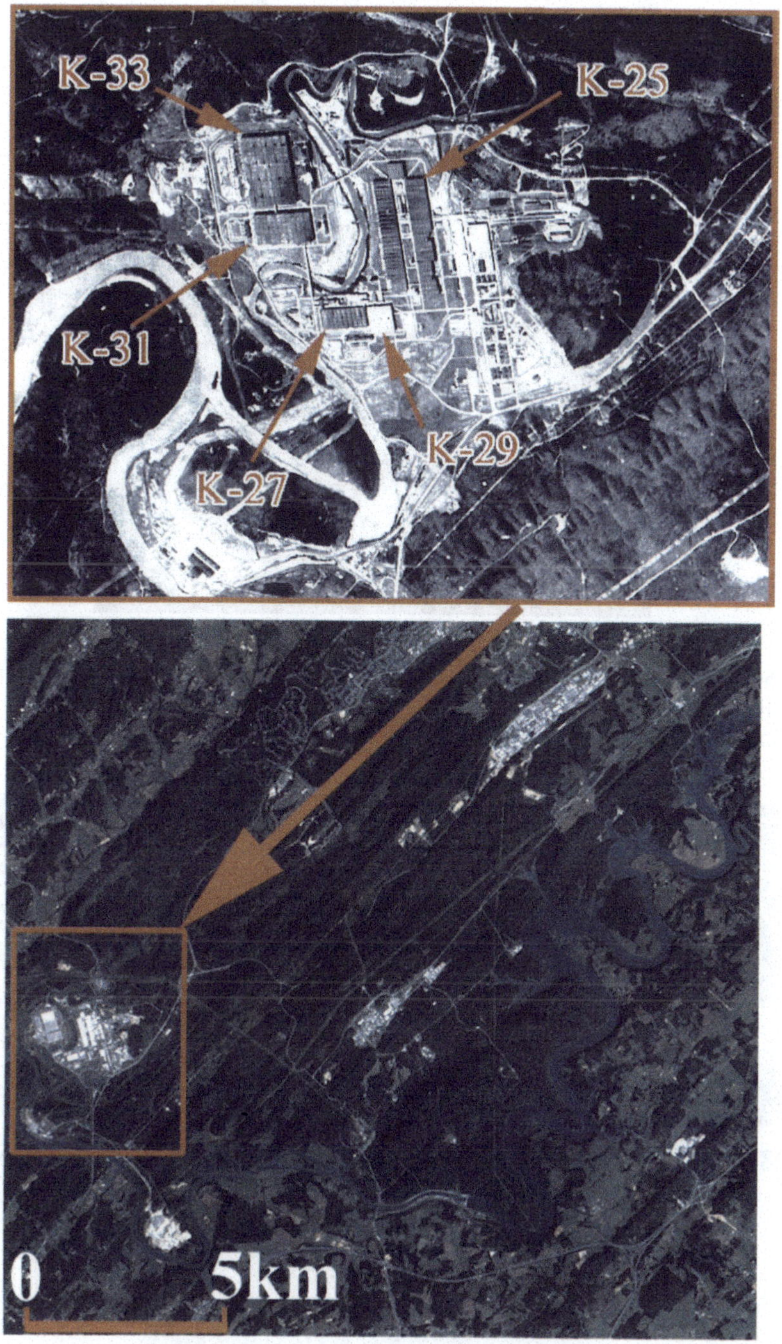

Plate 4 Extract from a Landsat 5 scene (resolution 28.5 m) acquired over the Oak Ridge facility on 8 September 1984 (lower image). The image is a combination of bands 1 (blue), 2 (green) and 3 (red). The enrichment facility is in the area marked by red rectangle. The top image is from the US early reconnaissance satellite KH4 showing roughly the same area as the rectangle in the Landsat image. The image was acquired on 8 March 1970. Various facilities are identified in the KH4 image, K-25 being the original diffusion plant [source: Landsat/EOSAT and Corona image].

Plate 5 Composite image of the central part of Rome. Colour is taken from the KFA-1000 image and 2-m resolution is taken from the KVR-1000 image (courtesy of Sovinformsputnik).

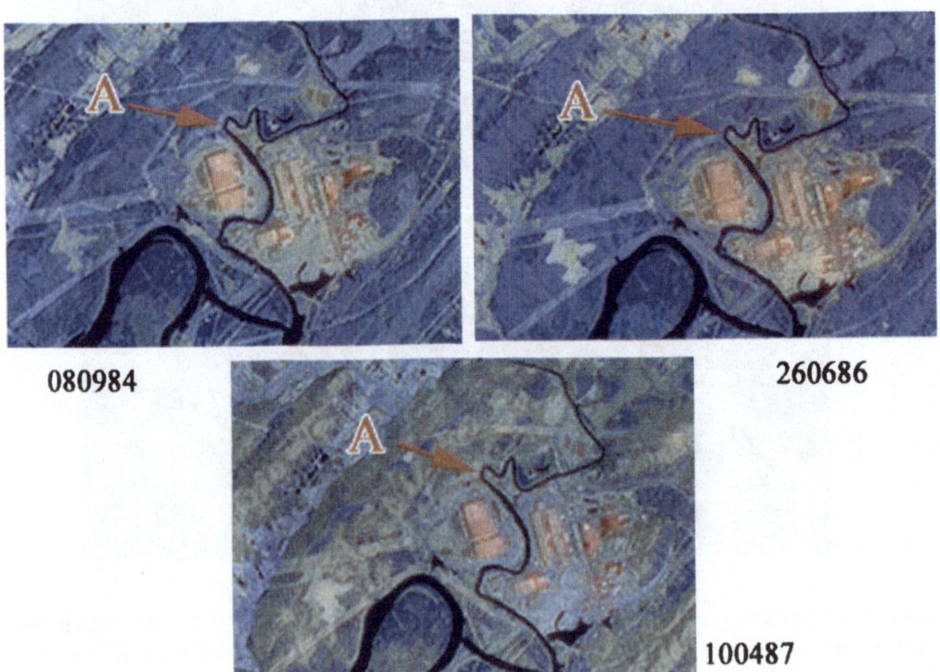

Plate 6 Extracts from Landsat images over the K-25 facility on three dates indicated under the images that are combination of bands 6 (red), 5 (green) and 4 (blue). The temperature of the water is indicated by the red colour **(A)**. This was found to be highest on 26 June 1986 and lowest on 10 April 1987 suggesting that the plants were operating in 1984 and 1986 but ceased operating by 1987.

Source: SPOT/CNES **Panchromatic image acquired by the French SPOT satellite (10m resolution) on 12 May 1991**

Source: KVR1000 **Panchromatic image acquired by Russian KVR1000 sensor (resolution 2m) on 27 June 1990**

Possible new sites deveoped **Sites damaged during the 1991 Gulf War or dismantled**

Plate 7 Extracts from two scenes acquired from SPOT (resolution 10 m) on 12 May 1991 (top image) and from a Russian KVR100 sensor (resolution 2 m) acquired on 27 June 1990 (lower image). Perimeter fences can be identified. Also some of the changes that have occurred on the Iraqi reactor site during a period of one year can be identified. The red boxes show new constructions and the blue boxes indicate changes in the site possibly due to bombing during the Gulf War or due to later dismantlement process. This is an example of visual change detection.

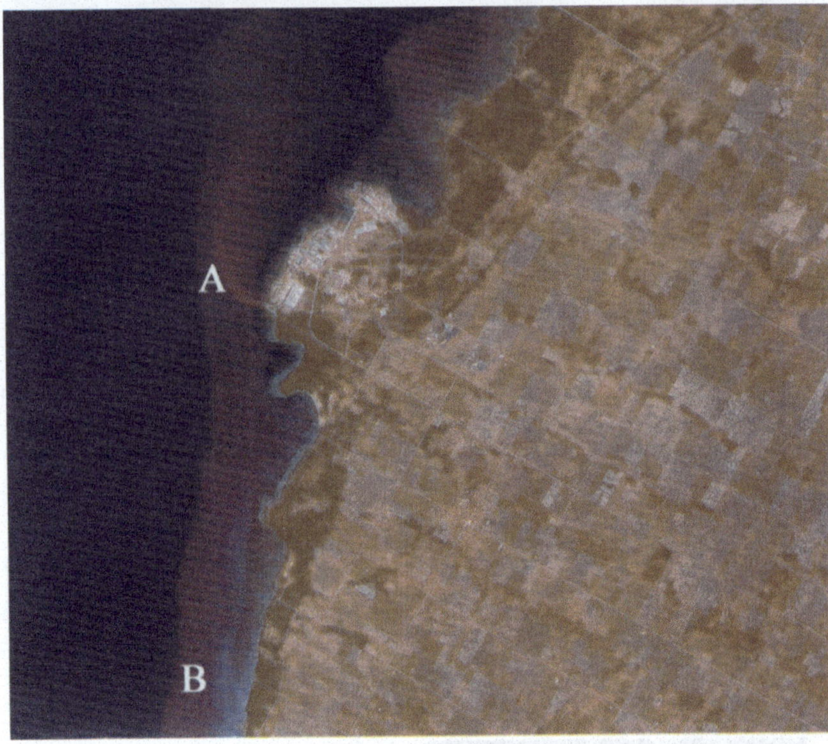

Plate 8 A combination of bands 1 (blue), 2 (green) and 6 (red) of the Landsat 5 image acquired on 7 May 1997. The thermal plume from the reactor can be seen at **A** while at **B** possible sedimentation from rivers can be detected.

Plate 9 A combination of three principal components, C4 (red), C1 (green), and C2 (blue), of the Landsat 5 image.

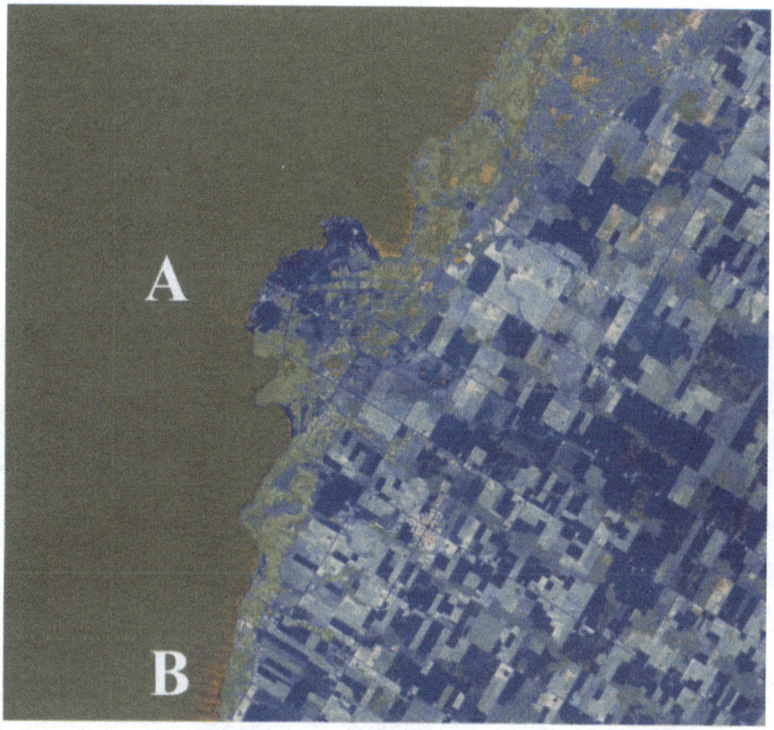

Plate 10 The result of combining the difference between the principal components C2 and C4 (red) with component C2 (green) and component C1 (blue).

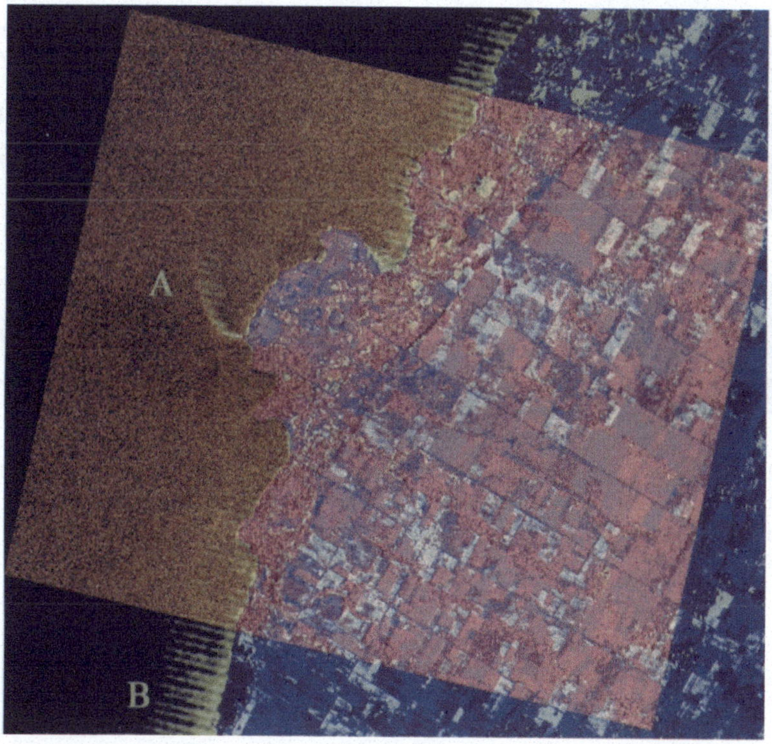

Plate 11 A combination of the co-registered ERS-2 image (as red) and the principal components C4–C2 (as green) and the component C1 (blue) as an RGB image. It can be seen that the warm water plumes coincide in all the images.

Plate 12 A USGS Landsat 7 Image from 26 August 1999. False colour composite of the BNPD site using the first, second and third principal components (30 m) fused with panchromatic image (15 m) [courtesy of M. Canty].

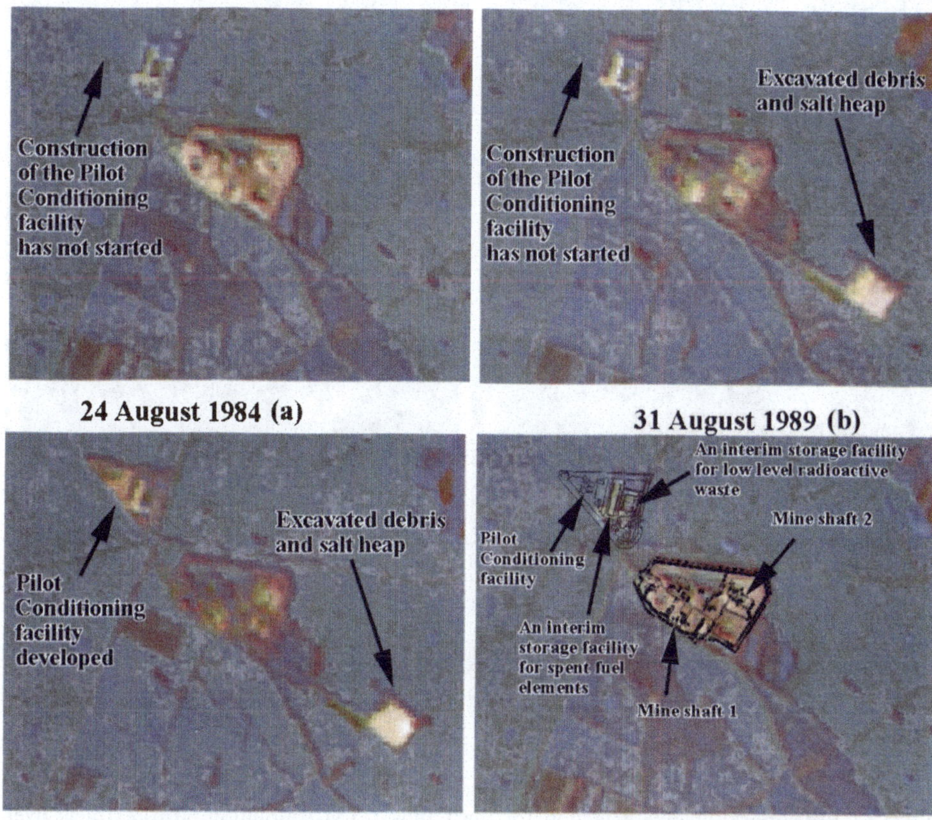

Plate 13 Extracts from the Landsat TM images (resolution 30 m) acquired at different times over Gorleben, shown as combinations of the principal components 1 (red), 2 (green) and 3 (blue) (a–c) and the site map superimposed over the image (d). The dates are indicated under each image.

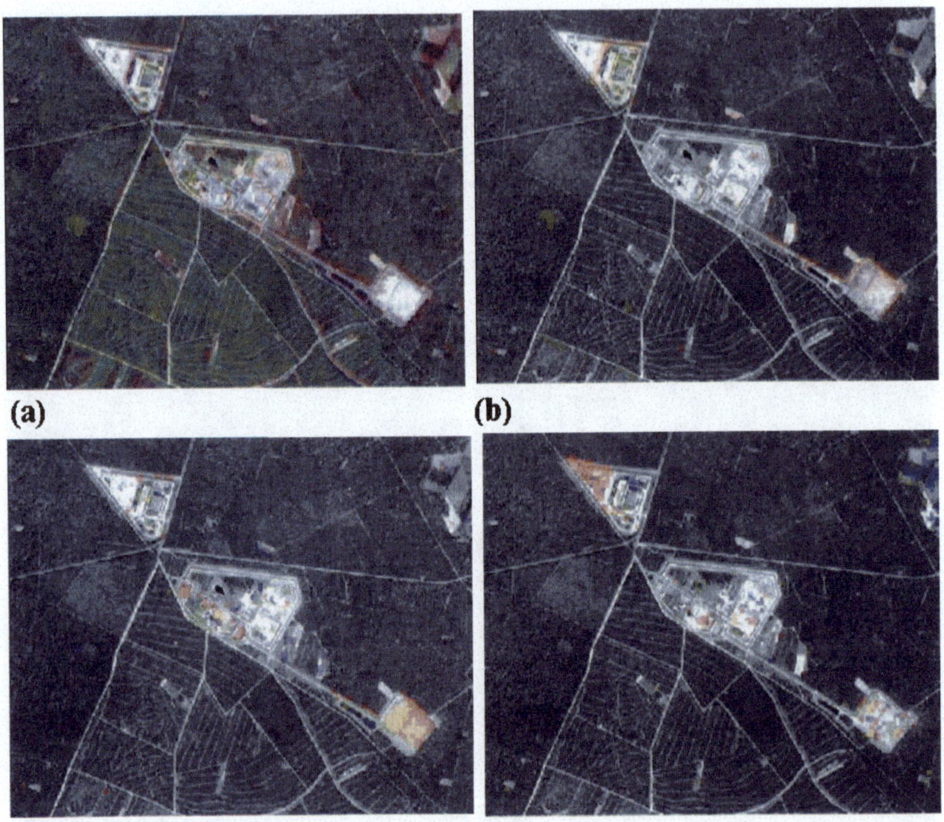

(a) (b) (c) (d)

Plate 14 Bi-temporal change detection for the Landsat TM (resolution 30 m) extracts over Gorleben acquired on 24 August 1984 and 31 August 1989 fused with a high-resolution KVR-1000 image. (a) PCA: Shown is the second principal component of the channels 5 (red), 4 (green) and 3 (blue) exceeding three standard deviations. (b) IPCA: Shown is the second iterated principal component of the channels 5 (red), 4 (green) and 3 (blue) exceeding three standard deviations. (c) MAD: Combinations of the MAD Components 6 (red), 5 (green) and 4 (blue) exceeding five standard deviations. (d) Bi-temporal change detection for the Landsat TM extracts over Gorleben acquired on 31 August 1989 and 21 August 1997, fused with a high resolution KVR-1000 image: Combinations of the MAD components 6 (red), 5 (green) and 4 (blue) exceeding five standard deviations.

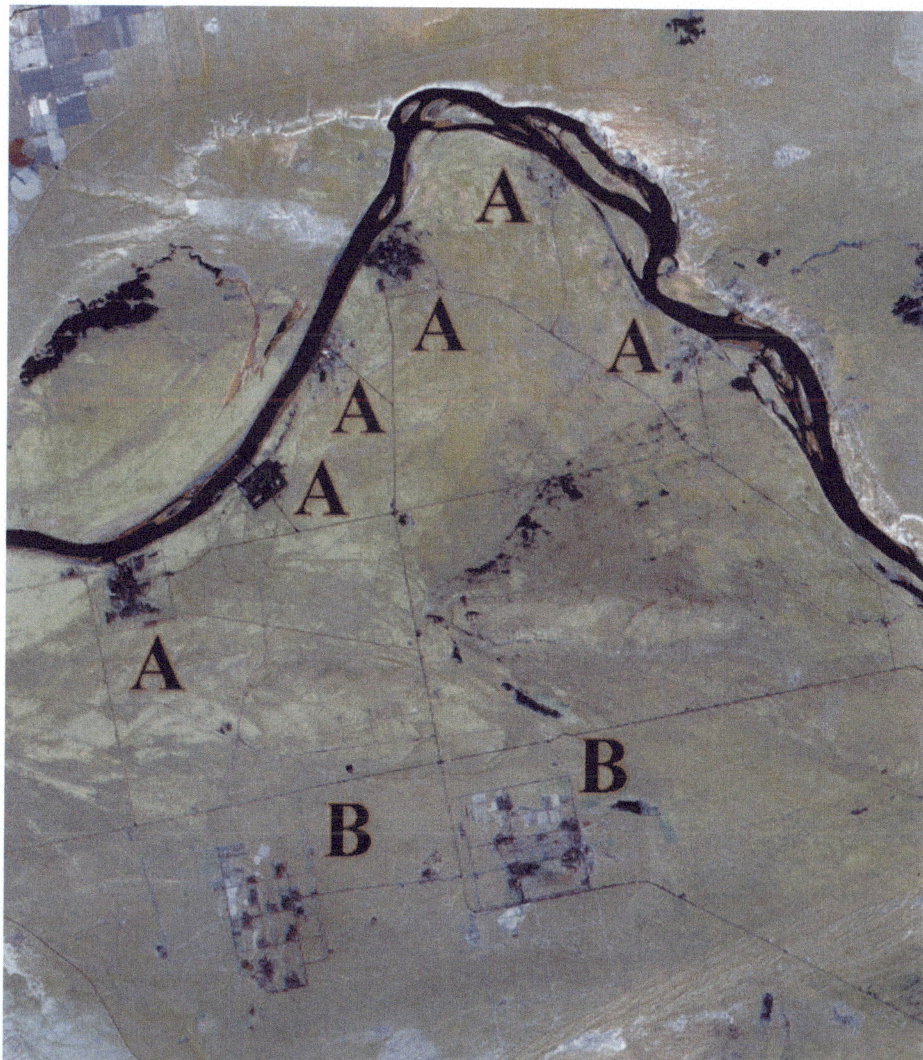

Plate 15 In this figure the images were combined as band five (red) of 1987 with bands five (green) and four (blue) of 1983. Sites marked **A** and **B** are production reactors and reprocessing facilities. In the image red and green colours represent changed areas. Compare this with the image in Figure 7.26.

Plate 16 This shows a 3 linear combination of the twelve bands of the bi-temporal image in red, green and blue resulting from the analysis using MAD. Twelve infrared bands of Landsat images dated 12 August 1983, 27 July 1986 and 28 June 1987 were used. The changes represented by colours were then superimposed on a single band from the image dated 12 August 1983. Compare this with the image in Figure 7.25.

Bruce NPD – Convair 580 Multi-polarised SAR

Bruce 'B' NGS cooling water discharge with southwards flow due to wind effect caused by Hurricane Floyd

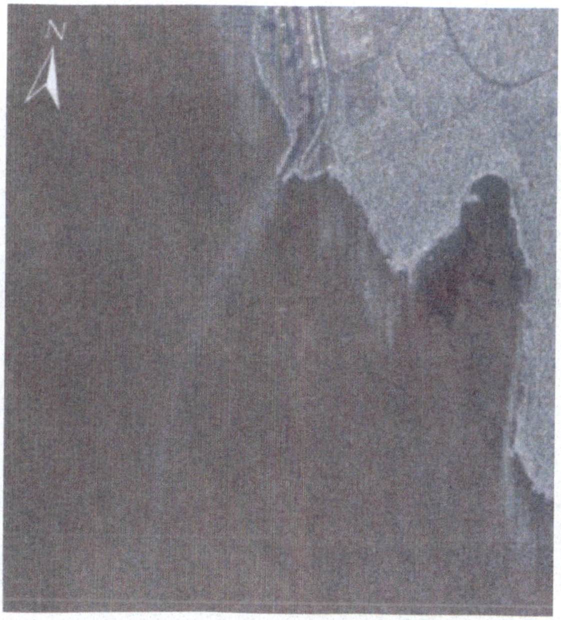

Blue Channel – HH
Green Channel – HV/VH
Red Channel – VV

16 September 1999

Plate 17 SAR airborne image showing the discharge at Bruce.

Plate 18 This is an image over south-east India and Sri Lanka acquired by the Ocean Colour Monitor (OCM) on board the IRS-P4 satellite.

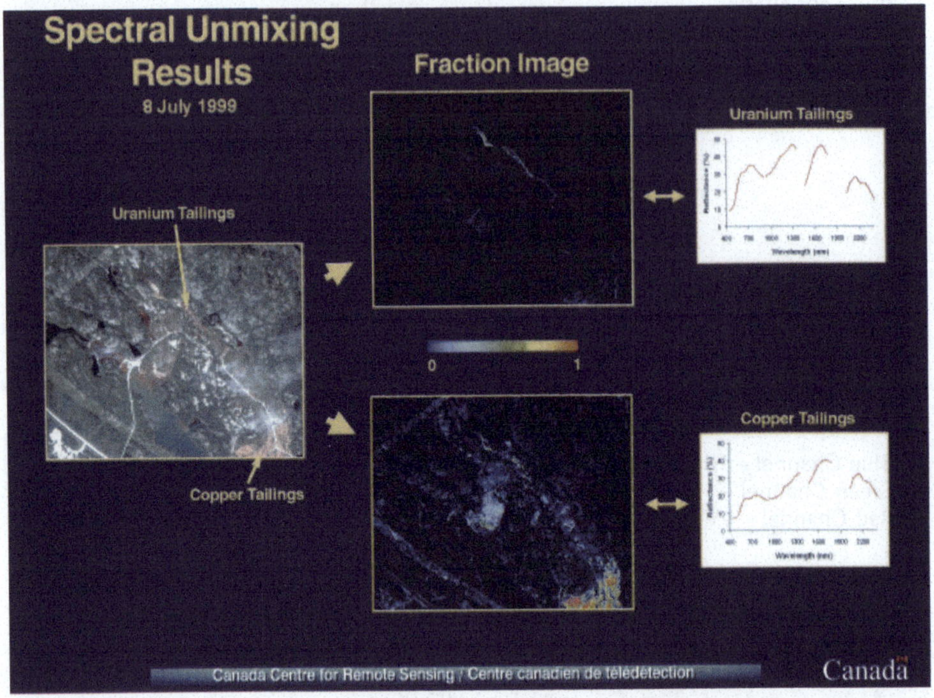

Plate 19 Spectral unmixing results – uranium and copper tailings.

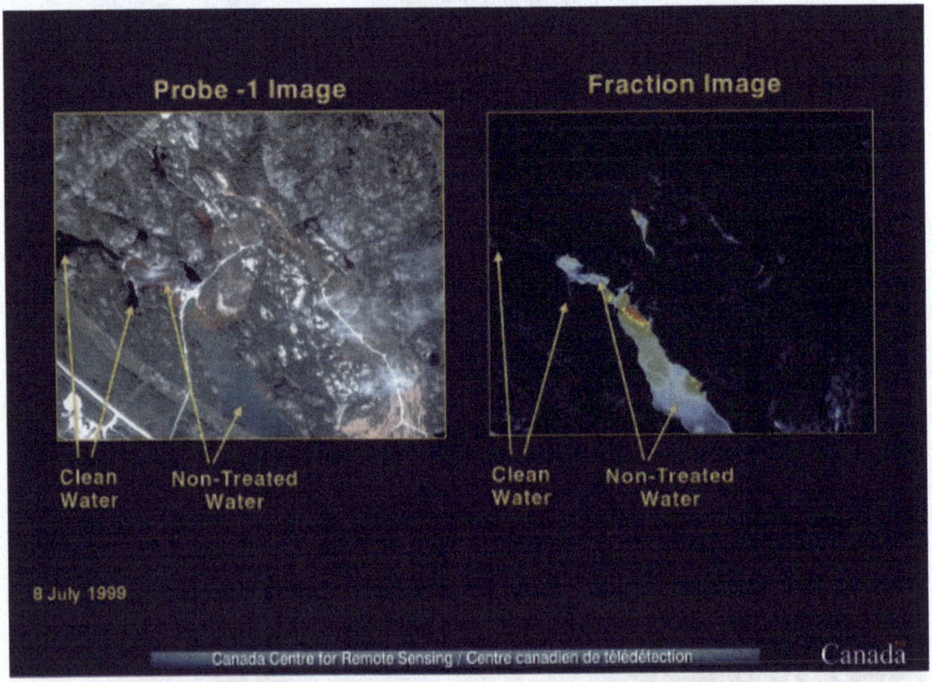

Plate 20 Spectral unmixing results – environmental effects in water.

Rotem, M. (1994) Spy Satellite for Arab Emirates 'Serious Threat', *Jerusalem Post*, 19 November, 1992; Sergey Koulik and Richard Kokoski, 'Verification Lessons of the Persian Gulf War,' *Conventional Arms Control* (New York: Oxford University Press, 1994), p. 199.

Schiff, Z. (1998) A serious satellite situation. *Haaretz*, 25 January.

Simpson, P., Acton, P., and Crowe, S. (1989) The Israeli Satellite Launch. *Space Policy*, May.

Sokoloski, H. (1994) Non-apocalyptic proliferation: a new strategic threat? *The Washington Quarterly*, Vol. 17, No. 2 (Spring).

Steinberg, G. M. (1998) Dual-use aspects of commercial high-resolution imaging satellites. Security and Policy Series Paper No. 17, BESA Center for Strategic Studies, Bar Ilan University.

Steinberg, G. M. (2000) Arms Control and Non-Proliferation Developments in the Middle East: 1998/9, Security and Policy Series Paper #44, BESA Center for Strategic Studies, Bar Ilan University.

Steinberg, G. M. (2001a) The Security Impact of Commercial Observation Satellites in the Middle East and the Persian Gulf. Baker, J., O'Connell, K. and Williamson, R. (eds), In: *Commercial Observation Satellites: At the Leading Edge of Global Transparency*, American Society for Photogrammetry and Remote Sensing (ASPRS) and RAND. Bethesada, Maryland.

Steinberg, G. M. (2001b) Starting over: the prospect for regional security and arms control in the Middle East in the next decade, Disarmament Forum, United Nations Institute for Disarmament Research.

Williamson, R. (2000) who was recently quoted as predicting that such ultra high resolution (UHR) (down to 0.5 m resolution) would 'make the world a safer and more transparent place, even with high-resolution satellite photographs in the hands of potential adversaries.' In: Loeb, V. (2000) US is relaxing rules of satellite photos after a year-long policy review, far greater detail being allowed. *Washington Post*, 16 December. p. A03.

Wollsey, R. J., cited in Charles Lane, 'The Satellite Revolution', *The New Republic*, 12 August 1996 and also in Marry Graham, 'High Resolution, Unresolved', *Atlantic Monthly* (July 1996), p. 26.

6.5 JAPANESE REMOTE SENSING SATELLITE AND DATA POLICY *(T. Hamazaki and M. Onoda)*

6.5.1 Introduction

Earth observation activities in Japan started in the mid-1970s, as the National Space Development Agency of Japan (NASDA) established the earth Observation Center (EOC) and started receiving and distributing earth observation satellite data. Since then, a number of countries have launched various earth observation satellites. Data are widely distributed and commercially available in the world market. Over these years, several private entities have launched operational satellites such as the IKONOS, OrbView and Quickbird satellites, now reaching resolutions greater than 1 m.

So, it is timely that the International Atomic Energy Agency (IAEA), in cooperation with its member states, has initiated an undertaking to use satellite imagery for safeguards procedures. It is the authors' view that remote sensing data are particularly useful in enhancing IAEA's safeguards capability. The most effective way to go about this would be to combine the many sources of satellite imagery available in the world today – those of government-owned satellite programmes, international organizations, commercial satellites and others. This is due to the fact not only that each satellite has different technical characteristics (i.e. resolution, frequency, etc.) but also that they are operated under different data policies, and it should therefore be best to explore the optimum combination of data use from various satellites to acquire information that will provide the basis for precise judgement.

In order to assess Japan's contribution to this undertaking and to present the range of earth observation activities Japan is conducting, this chapter begins with a review of the Japanese civil earth observation satellite programme, and discusses its potential contribution to IAEA's safeguard activities. An overview of the Advanced Land Observing Satellite (ALOS) will follow, and the data policy including the legal aspects will be discussed. Finally, some future prospects for commercialization and Japanese remote sensing activities will be introduced.

6.5.2 Japanese earth observation satellite programme

Overview

At the outset, note that in Japan the main body that carries out earth observation activities is still the government. To date, there is no Japan-based private entity that has launched and commercially operated its own remote sensing satellite. NASDA developed its earth observation satellite programme under the Japanese annual Space Development Plan. The data acquired from NASDA satellites are commercially available to the public.

NASDA earth observation satellites can be divided into two categories: for land observation and for atmospheric and oceanic observation. The former is mainly for practical use and high-resolution observation is one of its most important characteristics. The latter is mainly for scientific use and is characterised by multi-channel observation. The history of the Japanese earth observation programme is shown in Figure 6.5.1 (Hamazaki, 1998).

The initial step in Japanese earth observation satellite development was the Geostationary Meteorological Satellite (GMS), which NASDA developed for the Japan Meteorology Agency (JMA). GMS was launched in 1977. GMS satellites have been used in the World Meteorological Organisation's World Weather Watch (WWW) Programme. Many Asian countries use data from the GMS series for their meteorological services and research programmes. Today, the meteorological satellite programme has been handed over to JMA. As a successor to the currently operational GMS-5, a multi-functional transport satellite (MTSAT-1R) is scheduled for launch in early 2003.

NASDA developed the Marine Observing Satellite-1 (MOS-1), launched in 1987. Its successor, the MOS-1b, was launched in 1990.

The Japan Earth Resources Satellite-1 (JERS-1) was a joint development project

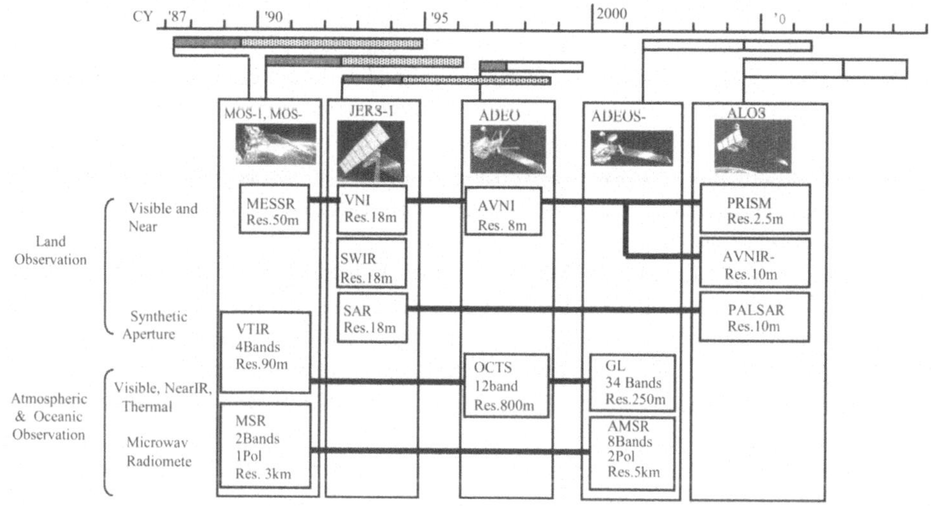

Figure 6.5.1. History of the Japanese earth observation satellite programme.

between NASDA and the former Ministry of International Trade and Industry, or MITI (now the Ministry of Economy, Trade and Industry: METI). NASDA was in charge of satellite development and MITI was responsible for development of the mission instruments, including the Synthetic Aperture Radar (SAR). JERS-1 was launched in February 1992 and was in operation until October 1998. The satellite carried an 18-m resolution Optical Sensor (OPS) and an 18-m resolution L-band SAR. The data acquired by JERS-1 were used to monitor earth resources and the environment, SAR in particular acquiring valuable forest monitoring data. NASDA has been producing global forest data sets in cooperation with JPL and the Joint Research Centre (JRC) of the European Commission.

The Advanced Earth Observing Satellite (ADEOS) was developed with the aim of establishing platform technology for future spacecraft and inter-orbit communication technology for transmission of earth observation data. The satellite carried eight sensors: two core sensors developed by NASDA – the Advanced Visible and Near Infrared Radiometer (AVNIR) and the Ocean Color and Temperature Scanner (OCTS) – and six sensors provided by other organisations, including NASA's Scatterometer (NSCAT) and NASA's Total Ozone Mapping Spectrometer (TOMS) and a radiometer called the Polarisation and Directionality of the earth's Reflectance (POLDER) of the French Centre National d'Etudes Spaciales (CNES). ADEOS was launched by the H-II rocket in August 1996. The satellite ceased operation in June 1997, due to breakdown of the solar array paddle. ADEOS marked a significant increase in capability and provided 8-m resolution panchromatic and 16-m resolution multi-spectral optical data in addition to data from six other scientific sensors.

ADEOS-II will monitor global environmental changes through international cooperation. It will carry six sensors: the Global Imager (GLI) and the Advanced

Microwave Scanning Radiometer (AMSR), both developed by NASDA, and instrumentation developed by NASA, CNES and the Japanese Ministry of Environment. ADEOS-II is scheduled for launch in 2002 by the H-IIA rocket. The resolution of its sensors is limited to 250 m as it focuses on global environmental change.

The Advanced Land Observing Satellite (ALOS) is planned for launch in June 2004. The panchromatic optical resolution of ALOS is improved to 2.5 m and its multi-spectral and L-band SAR resolution is 10 m. Details for ALOS are provided in the next section.

There is a plan to launch the Information Gathering Satellite (IGS), a high-resolution satellite constellation for purposes including national security. The satellites are being developed by NASDA, under contract with the Japanese government. The Japanese Cabinet Office will own, operate and use these satellites. While its data policy has yet to be announced, it is unlikely that the data will be openly disseminated to the public.

NASDA established the earth Observation Center (EOC) in 1978 to receive, process, distribute and archive data from the US Landsat and the French SPOT satellites. NASDA distributes data acquired at EOC to the general public through the Remote Sensing Technology Center of Japan (RESTEC), a non-profit foundation. NASDA/RESTEC have received and distributed to the public satellite data from various sources including Landsat, SPOT, ERS, RADARSAT and IRS, as well as Japanese satellite data.

6.5.3 Advanced Land Observing Satellite (ALOS)

NASDA plans to launch ALOS in 2004 as a follow-on satellite to JERS-1 and ADEOS. The mission objective of ALOS is to advance land-observing technology and to contribute to cartography, regional observation, disaster monitoring and earth resource survey. ALOS has three major sensors: the Panchromatic Remote Sensing Instruments for Stereo Mapping (PRISM), the Advanced Visible and Near Infrared Radiometer type 2 (AVNIR-2) and the Phased Array type L-band Synthetic Aperture Radar (PALSAR). ALOS is a large satellite that weighs 4,000 kg (Figure 6.5.2).

PRISM provides 2.5-m resolution panchromatic images and 3–5 m accuracy digital altitude data with triplet stereo mapping capability. Observation swathe width is 35 km in triplet stereo observation mode and 70 km in nadir observation mode. The data are mainly for cartography for generating and revising 1:25,000 scale maps and digital data for Geographical Information System (GIS). Simulated 2.5-m and 10-m resolution panchromatic images are shown in Figure 6.5.3.

AVNIR-2 is an improved version of ADEOS/AVNIR. AVNIR-2 provides multi-spectral 10-m resolution images. PALSAR is an improved version of JERS-1/SAR. NASDA and METI are jointly developing PALSAR. AVNIR-2 and PALSAR are used mainly for regional observation, disaster monitoring and earth resource survey. For the disaster monitoring mission, AVNIR-2 and PALSAR have a pointing capability and can provide images from anywhere in the world within 48 hours.

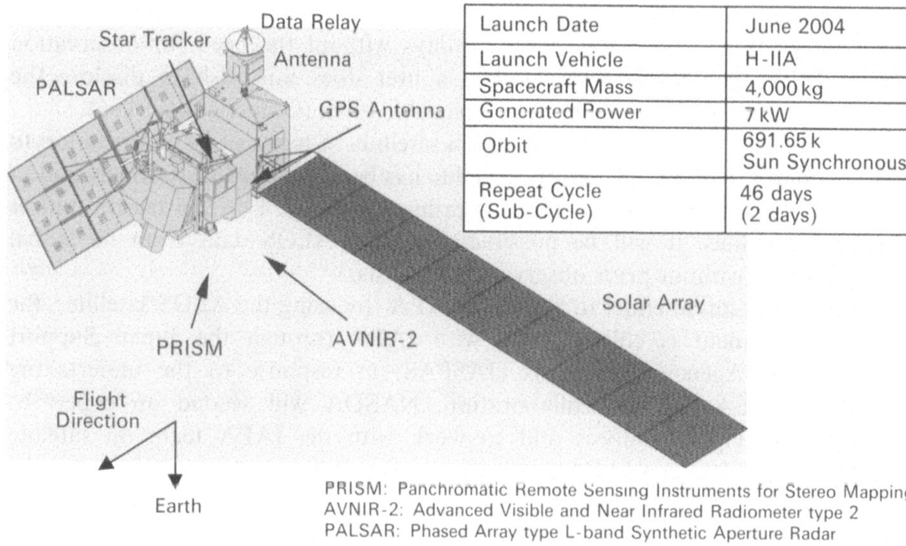

Launch Date	June 2004
Launch Vehicle	H-IIA
Spacecraft Mass	4,000 kg
Generated Power	7 kW
Orbit	691.65 k Sun Synchronous
Repeat Cycle (Sub-Cycle)	46 days (2 days)

Figure 6.5.2. ALOS satellite system.

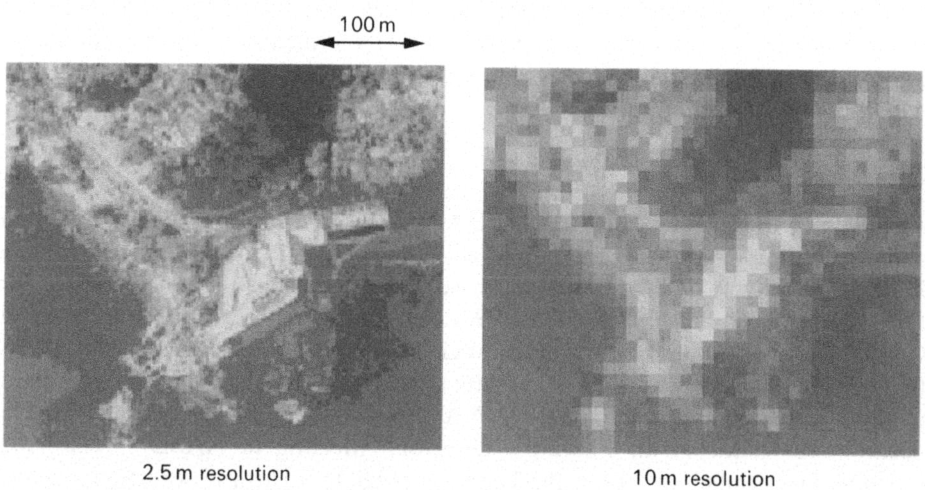

Figure 6.5.3. Simulated 2.5-m and 10-m resolution panchromatic image.

ALOS has some unique characteristics. Its inter-satellite communication capability allows real-time observation and data acquisition via Japanese and European data relay satellites. Its high-precision satellite position and attitude determination system enables users to locate the exact position of any ground object without any ground reference points.

The resolution of ALOS sensors is not as high as planned commercial high-resolution satellites; however, ALOS is able to provide imagery with many other special qualities. In particular, the wide swath width (70 km) of ALOS enables regular observation of any area every 46 days without the need for observation requests. This capability is critical when a user does not wish to disclose the reason for observing a particular site (Hamazaki, 1998; (Osawa *et al.*, 1996).

Therefore, the significance of the ALOS satellite, in terms of its contribution to IAEA safeguards activities, is that the satellite has been designed to routinely collect global coverage data in order to fulfil its primary mission. For this reason, unlike commercial satellites, it will be possible to obtain ALOS data from its global coverage archive without prior observation requests.

As part of its initial effort to assist the IAEA (by using the ALOS satellite), the Japanese government is collaborating with IAEA through the Japan Support Programme for Agency Safeguards (JASPAS) in response to the undertaking proposal by IAEA. In this collaboration, NASDA will second an expert to conduct basic imagery analysis and to work with the IAEA team on satellite imagery use for enhancing IAEAs, safeguards capabilities.

6.5.4 Data policy

In this section NASDA's Data Policy will be introduced, mainly in regard to high-resolution satellite imagery policies such as the JERS-1, ADEOS, and ALOS satellites. The legal aspects concerning the United Nations Principles Relating to Remote Sensing of the earth from Space (the UN Principles), Japanese domestic legislation and intellectual property rights, and future data policy (including that of ALOS) will also be examined.

Background

At the time of launching JERS-1, NASDA introduced its own policy for reception and distribution of earth observation satellites in July 1992 (the 1992 NASDA Data Policy). It was a two-tier data policy, based on discussions at the Committee for Earth Observation Satellites (CEOS, 1992) and the earth Observation International Co-ordination Working Group (EO-ICWG, 1994), where it was agreed that access to data should be open to all and that data should be made available for researchers at the lowest possible cost. The 1992 NASDA Data Policy was adopted after a decision was made at an Executive Directors' Meeting in NASDA. The 1992 NASDA Data Policy had been the basis for NASDA's earth observation satellite data policy from MOS-1 to TRMM, until it was revised in 2001.

During this time, we have seen a changing world of remote sensing activities, with policies diversifying between countries and regions and various forms of involvement of the private sector. For instance, the US Landsat and the French SPOT, the Canadian RADARSAT and the Indian IRS all represent forms of 'traditional' land remote sensing satellite operation by government–private sector cooperation. Private services such as those provided by IKONOS and OrbView satellites are emerging, while there are government-owned programmes such as NASA's EOS series and the European earth Explorer and earth Watch series. In the USA, it has been decided that low-level data from government-owned satellites will be distributed at low cost for fulfilling user requests, while value-added (higher level) data are reserved for the private sector (United States, 1992). On the other hand, in Europe, prices are defined by the categories of use, while at the same time data of higher levels are also put into the hands of the private sector (European Space Agency, 1998), Canada is also adopting a similar standpoint (Canadian Space Agency, 1994).

The domestic situation has also made great changes. NASDA now has a wide range of programmes from land observation such as ALOS to global change research with the Global Change Observation Mission (GCOM) series. The public's view of NASDA's space activities has charged significantly too, not only in that there is more awareness among the public, but also that it has become a requirement that NASDA show the benefit of public investment in space activities (i.e. effective use of data acquired, including widespread data dissemination and utilisation).

So, we can see today the striking diversity of remote sensing satellite activities worldwide. Based on several recent issues that have implied the need for improvement of the current data policy (Onoda, 2000), internal discussions have taken place in NASDA to fundamentally examine the existing policy. The policy was revised in 2001, and further discussions are ongoing about the next generation of earth observation satellites.

Current data policy

At present, the ADEOS Data Policy basically covers all of NASDA's earth observation satellites. The Policy was based on the 1992 Data Policy and became more detailed for ADEOS, incorporating some ideas from the IEOS Data Exchange Principles adopted in EO-ICWG, such as setting up a category for public purposes (EO-ICWG, 1994) and some technical changes, such as the rapid development of an information network.

The original ADEOS Data Policy set up different categories for data distribution. This Policy was amended in April 2001, so that currently all NASDA satellite data are available to all users at a low cost. This was done to facilitate the use of NASDA R&D satellites by the research community and the general public and setting a price that meets the user demand. The data from Japanese earth observation satellites, as well as data of foreign satellites received by NASDA, are open to the public and available on a non-discriminatory basis and in a timely manner. Data are distributed by the Remote Sensing Technology Center of Japan (RESTEC). For instance, the price of ADEOS/AVNIR panchromatic data (resolution 8 m, 80 km × 80 km scene, Level 1B processed data) is 2,700 yen (or US$20.76).

The data policy for ADEOS-II is expected to be similar to the framework currently valid for ADEOS data as described above. The ALOS Data Policy is also under discussion and will be described later in this chapter.

The UN Principles

With regard to the legal aspects of the Data Policy, since there is no specific legislation regarding remote sensing activities in Japan, the United Nations Principles Relating to Remote Sensing of the earth from Space (United Nations, 1987) currently provide the legal basis for remote sensing activities in Japan. The data policies of NASDA satellites are based on the provisions of the UN Principles, calling for open non-discriminatory access to remote sensing data.

The 15 general principles provided in the former part of the UN Principles along with some specific provisions such as avoiding invasion of the rights and interests of the sensed state (United Nations, 1987), promoting the protection of both the environment and humanity (United Nations, 1987) have been accepted as common practice.

The main issues, however, arise from Principle XII, which gives the right for a sensed state to gain access to data about its own territory, as soon as they are produced, on a non-discriminatory basis and on reasonable cost terms. This was a compromise to drop the condition of prior consent by the sensed state to distribute data to third parties. Issues of a similar nature also lie in the compromise that free access, or access without prior agreement of the sensed state, is acceptable as long as that information is made available (United Nations, 1987).

In the authors' view, there are several points for discussion concerning the UN Principles. First, although the UN Principles are contained in a UN General Assembly Resolution, they remain controversial despite the Principles being legally binding. Second, the term 'remote sensing' is defined in Principle I as 'the sensing of earth's surface from space...for the purpose of improving natural resources management, land use and the protection of the environment'. There is still room for discussion about whether this is a comprehensive notion that includes such areas as intelligence activities, commercial activities, or various other applications using particularly high-resolution satellite data information. Most states, including Japan, have not yet established domestic legal frameworks to secure compliance with the UN Principles.

National law

The law concerning NASDA (the NASDA Law) (Japan, 1969) is the only piece of legislation that governs the overall space activities of NASDA. In this law, however, there is no explicit provision about earth observation activities.

On the other hand, according to clauses in the 1992 NASDA Data Policy, its status in relation to the NASDA Law is:

- Reception of data from Landsat and SPOT satellites by NASDA and distribution of data for research purposes is pursuant to the NASDA Law and describes the scope of NASDA's activities as: 'The development of artificial satellites and rockets for the launching of artificial satellites (hereinafter referred to as 'artificial satellites, etc.') and development of facilities and equipment necessary therefor' (Art. 22, 1(1)).

- Data distribution for general purposes by the distributing agency is regarded pursuant to Art. 22, 1(4) of the NASDA Law, which allows NASDA to execute any activities that may be incidental to development activities.
- Note that in Art. 1 of the NASDA Law the activities of NASDA are clearly defined to be exclusively for peaceful purposes. This, in turn, limits NASDA's earth observation activities as well only to peaceful purposes.

As described above, legally, there is no restriction inside Japan, other than the UN Principles, about remote sensing activities. Commercial operators do not have to get licences for their activities in Japan, nor to report back to the government in any event. In contrast, in the USA, a commercial remote sensing satellite operator must use a data downlink format that allows the US government access to and use of the data, and the Government may limit data collection and/or distribution at certain times for national security and other purposes (1994 PDD 23).

Intellectual property rights

NASDA retains any applicable legal form of intellectual property rights to its earth observation data. Copyright in particular is not necessarily applicable to earth observation data, unless the data is so highly processed that copyright would obviously apply. Generally, data rights are protected under conditions agreed in respective contracts, in a way similar to copyright law. At present, it is not clearly defined, as it is in the USA, that governmentally owned earth observation data is not subject to copyright (Shaffer, 1992), nor is it indicated, as in Europe, that there should be strong protection for earth observation data rights including copyright (Harris, 1999).

Under the respective agreements, there are some specific restrictions for data use. Reproduction or redistribution of purchased or 'unenhanced' data to third parties is not allowed. Special arrangements and payment of royalties are required for this purpose. When the data is processed or 'enhanced' by the user (so that products no longer retain the original pixel structure and cannot by any means revert back to the original appearance of the standard product) the users have their own right for copy, reproduction and distribution of the enhanced data.

Note that the Law Concerning Access to Information Held by Administrative Organisations was passed by the Diet (a legislative assembly) in 1999 (Japan, 1999). This, in principle, also applies to public quasi-governmental organisations such as NASDA. In light of this legislation, the principle of open and non-discriminatory access to earth observation data is supported. Nevertheless, where commercial activities are involved, measures are to be taken so that the dissemination of data by NASDA will not interfere with those activities.

ALOS and future data policy

Based on NASDA's Data Policy, ALOS satellite data will be openly available to the public at an affordable cost. Its observation plan will be open and catalogue data will be accessible on its website. Any user may acquire ALOS data from the global archive.

As ALOS is capable of providing global coverage of land observation and to transmit the huge amount of data acquired, the volume of downlinked data is expected to be enormous. It is thus planned that the task of data processing and distribution be shared geographically. This framework is called the 'ALOS Data Node' concept. Under the current plan, the Data Node consists of one representative organisation in each region and cooperative organisations including data distributors. ALOS data is downlinked to NASDA's earth Observation Center (EOC) via a data relay satellite. NASDA will transfer the regional level-0 data it has obtained to the ALOS Data Node. The ALOS Data Node is responsible for providing regional level-1 data upon request from data users throughout ALOS's mission life. The plan is to divide the entire world into four regions: Asia, Europe and Africa, North and South America, and Australia and Oceania. The Data Nodes would serve both as regional ALOS data centres and scientific research centres.

The international policy of ALOS data distribution and exchange is actively discussed with the Data Nodes. Under the planned policy, each ALOS Data Node will be allowed to handle its regional data according to its own data policy. In this way, each region will be able to maximise data use by making the best of their existing infrastructure and user–provider community. As each region will carry out its activities under different policies, the most controversial issue is inter-node data exchange. At present, inter-node data exchange will be limited to direct reception data, while NASDA will provide the level-0 data of the required region. If these issues are resolved between international partners and this concept proves to function properly, it may well become a successful part of an internationally 'integrated' data policy.

6.5.5 Commercialisation and future prospects for Japanese remote sensing

We would like finally to put the authors' view on the role of NASDA and the private sector, and the future prospects of remote sensing activities in Japan. At present, we can see three areas of use in the world: operational (primarily weather and meteorology and partially land observation), R&D and science, and commercial.

If we look at Japan, except for the meteorological satellite that has been transferred to JMA, NASDA is the main body that develops, launches and operates earth observation satellites at the moment. This means that all three areas of use of earth observation are in practice still in the government's hands. As for data distribution, RESTEC has been the primary distributor in Japan, but it is a public non-profit foundation. There are chances that private companies selling foreign commercial satellite data in Japan might overtake the existing government-led framework in Japan, but the future is still unclear. Such a situation might suggest that the market for satellite data sales in Japan is still premature. It is so far unclear whether Japan is heading for earth observation by government or by private industry.

At this stage, NASDA should be fully involved at the political level in identifying space programmes and policies and in international dialogue such as within CEOS and the Integrated Global Observing Strategy (IGOS). It should then prepare an appropriate environment for developing new applications and commercial activities. Particularly for missions such as a future ALOS follow-on, we may eventually expect the private sector to be involved. With such prospects in view, the role of the government should be to define a simple, coherent policy that would allow easy access to satellite data, leading ultimately to the development of new applications and value-added use.

6.5.6 Conclusion

The effectiveness of the use of satellite imagery for IAEA safeguard activities has already been demonstrated in many studies (Jasani, 1998; Baines, 1998). Japanese earth observation satellites, especially ALOS, have enough potential to contribute to these safeguard activities. They have qualities that commercial high-resolution satellites do not provide. Efforts have already started for cooperation between Japan and IAEA, through the Japanese support programme. There are also possibilities that NASDA may provide satellite imagery to IAEA at a lower price, under certain arrangements.

To achieve the best results, IAEA should combine various sources of satellite imagery, including those of government-owned and private satellite systems. This should bring IAEA maximum benefits in terms of data quality, availability and policy. It is thus recommended that IAEA seek every possibility to obtain different earth observation satellite data, in a cost-effective manner. IAEA should clarify its requirements about issues such as copyright in order to negotiate with satellite data providers, since each provider has his own data policy.

Although satellite imagery cannot completely replace on-site inspection, it is a powerful aid for inspectors to acquire two-dimensional and multi-temporal information. The authors believe that analysis of satellite imagery and the basic technology to do so are now widely available and no longer unusual. The authors encourage the IAEA inspectors to familiarise themselves with and to utilise satellite imagery, thus considerably strengthening IAEA safeguards.

6.5.7 References

Anselmo, J. C. (2000) Competitors chasing Ikonos into orbit. *Aviation Week & Space Technology*, 31 January 2000.

Canadian Space Agency (1994) RADARSAT Data Policy, RSCA-PR0004, Canadian Space Agency, 13 July 1994.

CEOS (1992) Satellite data exchange principles in support of global change research, adopted at the sixth CEOS plenary meeting, December 1992, London.

EO-ICWG (1994) The International earth Observing System (IEOS) Data Exchange Principles, accepted by NASDA in February 1994.

European Space Agency (1998) ESA Envisat Data Policy, ESA/PB-EO (97), rev. 3, Paris, 19 February 1998.

Hamazaki, T. (1998) Current Status on Japanese earth Observation Satellite Programme: Data Availability and Restrictions, IAEA Technical Workshop on Safeguards: Sources and Applications of Commercial Satellite Imagery, 1998.

Harris, R. (1999) earth Observation Data Policy and Europe (EOPOLE). *Space Policy*, **15**, 175–177.

Japan (1969) Law Concerning the National Space Development Agency of Japan, Law No. 50, 23 June 1969.

Japan (1999) The Law Concerning Access to Information Held by Administrative Organisations, Law No. 42, 14 May 1999.

Jasani, B. (1998) IAEA Safeguards and Commercial Observation Satellite, IAEA Technical Workshop on Safeguards: Sources and Applications of Commercial Satellite Imagery, 1998.

Baines, P. J. (1998) Potential Applications of Commercial Satellite Imagery in International Safeguards, IAEA Technical Workshop on Safeguards: Sources and Applications of Commercial Satellite Imagery, 1998.

Onoda, M. (2000) Current Issues on earth Observation Data Policy in Japan, 22nd International Symposium on Space Technology and Science, Morioka, May 2000.

Osawa, Y. *et al.* (1996) High Resolution Remote Sensing Technology and the Advanced Land Observing Satellite (ALOS), IAF-96-B.2.01.

Shaffer, L. (1992) US data policy for earth observations from space. In: Space in the Service of the Changing earth, Vol. III, edited by T.D. Guyenne and J.J. Hunt, ESA SP-341, pp. 1477–1481, ESTEC, Noordwijk, Netherlands.

United Nations (1987) Principles Relating to Remote Sensing of the earth from Space, General Assembly Resolution 41/65, 42 UN GAOR Annex (95th plenary meeting, 3 December 1986), UN Doc A/RES/41/65, 22 January 1987.

United States (1992) Land Remote Sensing Policy Act, 15 U.S.C., Ch. 82, Sec. 5602.

6.6 RUSSIA *(O. V. Zhdanovich)*

6.6.1 Introduction

The Russian commercial remote sensing programme is based on the former Soviet Union earth observation activities, which, at the end of the 1980s, were very diversified and consisted of photographic observations from Resurs F and Kometa satellites, digital panchromatic and multispectral sensors on board the Resurs O satellites as well as the radar satellite Almaz. In addition, cosmonauts on the space station *Mir* performed observations using a multitude of instruments installed on board the station.

This chapter presents a brief historical background of the Soviet/Russian remote sensing programme, perhaps for the very first time. Today, there is still a lack of understanding of the real history of the development of remote sensing in the Soviet Union, because for many years Soviet space activities were classified. In the last 10 years, many books about the history of the space programme have been published in Russia, including books about the USSR military space programme. However, these books have not been translated into English.

Current Russian commercial remote sensing systems, applications, companies, prices and data distribution policies will be described in the chapter. A detailed overview of photographic and digital remote sensing satellites, representing the most successful Russian commercial remote sensing systems, is also given. A brief description of the data archive from the Modular Optoelectronic Multispectral Stereo Scanner (MOMS-2P, Deutsche Agentur für Raumfahrtangelegenleiten (DARA) remote sensing camera installed on board *Mir*) is given as an example of successful international cooperation while *Mir* was in orbit. An example of the application of Russian remote sensing data for geological and seismic engineering evaluation of an area selected for construction of a nuclear power station is presented as well.

6.6.2 Civil earth observation programme of the Soviet Union – historical overview

In an article on commercial high-resolution space imagery, published in *Geoinformatics* magazine in March 2001 (Gordon, 2001), the decision of the Soviet/Russian government to start selling civil remote sensing data worldwide and, subsequently, high-resolution data was presented as a 'totally surprising development' for the Western remote sensing community. In reality, it should not have been a surprise as the Soviet civil remote sensing programme was part of the national space programme (its basic architecture had already been designed and approved by the Soviet government by the mid-1960s).

The Soviet civil remote sensing programme, which was gradually developed from military reconnaissance and civil meteorological programmes, became fully operational in the 1980s.

In the USSR, the first launch in March 1964 of the Zenit-2 spacecraft was the beginning of regular launches of reconnaissance satellites. The design of this spacecraft was similar to Yuri Gagarin's spacecraft *Vostok* (Figure 6.6.1.). This was the period of the cold war and the space race between the USSR and the USA. This is why the space programmes of the Soviet Union and the United States had many common features (Cosmonautics and Armed Forces, 1997).

In the 1960s in the Soviet Union, the concept of intelligence from space developed in two major directions. There was earth observation for the USSR Ministry of Defence, which was performed by instruments on board satellites. The second direction was earth observation carried out by cosmonauts with the help of special optical cameras, installed on board manned space vehicles.

The first generation of Soviet military remote sensing satellites, called Zenit, used the same observation technology as their USA counterpart, the Corona spacecraft: photographic film was returned to earth in special capsules. One of the main purposes of reconnaissance satellites was to monitor compliance with the terms of the arms control treaties with special emphasis being put on control of the spread of nuclear weapons. Soviet National Technical Means (NTM) monitored nuclear explosions and closely observed the development of construction facilities for the production of nuclear weapons worldwide. The results were used not only for the government of the Soviet Union but were also shared with a small number of Eastern bloc states.

Figure 6.6.1. Zenit spacecraft at the assembly hall (source: Khrunichev company).

From 1982, the first generation of NTM was complemented by satellites with digital remote sensing cameras on board (Tarasenko, 1995b). However, up to the present films continue to be recovered for civil and dual use from photographic remote sensing satellites. Based of the Zenit technology with recoverable film capsules, a family of civil multispectral photographic satellites was developed and operated during the 1970s (Lantratov, 1996). This was followed by the development of a family of Resurs-F civil remote sensing satellites at the end of the 1970s, which began operations in 1981. The Central Specialised Design Bureau in Samara, better known as TsSKB, is the main industrial manufacturer of Soviet/Russian photographic satellites. The State Centre Priroda, which was created in 1973 within the USSR State Committee on Geodesy and Cartography (now this state agency is called Roskartographiya), was responsible for the distribution and processing of Fram and Resurs-F data.

Note that during the cold war era all Soviet satellites were known as Kosmos. They differed from each other by their specific numbers only. The actual names were known by a narrow circle of professionals only, while the real names of military satellites were classified.

For human military reconnaissance from space, the Soviet Union developed a military orbital station with special earth observation optical cameras, and this was tested in space in the first half of the 1970s. These space stations had common names like 'Salyut 2 and 3', but inside Russia they were better known as the Almaz military space station family. In the competition between automatic and human reconnaissance, which took place inside the Ministry of Defence in the USSR in the mid-1970s, the former approach took over, based on the experience gained with the Salyut 2 and 3 series. The explanation and reason was simple. Humans cannot make observations day and night and under all weather conditions, and they need very expensive life support systems and safety measures. Later Soviet cosmonauts would conduct earth observation mainly for peaceful scientific purposes (Development of Military Space Forces, 1997).

The company NPO Mashinostroenie, which developed the military space station Almaz, also developed a radar satellite with the same name Almaz. The first civil radar satellite Almaz, also known as Kosmos-1870, was launched into orbit in July 1987. The history of this civil satellite deserves some special consideration.

The origin of the civil radar satellite was directly connected with navy reconnaissance radar satellites with nuclear powered engines, one of which Kosmos-954 crashed in Canada in winter 1978. The satellites were given the name 'Radar Ocean Reconnaissance Satellites, RORSAT's (FAS *World Space Guide*, 1999) in the West. Because of their low orbit (250 km), these radar satellites were equipped with nuclear powered engines. They started regular operation in 1975. In 1978 they were joined by a passive navy reconnaissance spacecraft Electronic Ocean Reconnaissance Satellite, EORSAT, which orbited higher than the radar satellite. The RORSAT spacecraft was designed in such a way that in case of accident the nuclear powered engine was automatically shot from the spacecraft into a higher orbit. In the incident with Kosmos-954, the control centre failed to separate the nuclear powered engine and the whole satellite crashed and caused radioactive pollution. Fortunately, the impact area was remote and only one person witnessed this accident (Pieson, 1999). After this accident, the radar spacecraft's control system and the nuclear powered engine were redesigned, so that the engine was able to burn during any stage of its entrance into the upper atmosphere. In 1983 there was another accident with Kosmos-1402, but this time the nuclear powered engine burnt during reentry and the remnants of the satellite did not cause any radioactive pollution (Savin, *et al.*, 2001).

Since 1992, high-resolution photographic spy satellites were allowed for dual use. These satellites received the name 'Kometa'. Since that time, Russian 2-m resolution data has been successful in penetrating the world market of commercial remote sensing data. This commercial success was followed by the release of 1-m resolution imagery 6 years later.

The start of commercial use of Russian 2-m resolution satellite images from Kometa spacecraft initiated the worldwide use of civil high-resolution data. In 1995, in response to these Russian initiatives, the US government opened access to the archives of images from the Corona family of spy satellites (Editors, *Novosti Kosmonavtiki*, 1995). A few American companies began development of 1-m resolution commercial remote sensing systems at this time as well.

During the 1980s, only one state company Soyuzcarta was allowed, as a foreign representative of the State Committee of Geodesy and Cartography, to sell data

Table 6.6.1. Russian satellite imagery provided by different Russian state and commercial remote sensing companies.

Name of the company	Data from Resurs-F satellites	Data from Resurs-O satellites	Data from Kometa satellite and dual use satellites
State companies			
State Centre Priroda	✓		
State and Research Centre Planeta		✓	
Commercial companies			
Sovinformsputnik	✓	✓	✓
Geo Spectrum	✓	✓	✓
Northern Geographical Company	✓	✓	✓
ScanEx		✓	

from civil Soviet remote sensing satellites abroad. In 1991, still at the time of the Soviet Union, state companies, as owners of specific remote sensing data, were allowed independently to sell and market satellite imagery at the international remote sensing data market. At that time, this involved only a few big state companies: State Centre Priroda was responsible for satellite imagery from Resurs-F satellites, SRC Planeta offered satellite data from Resurs-O, Okean, Meteor and Elektro satellites, while NPO Mashinostroenie represented radar data and the Earth Observation Department from Russia Space Corporation Energia (RSC) marketed data from the photographic complex Priroda on the *Mir* space station, equivalent to that of the Resurs-F1 satellite (the Priroda and Spektr modules were added to the *Mir* space station in the mid-1990s).

During the 1990s, pure commercial remote sensing companies emerged in Russia. Nowadays, these companies, which are usually not large in size, offer their services in addition to the larger mainly still state-owned companies mentioned above. In Russia, a significant variety of applications and methodologies of remote sensing data processing was developed during almost three decades of operating the different types of civil remote sensing satellites. Commercial Russian remote sensing companies are now marketing satellite data as well as data processing techniques and applications. It is possible to order, through private companies, all types of Russian remote sensing data, applications developed on the basis of Russian imagery including archive data and ordering of new missions (Table 6.6.1). However, there are a few companies (e.g. ScanEx) that specialise only in digital remote sensing systems, data and applications with spatial resolution of less than 10 m.

6.6.3 Commercial Russian remote sensing programme

At the start of the 1990s, Russia inherited a well-developed remote sensing programme from the Soviet Union. In this programme, both the space segment and the ground segment had been developed excellently. Technology worked very well. All that was lacking was a commercial satellite imagery distribution network, both inside as well as outside Russia. There were two main reasons for this. The first was that, even inside the USSR, civil satellite imagery as well as the whole space programme was classified. Even to access and process (ivi) space imagery, very strict regulations were developed. The second reason was that the Soviet Union in its entirety was a state-owned country and any private initiatives or commercial business only started with Mikhail Gorbachev's perestroika in the mid-80s. At the time of the Soviet Union, manufacturers of earth observation satellites were in practice not able to benefit financially from the distribution of remote sensing data, because, as was shown in Section 6.6.2, the value chain, starting from the civil imagery satellites providers down to the data service providers, was very complicated: satellite manufacturers and data processing centres belonged to different state ministries and agencies. It was only in the 1990s that industrial enterprises, as producers of remote sensing satellites, started to be shareholders in remote sensing private commercial companies.

Since 1991, when the Russian space industry was allowed to market its products independently of state ministries, all Russian state enterprises as well as newly born commercial companies had a difficult time entering the international market. They didn't have any international marketing experience, and they were not ready for intense free market competition.

In the early 1990s, the commercial success of photographic and digital Russian remote sensing data on the international market was mainly due to two commercial companies: Sovinformsputnik, which was created in 1991, and ScanEx, created in 1992. Both these companies are based in Moscow and are now well known worldwide. These two companies gradually organised their business, based on international standards, and developed commercial satellite data distribution networks in Russia as well as abroad. However, the commercial success of these two companies was directly connected with their unique products offered to the market at that time. Sovinformsputnik offered high-resolution imagery from its Kometa satellite and ScanEx ground data receiving stations from Resurs-O satellite, based on personal computers IBM PC.

Commercial distribution of radar data from the Almaz satellite was also successful. NPO Mashinostroenie set a good example within Russia when it showed how an industrial company can distribute and market data successfully in addition to producing satellites. The Almaz-1A satellite was launched in March 1991 and acquired images with 15-m resolution. At that time, it was the only available civil radar system with such high resolution (the first RADARSAT was launched in November 1995). It was expected that the active lifetime of Almaz-1A would be about 30 months. However, because of fuel problems in October 1992, the satellite was de-orbited. Among the few Western companies that signed an agreement with NPO Mashinostroenie was the US company Hughes STX (Lanham, Maryland), which was responsible for the processing and commercial distribution of radar images outside Russia. The relationship was most successful. Total revenues from sales of Russian satellite radar data were around US$250,000 (but estimates of expected sales had been too high: at US$2 million). Today, there is no Russian radar satellite in space, although radar data from Almaz is still available from archives (Garbuk and Gershenzon, 1997). A new 1-m resolution radar spacecraft Kondor-E is under development.

Russian photographic imagery

Russian photographic remote sensing data differ in spatial resolution and type of satellite and comprise high-resolution and very high-resolution images acquired by dual use satellites, and other photographic data acquired by Resurs-F satellites. Photographic films are recovered at the end of missions. This is the best-known Russian remote sensing data on the international market.

Until 1995, Russian high-resolution photographic data dominated the high-resolution remote sensing international data market, but then in 1995 the Corona archives were declassified. Today, with the launch of the IKONOS satellite in 1999 and Quickbird-2 in 2001, although there is no real commercial competition between Russian remote sensing systems and satellites such as these, there are still a significant number of customers for Russian photographic imagery. Russian photographic remote sensing data have a few unique features that create the perfect basis for production of large-scale maps and for geo-information systems for a wide range of applications. The stereoscopic mode of survey, in particular, represents the clear advantage of Russian photographic remote sensing data for the development of digital terrain models (DTMs), digital elevation models (DEMs), 3-D models, animation, flight simulators, ortho photos and topographic maps. Another advantage of Russian photographic images is their data archives which go back to the 1980s. Recently, Russian remote sensing companies started to offer data from early remote sensing systems dating from the 1970s, but with spatial resolution less than 3 m. Archived data from Russian satellites is very useful for specific studies of given territories, when there is need to look into the dynamics of development of particular regions.

The excellent stereoscopic characteristics of Russian high-resolution imagery is borne out by the fact that the American company EarthWatch bought Russian high-resolution imagery for orthorectification of their products.

Kometa mapping system – high-resolution photographic satellite

High-resolution photographic remote sensing is performed by the series of Kometa spacecraft, which was specifically designed for production of large-scale topographic maps with/without additional ground control measurements and was used before 1992 only for military purposes. This feature makes this space system unique, especially for mapping of remote or mountain areas, where it is not possible to make ground surveys. Conducting a ground topographic survey can be very expensive; the cost of producing topographic maps and development of geo-applications based on Kometa satellite images can be much cheaper.

The first mission of the Kometa satellite took place in 1981. Dual use started in 1991. The active lifetime of a spacecraft is 45 days, in a 220-km near-circular orbit with inclination of 65 or 71°. The satellite has two on-board remote sensing cameras:

Figure 6.6.2. Kometa capsule that has landed in a forest (source: Sovinformsputnik).

a topographic camera TK-350 (10-m spatial resolution) and a high-resolution camera KVR-1000 (2-m spatial resolution); these two cameras work together. In addition, the on-board equipment includes two stellar cameras, a laser altimeter, navigation sensors and synchronising devices that provide determination of external parameters of satellite imagery orientation. Attitude information about the satellite itself is provided by on-board equipment as well as by three to four satellites of the Russian Global Navigation System (GLONASS). Satellite imagery is acquired on film, which is returned to earth in a capsule and then processed (Figure 6.6.2). Photographic images are digitised by precise scanners and orthorectified (Sovinformsputnik website, 2001). The main technical characteristics of cameras TK-350 and KVR-1000 are given in Table 6.6.2.

The Kometa mapping system is unique as the equipment installed on board acquires stereo-imagery in general with a 20–80% overlap (between 60 and 80% on average) resulting in high-resolution panoramic imagery. One frame of the TK-350 camera contains seven frames of the KVR-1000 camera. Even without ground control, the images from TK-350 provide very accurate photogrammetric measurements, with planimetric location accuracy between 20 and 25 m and accuracy in height definition of 10 m (Table 6.6.3). With ground control measurements, planimetric accuracy and height definition accuracy are even better. Images from KVR-1000 with 2-m resolution add even more detailed information to already accurate photo plans and maps developed on the basis of 10-m resolution stereo-images from TK-350. Enlargement of images does not cause loss of photogrammetric quality up

Table 6.6.2. Technical characteristics of the Kometa mapping system.

Camera	TK-350	KVR-1000
Focal length (mm)	350	1,000
Spatial resolution (m)	10	2
Frame size (cm)	30 × 45	18 × 72
Covered area (km)	200 × 300	40 × 160
Height of survey (km)	200–260	200–260
Scale	1:660,000	1:220,000
Longitudinal overlap	60-80%	8–10%
Spectral band (nm)	580–720	580–720
Years of survey	1981–1992, 1998, 2000	1984[1]–1992, 1998, 2000

[1] The KVR-1000 camera supplemented the TK-350 camera on board the Kometa satellite 3 years later.

Table 6.6.3. Mapping accuracy of the Kometa mapping system without and with ground control.

Camera type	TK-350	KVR-1000
Mapping accuracy without ground control		
Map scale	1:50,000	–
Accuracy in planimetric location	20–25 m	–
Accuracy in height definition	10 m	–
Mapping accuracy with ground control		
Map scale	1:50,000	1:10,000
Accuracy in planimetric location	10–15 m	4–5 m
Accuracy in height definition	5–7 m	–

to 1:40,000 scale for the TK-350 camera and up to 1:10,000 scale for the KVR-1000 camera (Sovinformsputnik website, 2001).

Figure 6.6.3 shows an image of part of Washington, DC made by camera TK-350. An enlarged KVR-1000 single image of the central part of Rome with 2-m resolution is shown in Figure 6.6.4.

The main advantages of images acquired by TK-350 and KVR-1000 cameras are their high resolution, large size of area covered, stereoscopic coverage within one shooting route and their high photogrammetric accuracy. Of course, these images have some disadvantages: the original images are on film (i.e. they are not digital) and the data cannot be acquired in real time.

Resurs-F[6] *satellites*

The civil remote sensing satellites Resurs-F were developed from the military satellite Zenit and the civil satellite Fram. Films are returned to earth in special capsules. The

[6] F stands for photographic mode of data acquisition.

Figure 6.6.3. TK-350 image of Washington, DC (courtesy of Sovinformsputnik).

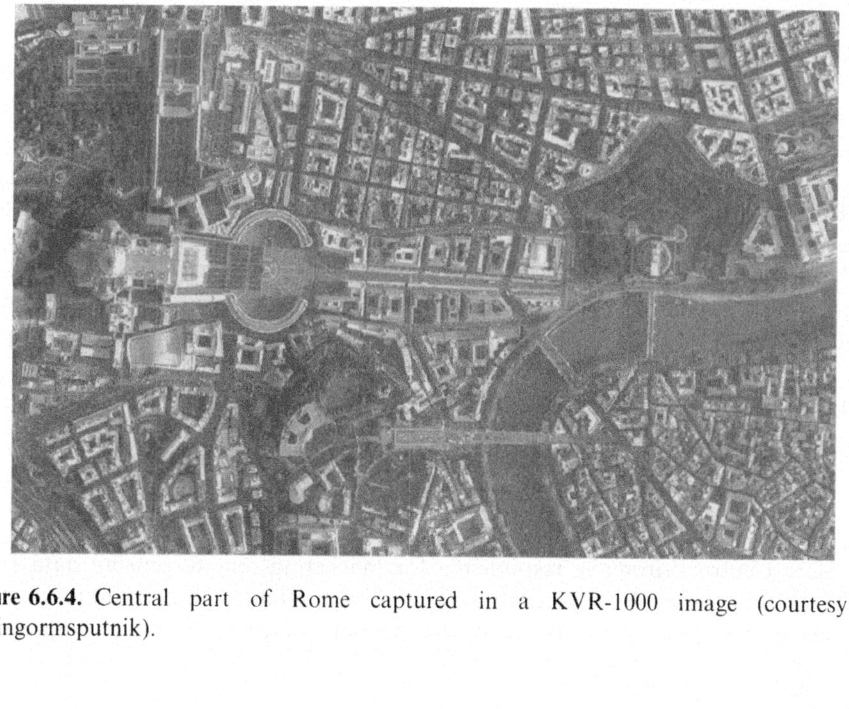

Figure 6.6.4. Central part of Rome captured in a KVR-1000 image (courtesy of Sovingormsputnik).

Table 6.6.4. KFA-1000, KATE-200 and MK-4 technical characteristics (Resurs-F).

Satellite	Resurs-F1	Resurs-F1	Resurs-F2
Camera	KFA-1000	KATE-200	MK-4
Height of survey (km)	190–360	190–360	180–360
Focal length (mm)	1,000	200	300
Scale	1:270,000	1:1,250,000	1:900,000
Covered area (km^2)	80 × 80	230 × 230	160 × 160
Type of survey	Colour	Multi-spectral panchromatic	Multi-spectral colour
Number of bands	1	4 from 6 possible	3 or 1
Spectral bands (nm)	570–680 680–810	635–690 810–900 515–565 460–505 580–800 400–700	510–600 600–700 700–850 or 450–700 (b/w)
Frame size (cm)	30 × 30	18 × 18	18 × 18
Number of spectral bands	1	3 or 1	4
Overlap (%)	60	60	60
Ground resolution (m)	4 (b/w film) 5 (colour film)	15–30	6–8 12–20

Resurs-F family consists of two types of spacecraft – Resurs-F1 and Resurs-F2. The active lifetime of these remote sensing satellites is 16 days for Resurs-F1 and 30 days for Resurs-F2, respectively. The average height of near-circular orbit is 270 k for Resurs-F1 and 300 km for Resurs-F2, with an inclination of 82.3° for both types of satellite. The very first test flight of Resurs-F1 was made in 1979 (Tarasenko, 1995a). Regular launches of the satellite started in 1981. Resurs-F1 has two types of remote sensing cameras on board: two or three KFA-1000 cameras and three KATE-200 cameras. The KFA-1000 camera provides colour images with 5-m resolution and the multispectral camera KATE-200 acquires images with 15–30-m resolution. Resurs-F2 is equipped with a multizonal camera MK-4. The first flight of the Resurs-F2 satellite took place in 1988. Its MK-4 camera provides three black and white images in three spectral bands with 6–12-m resolution and one image on colour film with 12–20-m resolution (Sovinformsputnik website, 2001). MK-4 is a topographic camera and has good geometric accuracy and good resolution (Table 6.6.4).

Examples of satellite images from cameras MK-4 and KFA-1000 are shown in Figures 6.5.5 and 6.5.6.

Companies

The State Centre Priroda is responsible for marketing remote sensing data from Resurs-F1 and Resurs-F2 spacecraft as well as from military photographic satellites that flew in the 1970s. In the 1980s, it was the main state centre and developed new

Figure 6.6.5. MK-4 image of Bolzano, Italy (band 635–690 nm; courtesy of Sovingormsputnik).

Figure 6.6.6. Beijing captured in a KFA-1000 image (courtesy of Sovinformsputnik).

techniques for various applications of remote sensing data from the KFA-1000, KATE-200 and MK-4 cameras as well as from cameras on board the Fram satellite family.

From the 1960s until the early 1990s, Soviet photographic high-resolution (2 m) and very high-resolution (1 m) images were used for military purposes only and were not commercially available on national or international markets. It was known that since the 1980s Soviet NTMs could obtain data with resolution better than 1 m, but these satellites were classified. In 1991, Sovinformsputnik was founded (its name indicates that it was created at the time of the Soviet Union) for commercial distribution of data from dual use satellites, mainly TK-350 data. A number of Russian defence organisations were shareholders in this company. Sovinformsputnik pioneered the international market of high-resolution data and was the first Russian company to start selling such data. In 1992, the government of the

Russian Federation allowed commercial use of 2-m resolution images and, in 1998, 1-m resolution images were declassified. Sovinformsputnik began to sell 1-m resolution images in June 2000. The release of 2-m resolution data in 1992 followed by 1-m resolution data was the result of intensive efforts by the founders of Sovinformsputnik, who spent practically the whole year of 1991–1992 lobbying the Russian government. Sovinformsputnik not only markets all Russian satellite imagery data, but also organises new satellite missions and develops various applications such as orthophotos, GIS layers, DEMs, and produces data processing software, etc.

Since the end of the 1990s, there have been other companies in Russia that offer similar services and products, including high resolution and very high-resolution data from dual use satellites. This includes companies such as Geo Spectrum and the Northern Geographical Company. Besides high-resolution imagery, these small companies also offer their own unique services and geo-products commercially. For example, the Northern Geographical company markets radar interferometry and shelf zone mapping, while Geo Spectrum offers navigation devices.

Products and services based on Russian space photos

The unique information acquired by the Kometa mapping system and Resurs-F satellites can serve a variety of applications, such as the production and updating of topographic maps, production of larger scale maps, digital maps, DEMs, and GIS layers. All of these are relevant to safeguard applications.

Products and services, developed on the basis of photographic remote sensing imagery, offered by Russian State and commercial companies include:

- Raw photographic and digital photo images for topographic reasons from Kometa spacecraft. Photographic products: film positive, film negative, contact paper prints, enlarged paper prints and ortho-photos. Images can be enlarged up to 1:40,000 for TK-350 camera and up to 1:10,000 for KVR-1000 camera. Digital products: raw digital images, digital ortho photos, and geocoded images.
- Raw photographic and digital images from Resurs-F spacecraft. Photographic products: film positive, film negative, contact paper prints, enlarged paper prints up to 9–10 times, colour and synthesised prints. Digital products: raw digital images, colour, and synthesised data.
- Production of colour composite images produced by combining and merging remote sensing data acquired in different spectral bands and/or by different Russian satellite systems (e.g. the very popular composite images produced on the basis of MK-4 and KVR-1000 images: (Figure 6.6.7).
- Production of mosaics from a number of satellite images for given areas.
- Production of topographic maps and plans at different scales based on digital processing of space images and topographic map updating.
- Production of thematic maps and GIS layers, including different geo-products up to scales of 1:2,000.
- Production of DEMs and DTMs using space stereo-data and radar interferometry.

Figure 6.6.7. Composite image of the central part of Rome. Colour is taken from the KFA-1000 image and 2-m resolution is taken from the KVR-1000 image (courtesy of Sovinformsputnik) (see Plate 5).

- Production of 3-D terrain models (Figure 6.6.8) with high accuracy showing separate buildings (Figure 6.6.9).
- Special Russian software for space and aerial image processing and data interpretation (Figure 6.6.10).
- Training and educational programmes.
- New data-acquisition missions.

The price of Russian photographic data

A few examples of prices for digitised photographic Russian satellite images offered on the international data market (from company price lists of November 2001) are given below. While prices for Russian products for foreign customers are traditionally given in US dollars, in view of the large variation in exchange rate of the rouble in recent years, the price list for Russian customers is different.

Tables 6.6.5 and 6.6.6 show examples of prices for high-resolution imagery charged by the Sovinformsputnik Company, who pioneered the data. Prices for 2-m resolution imagery from KVR-1000 depend on date of acquisition, quantity of ordered data (expressed in km^2, starting from areas larger than $100\,km^2$). In Table

Figure 6.6.8. Kamchatka 3-D model (courtesy of Sovinformsputnik).

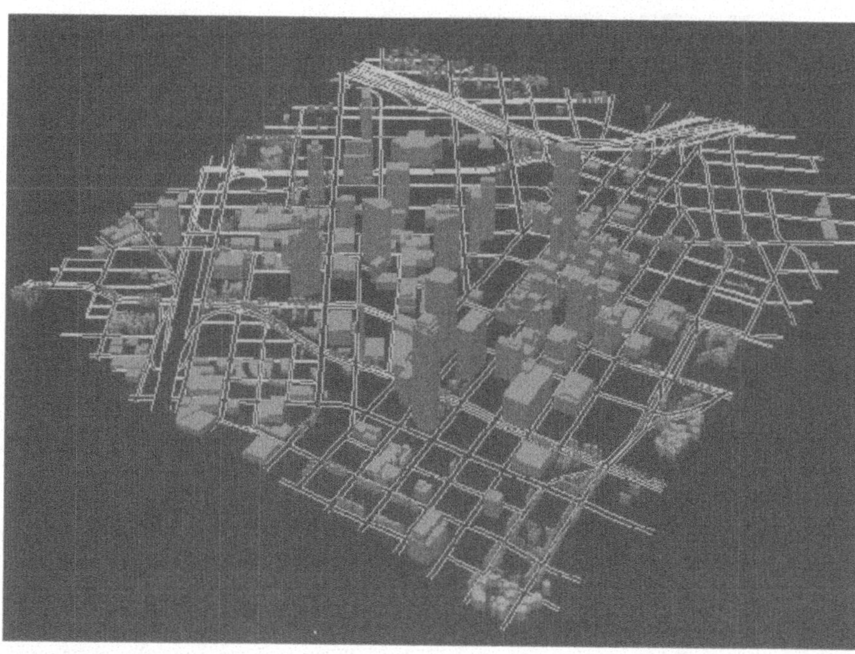

Figure 6.6.9. Building height extraction for Dallas, Texas showing building information joined with road network in a GIS layer (courtesy of Sovinformsputnik).

Figure 6.6.10. Example of ORTHOSPACE version 1.2 software developed by Sovinformsputnik. This imagery ortho-rectified toolkit has been specially developed for Russian high-resolution satellite imagery and Landsat TM data and aerial photos (courtesy of Sovinformsputnik).

Table 6.6.5. Examples of prices (as at November 2001) for raw digitised images from KVR-1000.

Area (km^2)	Years of acquisition		
	2000–1999	1998–1993	1992 and earlier
1–100	US$1,500	US$1,200	US$900
US dollars per square kilometre			
101–250	15.0	12.0	9.0
1,001–1,600	8.0	5.0	2.0
More than 1,600	Price negotiated		

Table 6.6.6. Examples of prices for ortho-rectified images from KVR-1000.

Area (km^2)	Years of acquisition		
	2000–1999	1998–1993	1992 and earlier
1–100	US$2,000	US$1,700	US$1,400
US dollars per square kilometre			
101–250	20	17	14
1,001–1,600	13	10	7
More than 1,600	Price negotiated		

6.6.5, examples of prices for raw digitised data for KVR-1000 are given. The cost of ortho-rectified images is approximately 10–15% more (see Table 6.6.6).

Because of the presence of several commercial companies on the Russian national market dealing with high-resolution data, prices for high-resolution imagery and geo-products and services may differ slightly from company to company.

The State Centre Priroda's price list is based on the cost of digitisation of 1 km^2 of an image. The Centre established the minimum quantity of data that can be ordered as corresponding to one-quarter of the square area of an image. For KFA-1000, the cost for black and white images per square kilometre is US$0.375 and US$0.411 for colour images. Prices for the MK-4 and KATE-200 cameras are given in Table 6.6.7.

For detailed information about prices, potential customers should contact the Russian State and commercial companies directly or visit their websites (see Appendix 6.6.I for the addresses of the main Russian remote sensing data providers). Remembered, though, that state or private companies do not necessarily keep their price lists on their websites and that not all companies have their own websites.

Table 6.6.7. Examples of prices for digitised satellite imagery from the MK-4 and KATE-200 cameras.

Images	MK-4 (US$/km^2)	KATE-200 (US$/km^2)
B/w, 1 channel	0.101	0.009
B/w, 3 channels	0.153	0.021
Colour	0.140	0.017
Synthesised	0.175	0.023
DEM	0.435	0.118

Russian digital remote sensing imagery

Russian digital remote sensing data originate mainly from Resurs-O satellites. Data are obtained in real time in several spectral bands. The Resurs-O system is similar to the Landsat system in that it acquires data with resolution of 34–45 m, as well as with resolution of 100 m. The data is utilised for global environmental studies, disaster monitoring, and natural resource monitoring. The moderate price of Resurs-O imagery is one of the main advantages of Russian digital data compared with other remote sensing data of similar spatial resolution.

Furthermore, archived data is available from the *Mir* space station by means of MOMS-2P.

Resurs-O [7] *satellite*

At present, there are two satellites of the Resurs-O1 family in orbit – Resurs-O1-3 and Resurs-O1-4. The former was launched in October 1994 and the latter in July 1998. Both satellites have polar sun synchronous orbits with an average altitude of 678 km for Resurs-O1-3 and 835 km for Resurs-O1-4. For both satellites the inclination is 97.97°. Each Resurs-O satellite is equipped with two MSU-E scanning radiometers and two MSU-SK radiometers. The MSU-E radiometer is a multi-spectral opto-electronic radiometer, containing three CCD linear arrays (one for each spectral band) and is considered the high-resolution scanner for the Resurs-O system with 34–45-m resolution (Table 6.6.8). The MSU-SK radiometer is a multi-spectral opto-mechanical radiometer with conical scanning of the Earth surface. This is why images obtained with the MSU-SK scanner are geometrically distorted, but spatial resolution remains constant within the images. The MSU-SK is considered as a low-resolution scanner, with 140-m resolution in the visible bands and 550-m in the infrared band (Table 6.6.9). Resurs-O1-3 and Resurs-O1-4 are equipped with the same sets of remote sensing scanners, but have different orbit heights. The different number of elements in the digital decomposition of the image lines result in slightly different spatial characteristics of the images (Garbuk and Gershenzon, 1997).

Russian digital remote sensing data acquisition has the advantage of multi-

[7] O stands for optical mode of data acquisition.

Table 6.6.8. Characteristics of the MSU-E scanner (Resurs-O).

Number of spectral channels	3
Wavelengths of the channels	0.5–0.6 μm 0.6–0.7 μm 0.8–0.9 μm
Field-of-view angular width	4.0°
Angular width of an instantaneous field of view of the CCD element in the along-track direction	5.9″
Angular width of an instantaneous field of view of the CCD element in the cross-track direction	7.1″
Angular distance between CCD elements (centre to centre)	7.1″
Scan rate	200 lps
Digital word size	8 bit/pixel/channel
Resolution	45 m across flight route 34 m in nadir along flight route
View width	45 km

Table 6.6.9. Characteristics of the MSU-SK scanner (Resurs-O).

Number of spectral channels	5
Wavelengths of the channels	0.5–0.6 μm 0.6–0.7 μm 0.7–0.8 μm 0.8–1.0 μm 10.4–12.6 μm
Angle at the vertex of the cone	39°
Angular width of the sector of view	66°
Angular width of an instantaneous field of view	42″
Scan rate	50 lps
Digital word size	8 bit/pixel/channel
Resolution	140 m in visible band 550 m in infrared band
View width	600 km

spectral global coverage in real time; however, its resolution is none too high (ScanEx website, 2001).

MOMS-2P data from Mir space station

The three-linear stereo camera of MOMS-2P was developed under the MOMS-2P/Priroda contract between the German space agency and the Russian space agency. In 1996, the instrument was installed on board the Priroda module of the *Mir* station.

Data acquisition, processing and archiving were performed by two ground centres DLR-DFD-FN (Neustrelitz, Germany) and RSC Energia (Mytishi, Moscow Region, Russia).

The combination of stereo and multispectral capabilities of this camera formed its unique features. The camera was able to perform: three-linear stereo imagery, along-track stereo imagery, high-resolution imagery, multispectral imaging, as well as combinations of stereo- and multispectral data acquisition. The MOMS-2P camera acquired data in the visible and near-infrared bands (four channels) with 18-m resolution and generated panchromatic, along-track stereo data using three linear scans, one for nadir with 6-m resolution and two off-nadir-looking modes with 18-m resolution (ScanEx website, 2001).

Companies

The State Scientific and Research Centre on Space Hydrometeorology Planeta (SRC Planeta) operates meteorological satellites (Meteor and Elektro), oceanographic satellites (Okean-O1 and Okean-O) and earth resource monitoring satellites (Resurs-O). In addition, the Centre acquires data from NOAA, METEOSAT, GMS, and SPOT satellites as well. The SRC Planeta ground segment has three regional data-receiving and processing centres over Russia: one in the European part in Obninsk, one in Siberia in Novosibirsk and one in the Far East in Khabarovsk. These stations are complemented by stationary and mobile stations (mostly on ships) located around Russia's shores, but mainly in Russia's Antarctic territories. The main customers of SRC Planeta are the Russian Federal Service on Hydrometeorology and Environmental Monitoring (ROSHYDROMET) and Emergency and Disaster Management Agency (EMERCOM) (see the SRC Planeta website).

ScanEx is one of the most successful private Russian commercial remote sensing companies. Having started in 1992 with the development of ground stations for Resurs-O data, this company now provides fully integrated solutions for satellite remote sensing applications.

RSC Energia and ScanEx market MOMS-2P data.

Products and services based on Russian digital satellite data

SRC Planeta provides a variety of services – from data acquisition to data processing – from a number of Russian as well as foreign satellites. Satellite imagery received by the Centre is mainly used for weather analysis and forecast, for ocean and terrestrial global observation studies, climate change, monitoring of forest fires, rivers, ice cover, and vegetation cover. The Centre provides sun/earth interaction data as well.

ScanEx offers the following services (ScanEx website, 2001):

- Development of personal ground stations for meteorological and environmental satellites: Liana, Scanex, ScanEr and EOScan. Liana acquires NOAA data in APT data format and Scanex receives NOAA HRPT data. ScanEr provides data from Resurs-O1 and Ocean-O satellites while EOScan acquires data from the EOS-AM1 (Terra) satellite. All these ground stations use personal computers

Table 6.6.10. Example of prices for raw MSU-SK data (Resurs-O).

Data volume (MB)	The price (US$/MB)	Discount (%)	
		First order	Educational
10–20	10.5	20	10
1,000–2,000	5.0	50	40
More than 1,000	3.0	50	50

Table 6.6.11. Example of prices for raw MSU-E data (Resurs-O).

Data volume (MB)	The price (US$/MB)	Discount (%)	
		First order	Educational
1–2	105	20	10
100–200	50	50	40
More than 1,000	30	50	50

provided with software running under MS Windows. Ground stations do not require a special shelter for the antenna system and can be handled by a single person without special training at any phase of operations.

- Satellite data archiving, distribution and processing. From May 1996 up to the present, the ScanEx archive of images, acquired from the Resurs-O1-3 and Resurs-O1-4 satellites, contains more than 600 GB of data covering territory from the Atlantic to Pacific Oceans.
- Distribution of Landsat-7, Radarsat-1 and MOMS-2P data.
- Remote sensing data applications such as development of software for satellite imagery processing.
- Training.

Prices for Russian digital data

At ScanEx, the price of Resurs-O data depends on data volume (ScanEx website, 2001). Examples of prices per megabyte for MSU-E and MSU-SK raw data are shown in Tables 6.6.10 and 6.6.11. Discounts are offered for first orders and to educational organisations.

These moderate prices for Russian digital data from Resurs-O satellites represent one of ScanEx's main advantages in the commercial market.

The geometric transformation service costs between 20 and 100% more than the cost of the raw data. A mosaic from several images (it may include both middle and high-resolution images) can be ordered. The price for such a mosaic can be up to 300% higher than the cost of the raw data.

ScanEx distributes data from the MOMS-2P scanner from the *Mir* space station. The price ranges from US$893 to US$1,785 per standard scene, depending on the acquisition mode selected. Discounts are available for first orders and for educational organisations. Depending on the volume of data ordered, discounts vary from 30 to 50%.

6.6.4 Example of analysis of a possible future construction site for a nuclear power station on the basis of KVR-1000 data

The Institute of Environmental Geoscience of the Russian Academy of Sciences has developed special techniques for application of Russian and international remote sensing data for geological and seismic engineering evaluation of an area selected for construction of a nuclear power station.

The Laboratory of the Institute, led by Dr A. Viktorov, developed a way of interpreting space and airborne images along with ground data to reveal hidden geological structures and processes (e.g. groundwater, karst, suffusion of land surface, etc.), based on image texture analysis, different landscape components, relationships and quantitative analysis of image patterns. The Laboratory developed a new theory of texture analysis: a classical interpretation augmented by mathematical modelling of morphological structures of genetically heterogeneous surfaces. The core of this theory, called 'mathematical landscape morphology', is that mathematical models have been created on various geological territories using the random process theory, which helps in defining later the most effective remote sensing data interpretation criteria.

For their recent studies, the Institute used KVR-1000 and Landsat-7 data. The following maps were compiled for geological and seismic engineering site evaluation (Viktorov *et al.*, 2001):

- Map of classification of the territory for micro-seismic zoning at a scale of 1:20,000. The map shows areas with different rock composition and rock subsidence, areas with different depths of groundwater deposition, dangerous geological processes, and geological structures important for territory seismicity.
- Map of geologic hazards and newest tectonic structures at a scale of 1:20,000. The map shows geological hazards, lithological differentiation (areas with different rock composition), and geologic structures important for geologic hazard identification.

Special attention was paid to identifying tectonic breaks.

The information extracted from these high-resolution photographic observations, combined with on-site observations, allowed the authorities concerned to give the go-ahead for construction of a new nuclear power plant.

6.6.5 Russian commercial remote sensing data distribution policies

Currently, there are no government restrictions on distribution of Russian civil remote sensing data. Any Russian commercial company that intends to sell or market Russian remote sensing data for national or international markets must obtain a licence from the Russian Aviation and Space Agency (ROSAVIACOSMOS). Russian original satellite photographic and digital images are the property of ROSAVIACOSMOS. Customers can obtain copies of Russian satellite imagery in any desired form: film, paper print or digital.

However, special regulations apply to the selling of Russian very high (1 m) and high-resolution (2 m) data from dual use satellites, because the original owner of such data is the Ministry of Defence of the Russian Federation. ROSAVIACOSMOS has a special agreement with the Ministry of Defence for national and international distribution of remote sensing data from dual use satellites. Under the current regulation, Russian satellite images, especially those with high and very high resolution, are released for peaceful purposes only. The Russian Ministry of Foreign Affairs regularly updates a special list of sensitive countries or of local conflicts worldwide. The release of Russian high and very high-resolution data on territories of countries or regions listed in this document is prohibited. Russian companies are not allowed to sell abroad 1-m resolution data of Russian territory. Data with 1-m resolution are available in only one format – as digital ortho-rectified images (see, e.g., Figure 6.6.11). Finally, the Russian space system involved in acquiring 1-m resolution images is classified (Orlovsky, 2000).

At present, it is just Russian State and commercial companies that are mainly given non-exclusive rights for distribution of Russian satellite imagery. Ten years ago, Western companies were often given exclusive distribution rights at particular geographic locations. This practice did not bring any significant benefits for Russian companies, since in several specific cases Western competitors bought exclusive rights without exploiting them. As mentioned above, in the past 10 years of operating in the international data market, Russian remote sensing companies have learned serious lessons. In some specific cases, non-Russian customers, ordering new, specific satellite mission surveys, can ask for exclusive rights and be given them for a fixed time period. For example, with project SPIN-2, when two American companies – Aerial Images, Inc. and Central Trading Systems, Inc. – ordered a 2-m resolution remote sensing survey for a particular USA territory, they got 2 years exclusive rights of high-resolution data distribution from the particular mission.

6.6.6 Concluding remarks

In recent years, the Russian space industry has undergone a dramatic development: from being completely state-financed to being really commercial. As a result, it has successfully penetrated the world (remote sensing) data market. The main advantages of Russian commercial satellite imagery are its highly accurate stereoscopic mode and its moderate prices. Russian high-resolution photographic data are still very successful in the international commercial data market, in spite of the presence in orbit of 1-m resolution commercial remote sensing systems.

As was illustrated in Section 6.6.4, utilisation of Russian high-resolution panchromatic data with data from other appropriate systems can give critical

Figure 6.6.11. Russian 1-m resolution image of Los Angeles (courtesy of the Northern Geographical Company).

information about selection and approval of sites for the construction of nuclear power plants. Utilisation of high-resolution, archived data should help in detecting changes in such sites, whether fast or slow, thus contributing to the planning of on-site inspections. Finally, it is known from experience described elsewhere that multi-spectral observations provide precise information about the level of activity of nuclear power plants.

Since the mid-1980s, the Soviet Union/Russia has pioneered the international remote sensing data market with high-resolution optical and radar data. Today, the Russian remote sensing programme is not as big and ambitious as in previous years, but there are a few high-resolution civil systems under development: optical Monitor and Resurs-DK, and radar Kondor-E. Experience gained from Soviet Union times shows that technology is not the only basis for successful operation in the international satellite imagery data market; a well-developed distribution network of value-added products at moderate prices is an essential part of such success.

6.6.6 References

Cosmonautics and Armed Forces (1997) *Russian Military Space Forces* (Vol. 1, 286 pp., in Russian).

Development of the Military Space Forces (1997) *Russian Military Space Forces* (Vol. 2, 431 pp., in Russian).

Editors (1995) USA declassifies secret satellite imagery. *Novosti Kosmonavtiki*, **4**, pp. 37–39 (in Russian).

FAS Space Policy Project (1999) *World Space Guide* (on-line at http://www.fas.org/spp/guide/russia/military/sigint/rorsat.htm)

Garbuk, S. and Gershenzon, V. (1997) *Space Earth Observation Systems* (280 pp.). Moscow (in Russian).

Gordon, P. (2001) Commercial high resolution space imagery, a very long gestation and a troubled birth. Geoinformatics, March, http://www.geoinformatics.com

Lantratov, K. (1996) TsSKB satellites for earth natural resources monitoring (a review). Novosti Kosmonavtiki, **26**, 40–46 (in Russian).

Orlovsky, A. (2000) Russian high resolution imagery. Lecture presented at the ISU Master of Space Studies Program, January, 2000.

Pieson, D. (1999) 'Navy Space Wings', *Delovoy Vtornik* (supplement). Tribuna, **26** (in Russian).

Savin, A., Zotov, G. and Petrustchenko, Yu. (2001) System of navy space reconnaissance and target destination. On-line at http://www.navy.ru/science/sor7.htm in Russian

ScanEx website http://www.scanex.ru

Sovinformsputnik website http://www.sovinformsputnik.com

SRC Planeta website http://sputnik1.infospace.ru

Tarasenko, M. (1995a) Finishing of the Program Resurs-F. *Novosti Kosmonavtiki*, **20**, 29–31 (in Russian).

Tarasenko, M. (1995b) Launch of satellite Kosmos-2320. *Novosti Kosmonavtiki*, **20**, 32–33 (in Russian).

Victorov, A., Trapesnikova, O., Manohin, A. and Sobchuk, T. (2001) Remote sensing data applications to hydro-geological and engineering geological tasks (Report, 10 pp).

Appendix 6.6.I List of Russian state and private companies, who supply Russian commercial remote sensing data

Geo Spectrum
Volgogradsky prospect, 45, Office 508
Moscow, 109316, Russia,
Tel. 7 (095) 177-8182, 177-8243
Fax. 7 (095) 177-8182
E-mail: map@geo-spectrum.com
URL: http//www.geo-spectrum.com

Northern Geographical Company
Ul. Cherskogo 13/4
127410 Moscow, Russia
Tel. 7-095-904-82-27
Fax. 7-095-904-82-22
E-mail: ngeocomp@mtu-net.ru

State Scientific and Research Centre Planeta
B. Predtechenskiy per. 7
123242, Moscow, Russia
Tel. 7-095-252-37-17
Fax. 7-095-200-42-10
E-mail: asmus@planet.iitp.ru
URL: http://sputnik1.infospace.ru

State Centre Priroda
V. Pervomaiskaya str., 4b
105264 Moscow, Russia
Tel. 7-095-163-1141
Fax. 7-095-164-49-07
E-mail: priroda@dol.ru

ScanEx
Lva Tolstogo str., 22/5
Tel. 7-095-939-5640
Fax. 7-095-939-4284
E-mail: sales@scanex.ru
URL: http//www.scanex.ru

Sovinformsputnik
Leningradskiy Prospect, 47
125167 Moscow, Russia
Tel. 7-095-943-0757
Fax. 7-095-943-0585
URL: //www.sovinformsputnik.com

6.7 COMMERCIAL REMOTE SENSING SATELLITE PROGRAMMES IN THE UNITED STATES (by Robert Pearson)

6.7.1 Introduction

The use of remotely sensed data for clandestine activities detection can be done with several US-based satellite platforms. The United States has for many years been a

Table 6.7.1. An overview of US based satellites currently in orbit or planned. [Source data from Mitretek, 7/6/2001.]

Satellite	Government or Commercial	Launch year	Panchromatic Resolution (m)	Multi-spectral resolution (m)	Number of spectral bands	Swathe width (km)	Global coverage (days to revisit)
LANDSAT 5	G	1984	—	30	7	185	16
LANDSAT 7	G	1999	15	30	7	185	16
Space Imaging IKONOS 2	C	1999	1	4		12	1.5
Orbview 4	C	2001	1	4	4	8	1
Quickbird 1 Quickbird 2	C	2001	0.6	2.5	4	16	2
Orbview 3	C	2002	1	4	4	8	1
Space Imaging IKONOS 3	C	2004	0.5	2	4	12	1.5
TERRA (ASTER)	G	1999	—	15	14	60	7.5
NEMO	G	2002	5	—	210	30	3.7

world leader in the deployment of remote sensing satellites. The early efforts were mainly for military purposes and the imagery has been classified and will not be discussed here.

Both government and private commercial entities sponsor satellites launched by the United States. Both make the data available to the public on a cost per image basis. Therefore, both sources of data can be used for the purpose of detection of clandestine activities. Table 6.7.1 summarises the US government and commercial satellites currently in orbit or planned to be orbited.

6.7.2 Government satellites

The first series of government satellites launched were the Earth Resources Technology Satellite (ERTS) series. These were later renamed Landsat (Table 6.7.2). The launch of the Landsat satellites starting in 1972, made publicly accessible data available. While the early Landsat satellites provided coarse imagery (80 m in the early satellites and 15 m to 30 m data in the later satellites) the imagery is now of a quality to be useful in the detection of clandestine activities. In Table 6.7.2 a summary is given of all the Landsat satellites launched till the time of writing this chapter.

Table 6.7.2. The Landsat series of satellites.

System	Launch (End of service	I(s)	Resolution (m)	Communications	Alt. km	R Days	D Mbps
Landsat 1	7/23/72 (1/6/78)	RBV MSS	80 80	Direct downlink with recorders	917	18	15
Landsat 2	1/22/75 (2/25/82)	RBV MSS	80 80	Direct downlink with recorders	917	18	15
Landsat 3	3/5/78 (3/31/83)	RBV MSS	30 80	Direct downlink with recorders	917	18	15
Landsat 4*	7/16/82	MSS TM	80 30	Direct downlink TDRSS	705	16	85
Landsat 5	3/1/84	MSS TM	80 30	Direct downlink TDRSS**	705	16	85
Landsat 6	10/5/93 (10/5/93)	ETM	15 (pan) 30 (ms)	Direct downlink with recorders	705	16	85
Landsat 7	4/15/99	ETM+	15 (pan) 30 (ms)	Direct downlink with recorders (solid state)	705	16	150

I(s) = Instrument(s)
R = Revisit interval

The launch schedule of the Landsat series of satellites has resulted in at least one satellite being in operation since 1972 to the present time (Figure 6.7.1).

The latest in the series is Landsat 7, launched in 1999. The Landsat satellites operate in a sun synchronous polar orbit with an equatorial crossing line of about 9:00 local time and a 16 to 18 day orbit repeat cycle. Data collected by the Landsat satellites are commercially available from the US Geological Survey EROS Data Center in Sioux Falls, South Dakota. With an agreement with the Landsat authority, data receiving stations are also established in various parts of the world. This facilitates states within the footprint (see Figure 6.7.2) of the satellite transmitter antennae to receive images for distribution to their user community.

Another more recent US government sponsored satellite is the NASA TERRA satellite launched in December 1999. The satellite carries several sensors including the Advanced Space born Thermal Emission and Reflection Radiometer (ASTER). ASTER collects detailed maps of surface reflectance, temperature, emissivity and elevation. The ASTER instrument is jointly sponsored by the US and Japan and was constructed in Japan to be carried in the US TERRA satellite. ASTER has 14 spectral bands, three in the visible and near infrared, six in the short-wave infrared and five in the thermal infrared regions. The data from the various spectral bands can be combined into several types of images. The spatial resolution of ASTER is 15 m.

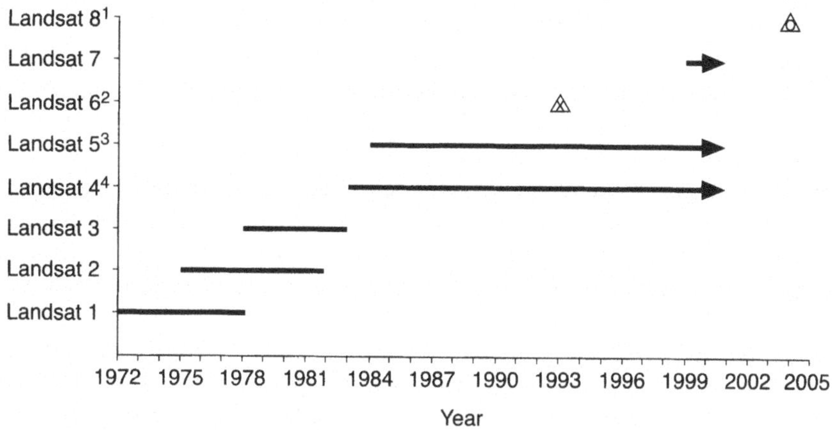

[1] ETM+ follow-on instrument on EOS AM-2; launch anticipated in 2004.
[2] Launched 5 October 1993 but failed to obtain orbit.
[3] Data transmission by direct downlink only.
[4] No longer transmitting TM data.
Arrows indicate that the satellites are currently in orbit transmitting data.

Figure 6.7.1. Launch schedule of the Landsat series of satellites. [Source: National Aeronautics and Space Administration (NASA).]

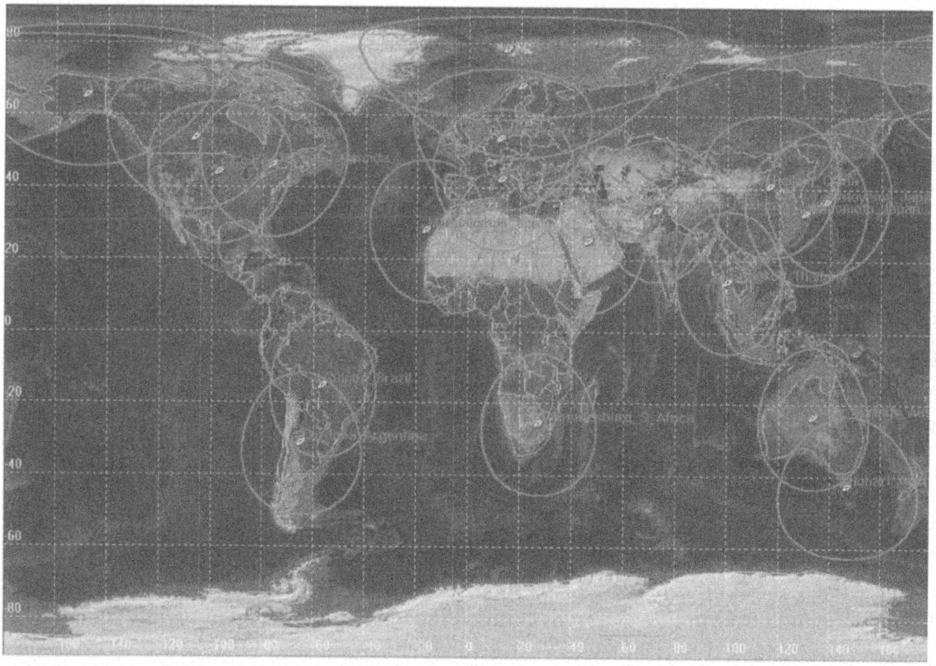

Figure 6.7.2. Distribution and coverage of the US Landsat data receiving stations [source: http://ltpwww.gsfc.nasa.gov/IAS/handbook/handbook_htmls/chapter4/htmls/igs.html].

Other sensors on the TERRA satellite are the MODerate-resolution Imaging Spectro-radiometer (MODIS), Measurements Of Pollution In The Troposphere (MOPITT), Multi-angle Imaging Spectro-Radiometer (MISR) and Clouds and the Earth's Radiant Energy System (CERES). These instruments operate with moderate to coarse resolution and are not suitable for observing clandestine activities.

The other US government satellite that can potentially be used for clandestine detection purposes is the Naval EarthMap Observer (NEMO) Hyperspectral Remote Sensing Technology Demonstration project. NEMO is a joint government and industry effort between the Space Technology Development Corporation and the Naval Research Laboratory.

The first NEMO satellite is planned for launch in 2002. NEMO is a space-based remote sensing system for collecting broad-area, synoptic hyperspectral imagery for the military and the private sector. It is designed to gather remote sensing data in the coastal regions and land-based data for land-use management, agriculture, environmental studies, and mineral exploration. The satellite will carry a panchromatic sensor with a 5 m resolution and multispectral scanners with 105 bands in the visible and near infrared regions and 105 bands in the short wave infrared regions.

6.7.3 Private commercial satellites

Several companies in the United States are launching commercial satellites for the purpose of collecting and selling remote sensing imagery on the commercial market. The United States government in 1994 established rules stipulating 3 m as the highest spatial resolution that commercial satellites could use. Companies immediately requested licenses for these satellites and began requesting permission for finer resolution. Three companies have now made plans for the launch of 1 m or finer resolution commercial satellites. Those companies are Space Imaging, Earth Watch and Orbimage.

Space Imaging is headquartered in the Denver area of Thornton, Colorado. The company was the first in the US to launch a commercial 1 m IKONOS satellite. The IKONOS 2 satellite carries a 1 m resolution panchromatic sensor and a four-band 4 m resolution multispectral scanner. The IKONOS 3 satellite planned for launch in 2004 will have a 0.5 m resolution panchromatic sensor and a four-band 2 m resolution multispectral scanner. The company's internet address is www.SpaceImaging.com.

Earth Watch is headquartered in Longmont, Colorado. The company has recently launched a less than 1 m resolution satellite called Quickbird. The Quickbird 2 satellite has a 0.6 m panchromatic resolution and a four-band 2.5 m multispectral scanner. The company's internet address is www.DigitalGlobe.com.

ORBIMAGE, an affiliate of Orbital Sciences Corporation, is headquartered in Dulles, Virginia and in St Louis, Missouri. Plans are in production for a five-satellite system. ORBIMAGE currently operates the OrbView-1 and OrbView-2 low-resolution satellites (10 m and 1 km) and plans to add two high-resolution satellites, OrbView-3 and OrbView-4. OrbView-3 and -4 will have a 1 m panchromatic resolution and a 4 m four-band multispectral scanner. The company's internet address is www.Orbimage.com. Imagery from all three companies is available commercially. They can provide data either via the internet for download or by copying to CD-ROM for direct shipping. There currently is no restriction on the dissemination of this imagery.

6.7.4 Data sale and cost policy

Details regarding the availability and cost for imagery from various satellites can change. The reader is directed to the internet sites listed in this chapter for current availability and cost information.

Data from the US-based satellites discussed above are generally available to the public and can be used without restriction, except that data from the commercial imagery providers generally is accompanied by a licensing agreement that the purchaser must sign to receive the data. The licensing agreement generally limits the data to be used in a certain processing centre or by a certain number of users and limits the ability of the purchaser to send the purchased data to others. The agreements generally do not limit who can purchase the data and what purpose the data can be used for (so long as any publication of the data is accompanied by an acknowledgement of the source of the data). See the terms of each licensing agreement for details.

The cost for the data varies widely depending on the satellite provider, the type of data requested and the level of processing the data has received. Also, the cost could change with time, particularly as more satellites are launched and the operators price their imagery to be more competitive. Some imagery from the US government can be obtained free if the delivery is via internet FTP data transfer. Otherwise the cost is nominal to cover the data medium (CD-ROM, 8 mm tape, etc.) that is sent to the purchaser and the copying cost. Data from the commercial satellites is generally higher again depending on the type and quality of data purchased, the processing it has received and the method of delivery of the data to the purchaser. For data from the following satellites, specific availability and pricing information can be obtained from the respective internet web site.

US government satellites, Landsat, Terra
http://edcdaac.usgs.gov/pricing.html

or for TERRA data specifically:
http://eosdatainfo.gsfc.nasa.gov/eosdata/terra/terra_dataprod.html
For data from the commercial suppliers the reader is directed to the respective Internet websites for details on pricing, data formats and licensing agreements:

Space Imaging
www.SpaceImaging.com,

Earth Watch:
http://www.digitalglobe.com/?goto=products/how_to_order

Orbimage:
www.Orbimage.com.

Another possibility for receipt of satellite imagery is for the user to set up a ground station to directly receive and process imagery from the satellite. Each satellite has ground stations around the world that receive the data and forward it to the respective data centres for processing and storage. Operating a ground station is likely to be more easily arranged with the US government for data from their satellites than it is for the commercial satellites. The commercial satellite companies are specifically in the business of collecting and selling the imagery to users whereas the US government generally is not. Therefore, the commercial operators would likely not be inclined to let someone directly receive the data without compensating the commercial operator. However, they may be willing to make specific arrangements and contractual agreements to allow someone to set up and operate a ground station to directly receive and pay for the data. Contact the satellite operator for further details.

6.7.5 Conclusions

With the availability of high-resolution imagery from three US-based commercial sources, the use of these data for detection of clandestine activities is greatly enhanced. The only barrier to the use of these data for detection purposes is the need to purchase these data on an ongoing basis and to set up and operate a facility to process and store the information. Anyone can perform this activity if they so desire and if the funding needed to carry out the work is available.

7

Applications of commercial satellite imagery: some case studies

B. Jasani and M. D. Ward

7.1 INTRODUCTION

Under the IAEA safeguards system, the Agency monitors a large number of declared facilities and nuclear materials in more than 50 countries. However, Iraq's non-compliance with the provisions of its safeguards agreement with the IAEA and the magnitude of its undeclared nuclear weapons related programme, indicated the inadequacy of the Agency's safeguards procedures as applied through nuclear material accountancy, surveillance and containment at declared facilities alone. They must be complemented and strengthened by other methods that would provide assurances that states party to the NPT do not possess undeclared nuclear facilities for the production of weapons usable nuclear materials.

The IAEA, under the SSS initiative, examined ways to strengthen its safeguards procedures by improving its ability to detect undeclared nuclear activities in states with which it has safeguards agreements. It is well recognised that some states developed some of their space assets in support of their earth observation activities. A by-product of this was the use of observation satellites for monitoring compliance with the terms of a number of arms control treaties. Thus, observation from space became a vital element of their NTM of verification system. The importance of satellites became evident when a number of bilateral agreements were reached. However, these states have not been willing to share widely either the technology or the information obtained by their NTMs. To some extent, this gave an impetus to the development of MTM verification. An example of this was the establishment, in 1991, of the Western European Union Satellite Centre in Madrid. Thus, commercial satellite imagery (CSI) is beginning to play an important role.

It is well recognised that 'there is a need to develop and test the procedures and techniques relevant to the potential utilisation of commercial satellite imagery in the support of safeguards.[1] Information gathered from space has played an important rle in many instances during the course of the IAEA's work. However, the nature of the source of information often did not allow the Agency to carry out detail analysis and assessments by itself. This is frequently important when dealing with non-compliance issues. In this context, CSI could be an important open source of information that the Agency could use. This is particularly so when the quality of satellite-based information is rapidly improving in its quality and several states are launching remote sensing satellites. Moreover, the data they generate are commercially available. It was suggested that civil earth-orbiting observation satellites with panchromatic, multispectral and radar sensors on board, could be used to monitor multilateral agreements such as the 1970 NPT (Jasani, 1990).

Also, the possible use of radar satellites for monitoring arms control treaties was previously discussed in 1986 (Jasani and Larsson, 1986). An advantage of such a sensor is that it can be used under all-weather conditions and in total darkness. This is because a radar sensor is an active one that sends its own signals and detects them when some of the radiation is reflected back from the targets to the antenna. The returning radiation can be reflected from a diffuse, specular or a corner reflector. The change in the surface structure caused by turbulence in water due to discharge of cooling water would be detected in the returned radar beam. Metallic objects, such as railways tracks and pylons carrying power cables, reflect radar beams efficiently.

The basic imageries from such sensors can be processed in various ways to enhance their interpretability. For example, images from panchromatic, multispectral and radar sensors could be processed and used either on their own or in combination to enable derivation of detailed information. Such data could also be used either to construct most up-to-date maps or they could be used in conjunction with existing maps to learn about the sites being observed.

In the following sections, the use of image enhancement techniques, maps and visual as well as computer aided detection of changes in a site illustrated the usefulness of commercial satellite imagery for the IAEA's safeguards procedures. The uses can range from the validation of information provided by a member state to the Agency to the detection of declared and undeclared facilities and sites.

7.2 EXAMPLES OF HOW CSI COULD BE USED FOR SAFEGUARDS PURPOSES

In a series of studies some 37 images over various sites were analysed (Jasani, *et al.*, 1996, 1998, 1999, and 2000). In this chapter, however, only some of the analyses will be presented as examples. Most of this work was carried out before higher resolution images (IKONOS with 1 m and QuickBird with 0.61 m) became available. Nevertheless, even the low-resolution data demonstrated the usefulness of CSI. With the

[1] Developing the fundamental elements of the SSS. IAEA document on **Safeguards Research and Development and implementation Support Programme for 1999–2000, Annex 3, Group I**, 14 June 1998, p. 9.

availability of higher resolution imagery, the potential of CSI is considerable. Optical data from commercial satellites in the visible range of the EMS tend to have higher resolution than that in the near and far infrared range. Such data are useful for identification of objects and activities on the earth's surface. Attempts were made to identify features that would help distinguish between a facility used for civil purposes from that used for defence.

No analysis has yet been carried out on images acquired by commercial satellites as stereo-pairs. The utility of such images is worthy of further examination.

7.2.1 Examples of some characteristics of nuclear facilities that can be observed

Security measures

Generally, civil facilities have less stringent security precautions. For example, no high-security fences or anti-aircraft guns are found around such facilities but these may be present at defence related sites. The Iraqi facility near Baghdad (Al Tuwaitha) is an example of the latter. Passive security measures such as a high wall and active measures such as anti-aircraft guns can be identified in the SPOT image of Al Tuwaitha facility shown in Figure 7.1.

Site locations

Civil facilities tend to be located near towns and cities so that they have access to industrial infrastructures while, on the whole, defence related activities are carried out in remote sites. An example of the latter is the US Oak Ridge National Laboratory.[2] This was established in 1942 as part of the Manhattan Project under which the US nuclear weapons programme began. It was at this site the first US weapons-grade uranium and plutonium were produced. Most of the major towns are some 20–30 km from the facility, which is enclosed in an area of about 245 km^2. The map, acquired over the internet of the location of the Oak Ridge facility, is shown in Figure 7.2. Other such examples are those of the US Hanford and Rocky Flats facilities and the Russian plutonium production plant at Kyshtym.

Power lines associated with some facilities

Nuclear power reactors have power lines associated with them. Under certain conditions, these can be detected in commercial satellite images. For example, if there is good contrast, even the actual power cables can be identified in optical low-resolution images. This can be seen in a KVR-1000 image from a Russian satellite over the UK Dungeness Nuclear Power plant, from which power cables can be detected when they are over fields (see Figure 7.3). A radar sensor is ideal for this as a radar beam is efficiently scattered back to the sensor by such metallic objects as pylons carrying

[2] **Oak Ridge National Laboratory**, internet address: http://www.ornl.gov/glance/glance.html: and **Oak Ridge National Laboratory – A Self-Guided Tour**, internet address: http://www.ornl.gov/tour/tour.html.

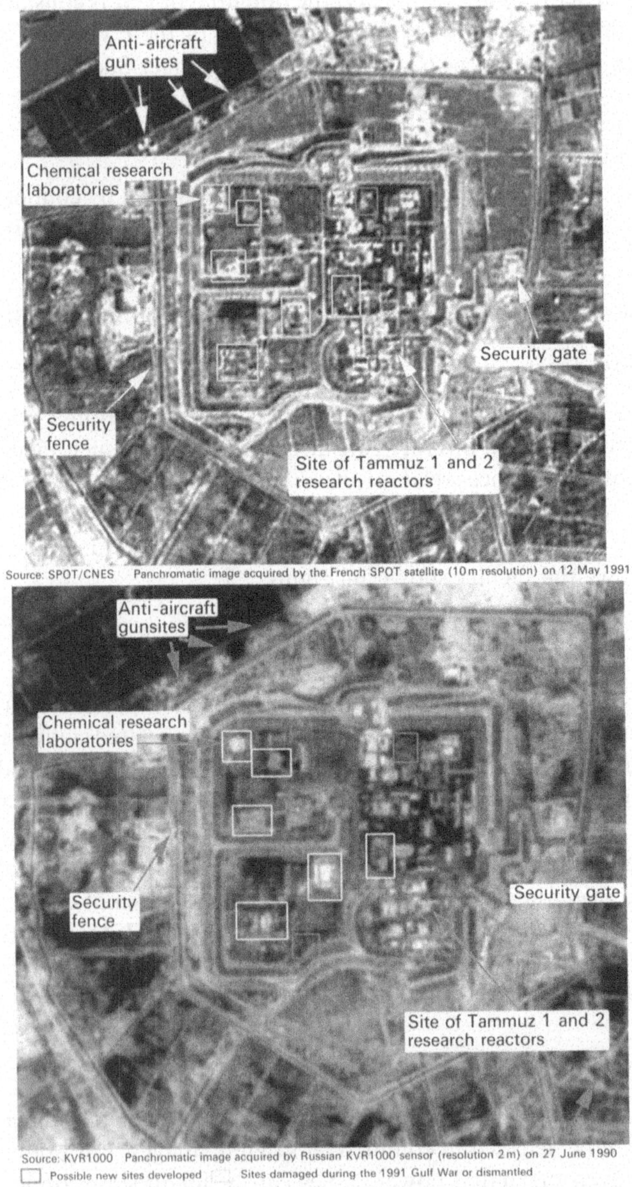

Figure 7.1. Extracts from two scenes acquired from SPOT (resolution 10 m) on 12 May 1991 (top image) and from a Russian KVR100 sensor (resolution 2 m) acquired on 27 June 1990 (lower image). Perimeter fences can be identified. Also some of the changes that have occurred on the Iraqi reactor site during a period of one year can be identified. The red boxes show new constructions and the blue boxes indicate changes in the site possibly due to bombing during the Gulf War or due to later dismantlement process. This is an example of visual change detection (see Plate 7).

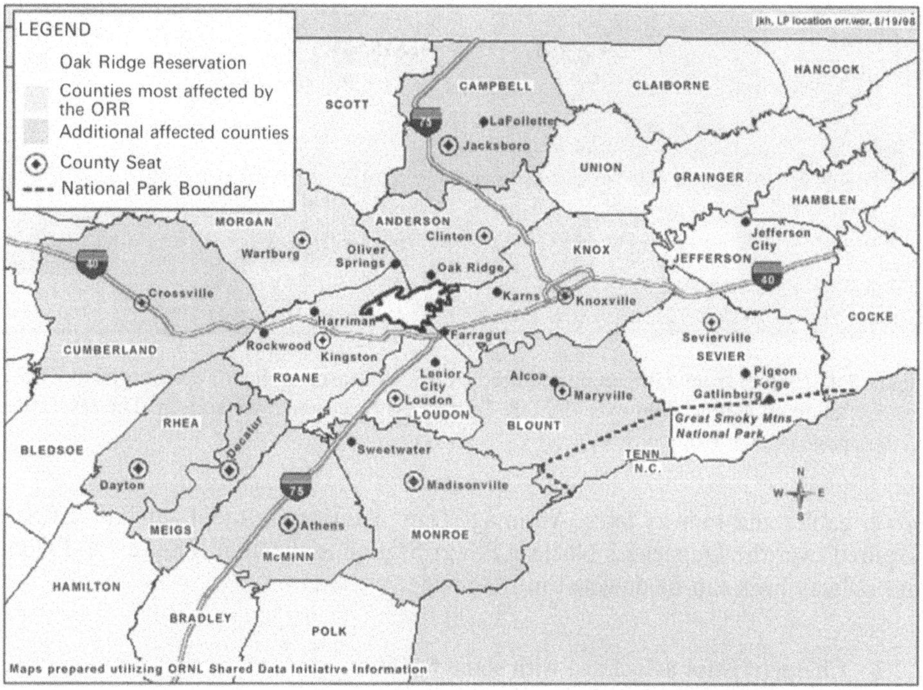

Figure 7.2. The map shows the location of the US Oak Ridge nuclear facility (source: internet address: http://www.ornl.gov/~dmsi/landUse).

Figure 7.3. This extract from the Russian KVR-1000 image with a resolution of at least 2 m over the Dungeness area shows the pylons and the power cables they carry from Dungeness Nuclear Power station (source: Russian KVR1000).

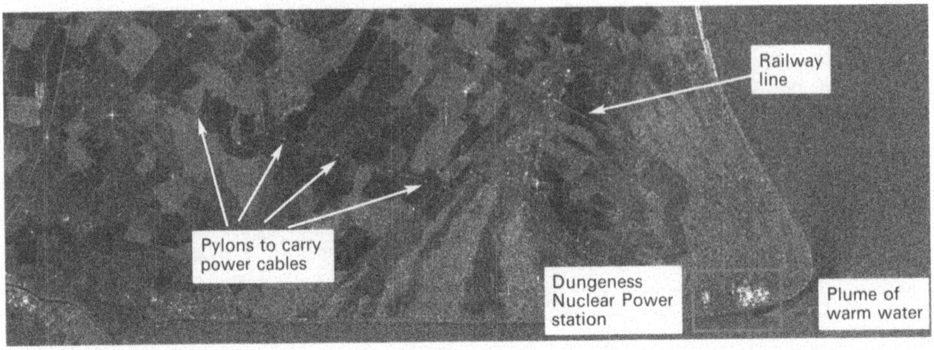

Figure 7.4. Extract from a full scene acquired by a SAR sensor on board the European ERS-2 radar satellite on 4 June 1996 over the UK Dungeness Nuclear Power station. The resolution of the image is 25 m (source: ESA).

power cables and railway lines. An image from a European radar satellite, ERS-2, acquired over the Dungeness Nuclear Power Station is shown in Figure 7.4. Pylons and railway lines can be detected in the image.

7.2.2 Characteristics associated with some facilities

Gaseous diffusion plants

Because of their large size and service requirements, these plants are most conspicuous; an example is the Oak Ridge diffusion plant shown in Figure 7.5. A typical diffusion plant producing weapons-grade uranium may contain several thousands of enrichment stages and can measure some 900 m in length and 150 m in breadth. These plants also consume a large amount of power. A diffusion plant producing, for example, 6,000 kg of weapons-grade uranium a year would consume an average of 400 MW of electrical power (Hipple and Levi, 1986). The Tennessee Valley Authority supplied the large amount of electrical power required for the Oak Ridge plant (Kiser, 1975). The pylons carrying electrical cables and relatively large switchyards can be detected at such facilities.

Often, in a diffusion plant, cooling towers are used to remove the excess heat. No such towers can be seen at the Oak Ridge facility (see Figure 7.5). As some components of the enrichment process require cooling, the water from the tributary of the Clinch River that runs through the K-25 complex may have been used for this purpose. It would then be possible to detect the discharged warm water into the river indicating the operation of K-29, K-31 and K-33 plants. Extracts from the three Landsat images over the K-25 facility were examined using bands 4, 5 and 6. The latter is the thermal band. These are shown in Figure 7.6. The temperature of the water is indicated by the red colour **(A)**. This was found to be at a maximum on 26 June 1986 and a minimum on 10 April 1987 suggesting that the plants were operating in 1984 and 1986 but not since 1987.

While waiting for decontamination and decommissioning, the present purpose

Figure 7.5. Extract from a Landsat 5 scene (resolution 28.5 m) acquired over the Oak Ridge facility on 8 September 1984 (lower image). The image is a combination of bands 1 (blue), 2 (green) and 3 (red). The enrichment facility is in the area marked by red rectangle. The top image is from the US early reconnaissance satellite KH4 showing roughly the same area as the rectangle in the Landsat image. The image was acquired on 8 March 1970. Various facilities are identified in the KH4 image, K-25 being the original diffusion plant [source: Landsat/EOSAT and Corona image] (see Plate 4).

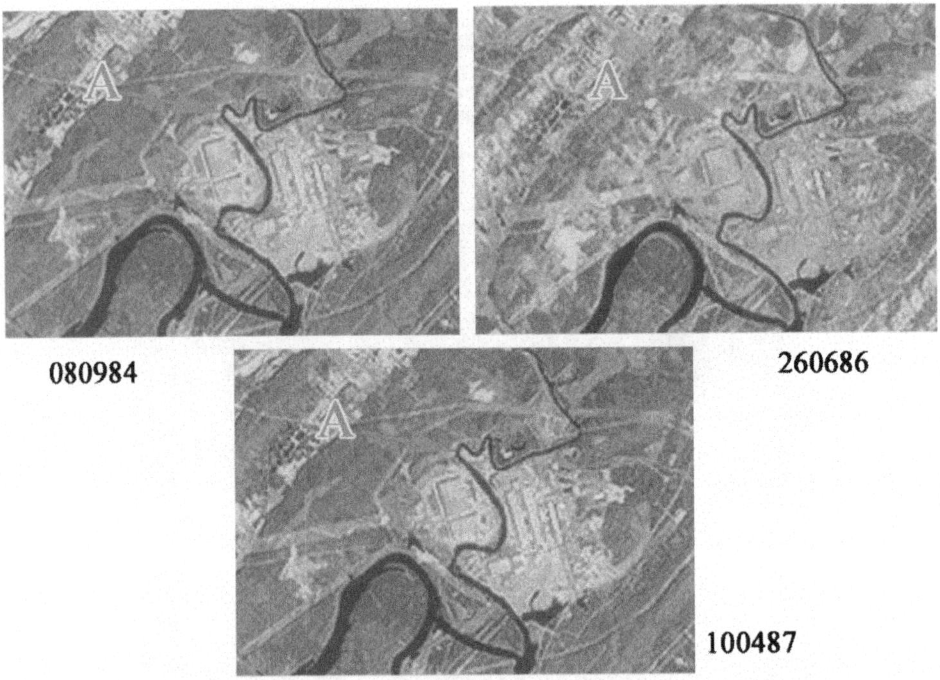

Figure 7.6. Extracts from Landsat images over the K-25 facility on three dates indicated under the images that are combination of bands 6 (red), 5 (green) and 4 (blue). The temperature of the water is indicated by the red colour **(A)**. This was found to be highest on 26 June 1986 and lowest on 10 April 1987 suggesting that the plants were operating in 1984 and 1986 but ceased operating by 1987 (see Plate 6).

of all the five buildings in the K-25 site (Figure 7.5) is to store large amounts of low-level radioactive and hazardous wastes (Federation of American Scientists). As the buildings have been adopted for storage, no outward changes seem to have occurred. Plants K-25 and K-27 stopped operation in 1964 while the remaining three were placed on standby at the end of 1985.

Gas centrifuge plants

In contrast with the plant discussed above, gas centrifuge plants, because of their greater efficiency, need less enrichment stages and, because of their compactness, occupy less space. Nevertheless, under certain conditions, power lines supplying electricity to such a facility and relatively small electrical switchyards can be detected. Moreover, the centrifuges and possibly other associated equipment may require cooling so that a source of cooling water is needed and the warm water may then be discharged into a river or a lake. Some of these features can be identified in an image acquired by a Russian satellite over the German centrifuge facility at Gronau shown in Figure 7.7. The image was then corrected against a site map

Figure 7.7. An image of the German gas centrifuge facility at Gronau acquired by a Russian satellite. The resolution is between 2m and 5m. The date is unknown.

and the two were combined in Figure 7.8 to enable identification of the facility and other features.

7.2.3 Sources and uses of maps

Under the New Model Protocol, the Agency should receive considerably more information about the nuclear activities within a member state.[3] This would

[3] 'IAEA Board of Governors Approves Strengthened Measures to Verify Nuclear Weapons Pact', IAEA Press Release, 97/9, 16 May 1997; and INFCIRC/540, Model Protocol Additional to the Agreement(s) between State(s) and the IAEA for the Application of Safeguards.

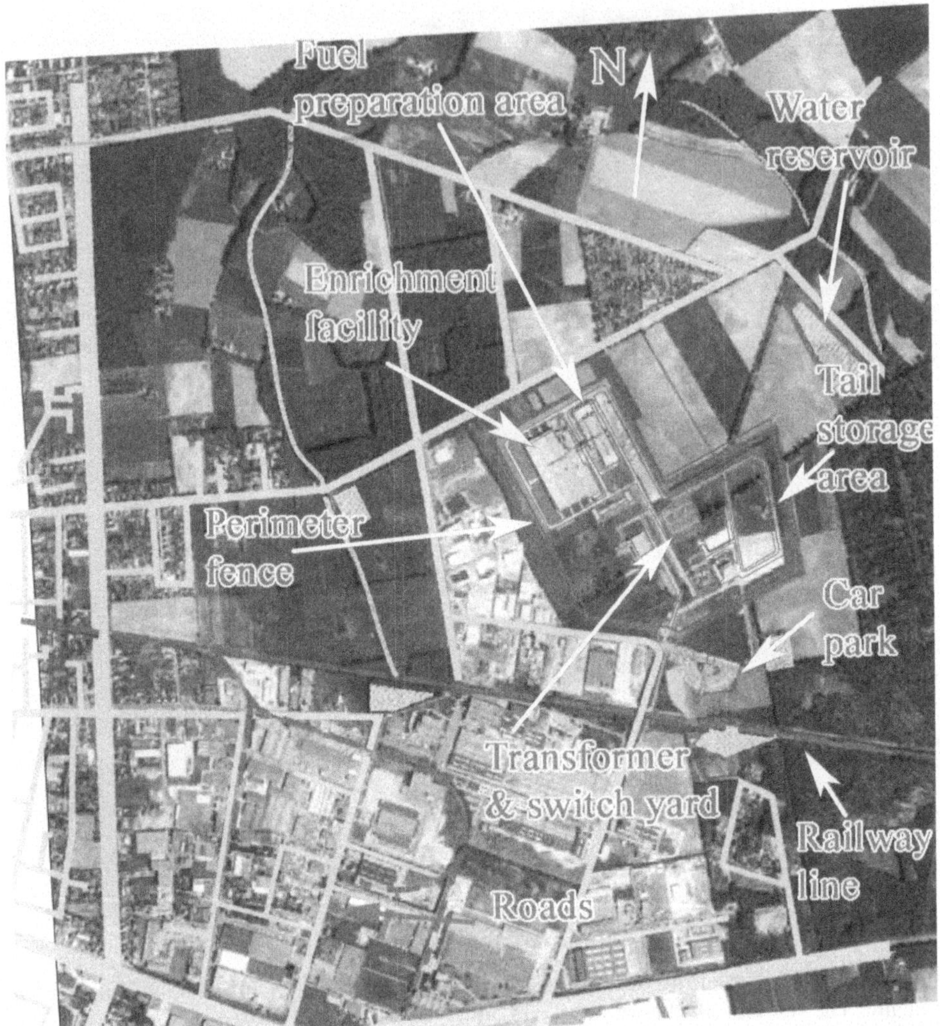

Figure 7.8. The image of Figure 7.7 as corrected against a map of the site area and then superimposed over the image.

include, for example, the coordinates of facilities, maps, site plans, the dimensions of the buildings and their descriptions. In order to illustrate how the Agency would use such information, a number of sources of maps and their potential for checking against images were examined.

US National Imagery and Mapping Agency (NIMA)

Maps of some of the sites studied were acquired from National Imagery and Mapping Agency (NIMA). The maps are provided on four CD-ROMs covering

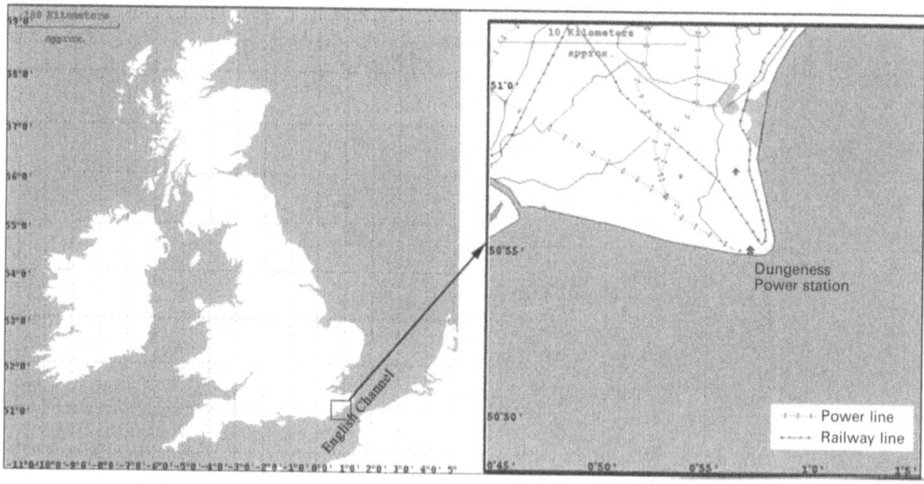

Figure 7.9. A full map of the UK in which the location of the Dungeness Nuclear Power Station is shown; the inset shows its immediate surrounding (source: US National Imagery and Mapping Agency (NIMA).

the whole of the earth (NIMA, 1997). They also contain such data as geographical, cultural and political information. From such data, maps like the one in Figure 7.9 can be constructed. The scale of the UK map is 1:4,000,000 and that of the inset is about 1:100,000. It is interesting to note that while the NIMA maps can be built up with a considerable amount of detail of all kinds, for the Dungeness Nuclear Power Station site the location of the power station itself is not given.

Ordnance Survey maps

A UK Ordnance Survey map with a 1:25,000 scale of the Dungeness area was also acquired for comparison. Compare the map in Figure 7.9 (in the inset) with that in Figure 7.10 (a section from the Ordnance Survey map). The latter shows more details than those given in the NIMA maps. For example, even the site diagram of the Dungeness Nuclear Power Station, the associated power lines, a railway line and roads are given.

Internet as a source of maps and site plans

A third source is the internet. In Figure 7.11, a map, acquired over the internet, of the Oak Ridge facility in the USA is shown. The gaseous diffusion and the electromagnetic isotope separation facilities together with the location of the main Oak Ridge National Laboratory are also given on the map.

Information from other sources

Often, detailed information on a state's nuclear programmes can be obtained from open literature. A map of the Iraqi facility at Al Tuwaitha acquired from an open source is shown in Figure 7.12.

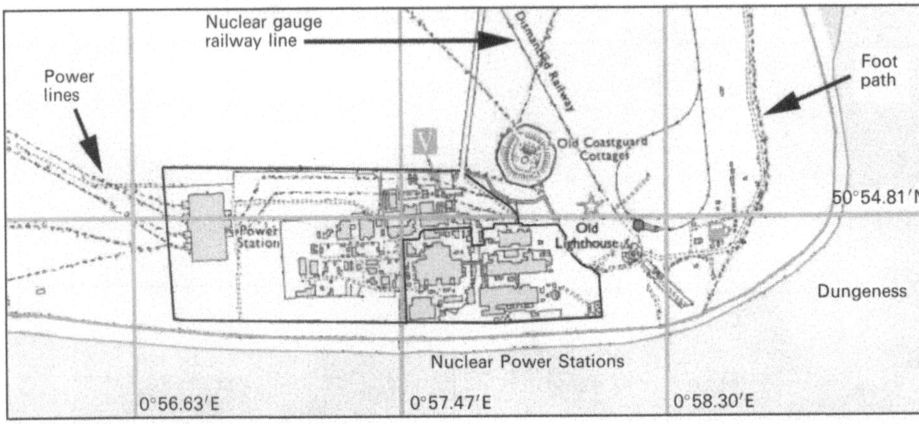

Figure 7.10. A map of the Dungeness area from an Ordnance Survey map. The scale of the map is 1:25,000.

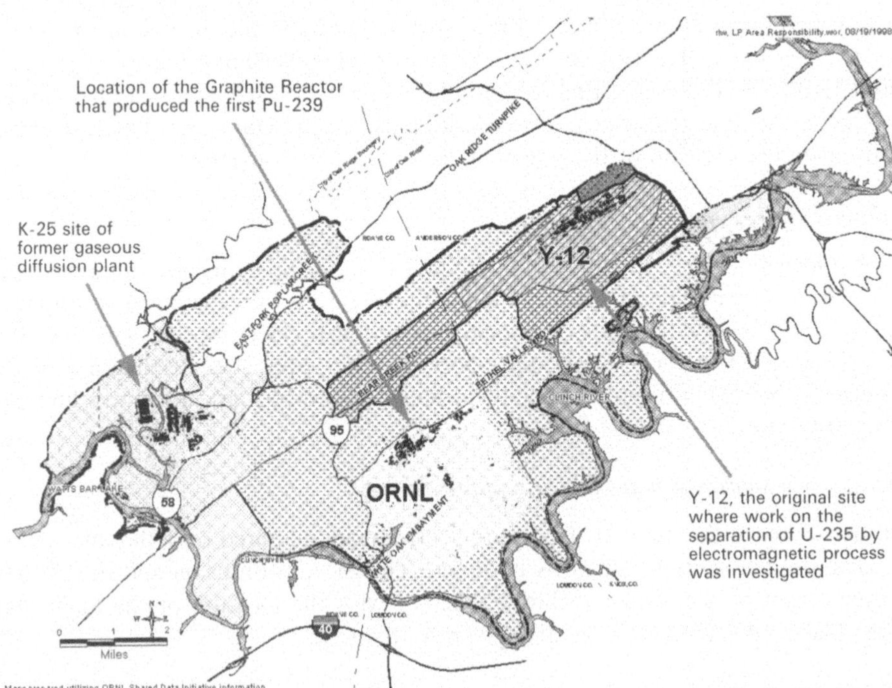

Figure 7.11. The map shows the locations of the Oak Ridge National Laboratory (ORNL) main facility and the site of the two isotope separation facilities (source: Internet address: http://www.ornl.gov/˜dmsi/landUse).

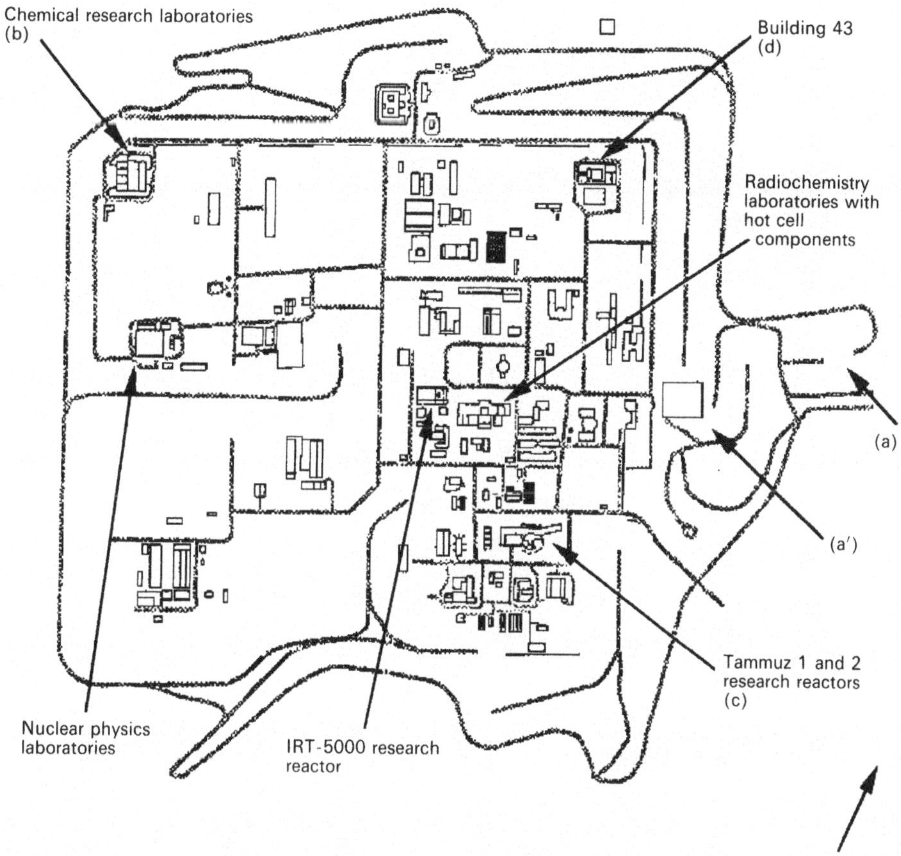

Figure 7.12. Site plan of the Iraqi nuclear facility at Al Tuwaitha in Baghdad (Kokoski, 1995).

Use of maps

Under the safeguards agreement with the IAEA, a state may provide a variety of information but not give such details as the network of roads and railway lines; nor, for that matter, the power lines, if any, associated with a nuclear facility. Even when a map of a site is provided to the Agency, the first task may be to verify the validity of the map. Here, a comparison between the map and a satellite image would be of great value. Such a comparison is made in Figure 7.13. Panchromatic images acquired by, for example, SPOT satellite are useful for mapping and identification of buildings. In Figure 7.13, an extract of an image acquired by the French SPOT satellite of the Iraqi facility at Al Tuwaitha is shown with the site map from Figure 7.12 superimposed over it. The image was corrected for size and orientation. It can be seen that a number of buildings which appear on the satellite image are not indicated on the map. While in this case the difference may be because the site diagram was simplified, it nonetheless does illustrate the potential usefulness.

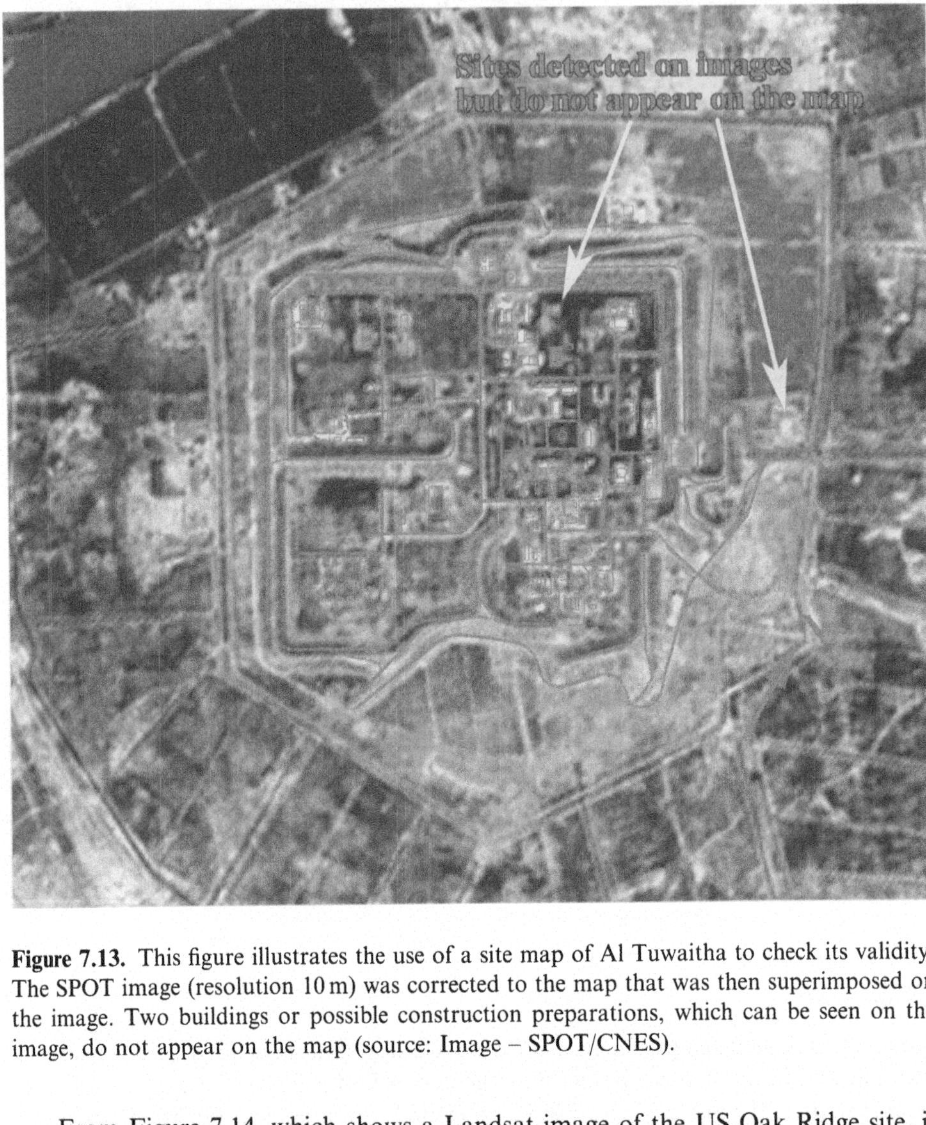

Figure 7.13. This figure illustrates the use of a site map of Al Tuwaitha to check its validity. The SPOT image (resolution 10 m) was corrected to the map that was then superimposed on the image. Two buildings or possible construction preparations, which can be seen on the image, do not appear on the map (source: Image – SPOT/CNES).

From Figure 7.14, which shows a Landsat image of the US Oak Ridge site, it can be seen that by comparing it with the site map (Figure 7.11) it is possible to identify not only various facilities but also any changes in the site that may have occurred. As before, the map is overlaid on the image. For example, a close examination of the image in Figure 7.14 indicates the development of new roads at **(A)** probably constructed after the publication of the map.

7.2.4 Determination of the operational status of nuclear facilities

The most significant indicator of the operation of nuclear facilities such as a reactor or an enrichment plant is heat loss. Heat is mostly discharged into large water bodies

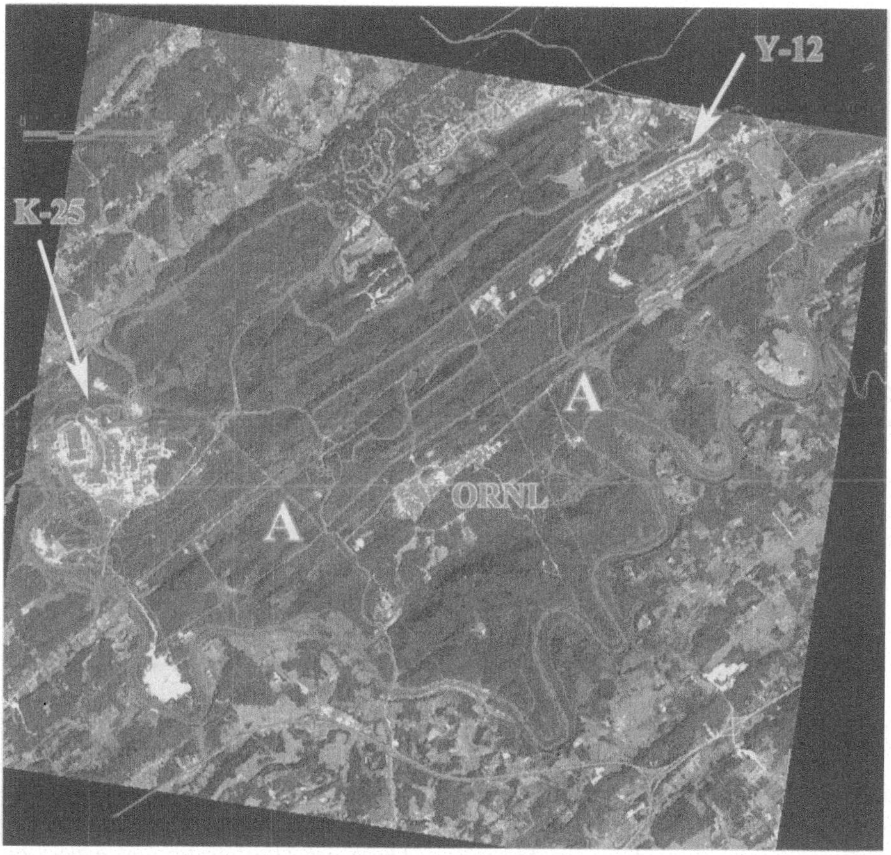

Figure 7.14. An extract of a Landsat 5 image (resolution 30 m) acquired on 8 September 1984 over the US Oak Ridge nuclear facility. The image of Figure 7.8 was corrected for size and orientation using the site map shown in Figure 7.11. A line drawing from this figure was then superimposed on the satellite image. Roads that can be identified on the image **(A)**, for example, are not given in the map suggesting that these were built after the map was drawn (source: Landsat/EOSAT).

such as rivers or lakes or even seas or via cooling towers or naturally from buildings into the environment. Water is often used to remove heat from a reactor core converting it into steam to drive turbines to generate electricity. In other facilities, such as an enrichment plant, water may be used to remove excess heat from components of centrifuges or various diffusion elements). The warm water is then discharged into a large water body, as in the case of the Oak Ridge plant (see Figure 7.6), or it may be converted into steam and discharged into the atmosphere via large cooling towers. Multispectral images acquired by the US Landsat satellite enable the detection of such discharged water. At present, there are two satellites, Landsat-5 and –7, in orbit that have thermal sensors on board. The use of the thermal band gives some indication on the operational characteristics of, for example, a reactor

Figure 7.15. An extract from a quarter scene acquired on 3 November 1997 by the US Landsat 5 satellite (resolution of 28.5 m) over the UK Dungeness Nuclear Power Reactor station. The warm water discharged into the sea can be seen in this image which is a combination of the thermal band 6 (red) and bands 5 (green) and 4 (blue). Some heat escaping the switchyard building can also be detected while clearly the reactor buildings are reasonably well insulated. Some of the buildings in the reactor complex can be distinguished as separate buildings (source: ESA/Eurimage) (see Plate 2).

(Figure 7.15). If the reactor, for instance, was shut down frequently, it might indicate that plutonium is being produced. This would then trigger further investigations. Such information would also be useful to confirm the operational status of declared closed down/decommissioned facilities.

Certain regions of the Earth suffer from prolonged periods of cloud cover or darkness. Under these circumstances, a radar sensor is very useful because of its ability to penetrate clouds and to observe sites even in darkness (Figure 7.16). There are four important applications for radar. These are (1) detection of excess heat discharged through cooling water or cooling towers; (2) use of its ability to detect heights; (3) possible detection of underground activities; and (4) the radar interferometry to detect movement of vehicle traffic.[4] The first two could soon be used in the

[4] A third application may be to detect if any underground activities are taking place. Under certain soil conditions, some radar beams may be able to penetrate the surface a few metres (McKinley, J. F., *et al.*, Surface valley and geoarcheology of the Eastern Sahara revealed by Shuttle radar (1982) *Science*, Vol. 218, no. 45–76, 3 December, pp. 1004–129; Blom, R. G., *et al.* 'Detection of subsurface features in Seasat radar images of Means Valley, Mojave Desert, California,' **Geology**, 1984, Vol. 12, pp. 346–349; Berlin, G. L., *et al.* 'SIR-B subsurface imaging of a sand-buried landscape, Al Labbah Plateau, Saudi Arabia', **IEE Trans. Geoscience Remote Sensing**, Vol. GE-24, no. 4, pp. 595–603; and Dabbagh, A. H., *et al.* 'Detection of sand covered geologic features in the Arabian Peninsula using SIR-C/X-SAR data', **Remote Sensing of Environment**, Vol. 59, no. 2, pp. 375–383.). The possible fourth application is the detection of vehicles and their movements near a facility, particularly those that may occur at night or under cloud cover. Some may use the cover of darkness to move, for example, nuclear fuel by trucks or railway in order to avoid detection by optical satellites. Such movements could be detected using radar interferometry.

Figure 7.16. An image from a weather satellite, reproduced here from the internet, shows Dungeness on 31 December 1996 obscured by clouds while in the ERS-2 radar image (resolution 25 m), acquired on the same day, the reactor complex and the surrounding can be clearly seen. Warm water plume can also be detected (source: ESA and weather satellite images from the Internet).

IAEA's safeguards procedures with two more uses for future consideration. In one, it may be possible to use it to detect the discharge of warm water from nuclear facilities to determine their operating status (Figure 7.16).

Often, in analysing an optical image, it becomes difficult to know what a particular signature represents. It may be due to a building or just disturbance of the soil. This could be resolved by the second application of radar. With a radar signal, it is possible to measure height very accurately and it is possible to construct a contour map of the site that would assist in determining the nature of the observed signature.

Moreover, power lines are not always easy to detect in an optical image unless there is good contrast. On the other hand, with radar, pylons carrying power cables and railway lines can be readily observed (Figure 7.16).

In another example, two images, one acquired on 7 May 1997 by the US Landsat 5 optical satellite and a second one acquired on 13 May 1997 by the European ERS-2 radar satellite, were used to observe the Canadian Bruce Nuclear Power Development facility on Lake Huron. It consists of nine reactor units, four (769 MWe each) at Bruce A, four (860 MWe each) at Bruce B and one a prototype (208 MWe) at Douglas Point sites (Figure 7.17). However, only the four units at Bruce B site are operating at present. The shut down dates for the Bruce A units are Unit 1 on 16 October 1997, Unit 2 on 8 October 1995, Unit 3 on 9 April 1998 and Unit 4 on 16 March 1998.[5] The Douglas Point reactor has been decommissioned.

[5] Ontario Power Generation – System Map, Internet address: http://www.ontariopowergeneration.com/newgen/powerprod.asp

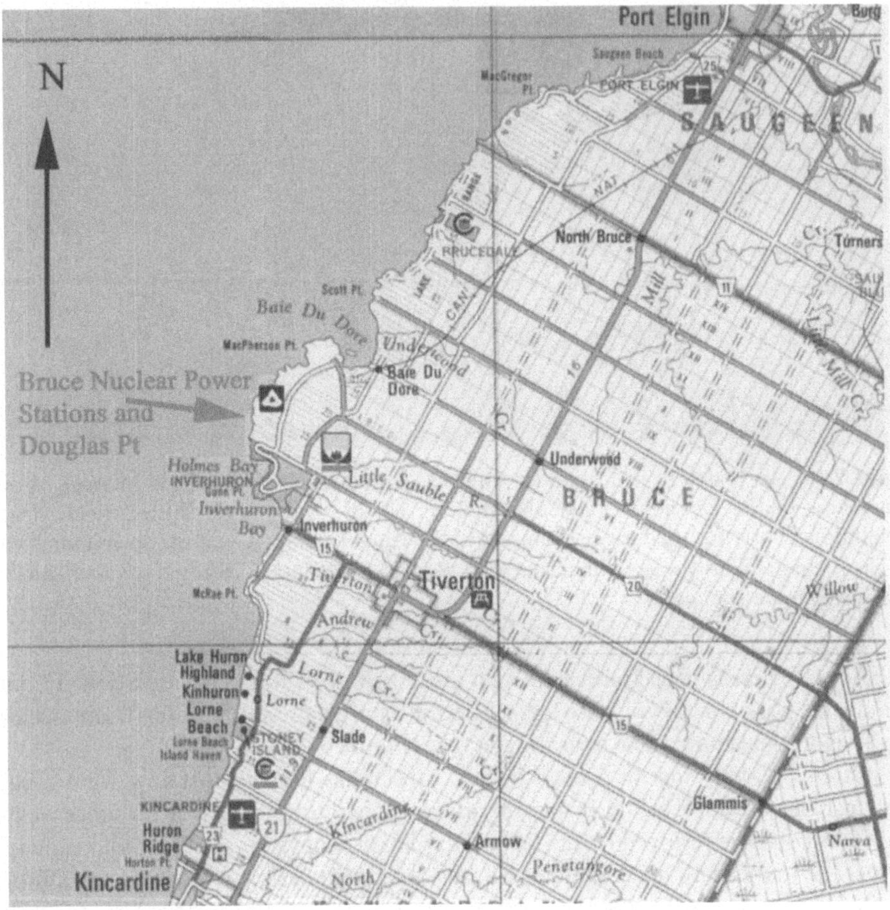

Figure 7.17. Map showing the Bruce Nuclear Power Development (NPG) facility on Lake Huron (source: MapArt Corporation, Ontario (1 April 1988).

These reactors use natural uranium and heavy water. The latter is produced in a heavy water production plant situated in the same complex.

Bands 1, 2 and 6 of the Landsat 5 sensor were combined to obtain the image shown in Figure 7.18. The thermal plume recorded in band 6 (red) can be seen **(A)** from the Bruce B complex. This is masked, to some extent, by what appears to be several small plumes coming from the Kincardine area, south of the reactor complex **(B)**. These could be caused by several small rivers which discharge into Lake Huron along its western bank bringing with them materials that then cause sedimentation all the way up to the reactor complex. Light with wavelengths between 0.45 and 0.52 microns (band 1, blue) and between 0.52 and 0.60 microns (band 2, green) can penetrate water. It is possible that the signature along the western coast of the lake (see Figure 7.17) could be due to the sedimentation.

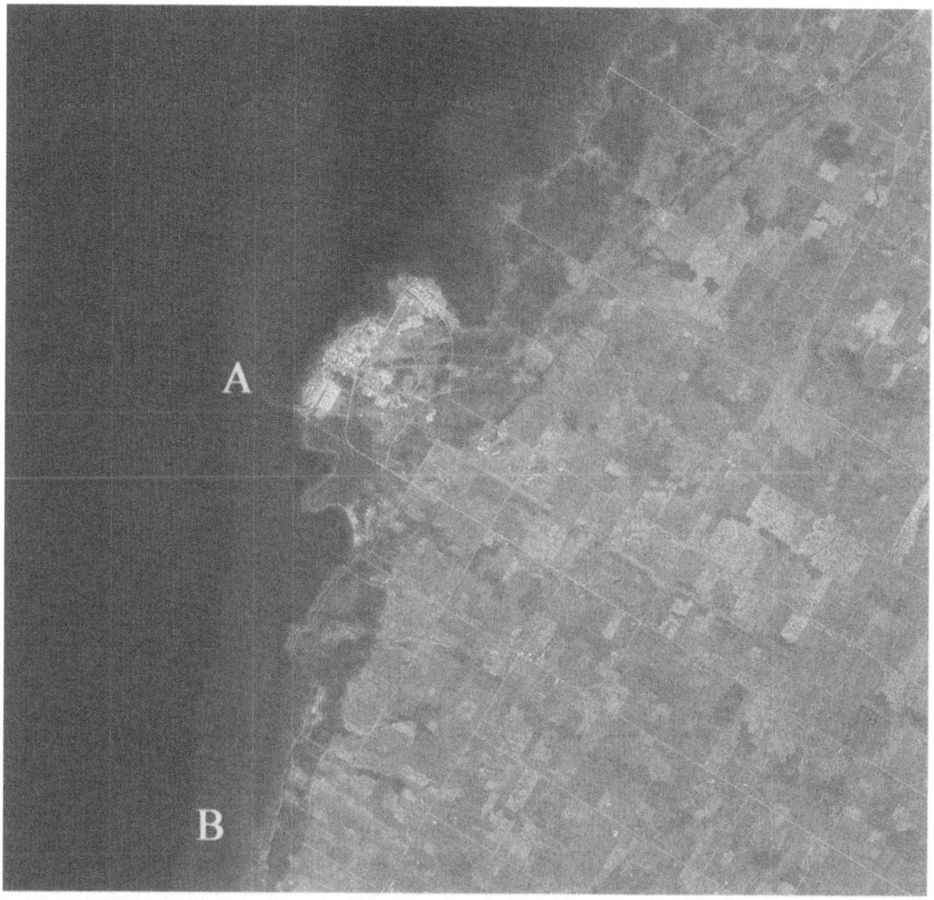

Figure 7.18. A combination of bands 1 (blue), 2 (green) and 6 (red) of the Landsat 5 image acquired on 7 May 1997. The thermal plume from the reactor can be seen at **A** while at **B** possible sedimentation from rivers can be detected (see Plate 8).

In order to improve the discrimination between the warm water plume and the possible sedimentation effect PCA was carried out. Figure 7.19 shows a combination of three components, C4 (red), C1 (green) and C2 (blue). It can be seen that the plume appears a little more prominently in this figure than in Figure 7.18. Moreover, in the former, the rivers can now be identified. Subtracting the PCA component C2 from C4 and combining the resulting image with components 1 and 2 (see Figure 7.20) resulted in an image with an enhanced warm water plum. It can be seen that the PCA has been useful in enhancing the image so that the identification and the locations of even smaller rivers became possible. This was important because it gave a possible explanation for the sedimentation along the bank of the lake. Of course, this may also be due in part to waste discharged from the townships lining the bank of Lake Huron.

Figure 7.19. A combination of three principal components, C4 (red), C1 (green), and C2 (blue), of the Landsat 5 image (see Plate 9).

It is interesting to note that no such signature can be seen from the Bruce A facility suggesting that Units 1, 3 and 4 were not in operation although the shut down dates for these were 16 October 1997, 9 April and 16 March 1998 respectively.

The European ERS-2 satellite acquired a radar image on 13 May 1997, six days after the acquisition of the Landsat image. This is shown in Figure 7.21. Again no plume is detected at the Bruce A site. It is interesting to note that the radar image does not show strong signature for the sedimentation which means that this effect observed in the Landsat image is not a surface effect but most likely to be that due to sedimentation under water. However, the warm water plume from the reactor facility can be seen clearly at **A**. The ERS-2 image was co-registered with the principal components of the Landsat images. It was then combined with an image obtained by subtracting component 2 from component 4 and component C1 (as an RGB image as shown in Figure 7.22).

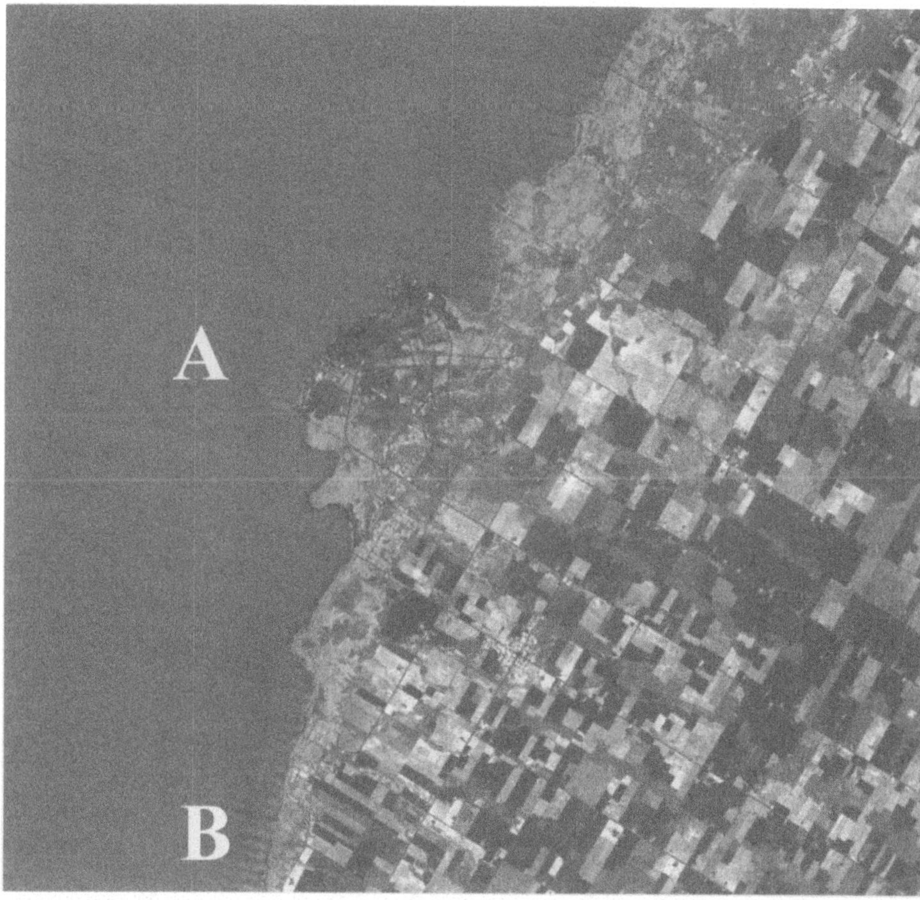

Figure 7.20. The result of combining the difference between the principal components C2 and C4 (red) with component C2 (green) and component C1 (blue) (see Plate 10).

7.2.5 Detection of changes in facilities

To detect changes in a state's nuclear-related activities, images could be acquired over a period of time. Construction of new buildings or alterations to the existing ones (i.e. changes in the geometrical patterns in a scene), can be detected by visual inspection of images. In order to facilitate further interpretation, multispectral images could be acquired and enhanced by such techniques as PCA. An image having undergone PCA generally gives a good contrast between man-made features and natural vegetation.

This was illustrated when PCA was performed separately on three Landsat TM images, acquired on 24 August 1984 (Figure 7.23a), 31 August 1989 (Figure 7.23b) and 21 August 1997 (Figure 7.23c) over an interim nuclear spent fuel storage plant in Germany near Gorleben. Various elements of the facility were identified with the aid

Figure 7.21. This shows an ERS-2 image over the BNPD. The image was acquired on 13 May 1997. The warm water plume can be clearly seen at **A**.

of a site map provided by the operator (Figure 7.23d). From the figure it can be seen that only the interim storage but not the pilot conditioning facilities near the top edge, left of the centre of the scene in (a) and (b), were present in August 1984 and 1989. The trapezium shape indicates this. Compare this with Figure 7.23d in which the site diagram was superimposed on the image acquired on 24 August 1984. However, by August 1997, this part of the site was further developed as indicated by its now triangular shape in Figure 7.23c. A further development had occurred by August 1989, as indicated by a rectangular shape in the lower right-hand corner of the images, which is indeed the excavated debris and salt heap.

The use of PCA can determine changes in a site using computer-aided technique. It was used to combine two multispectral images (24 August 1986 and 31 August 1989, and 31 August 1989 and 21 August 1997) to create the first and second

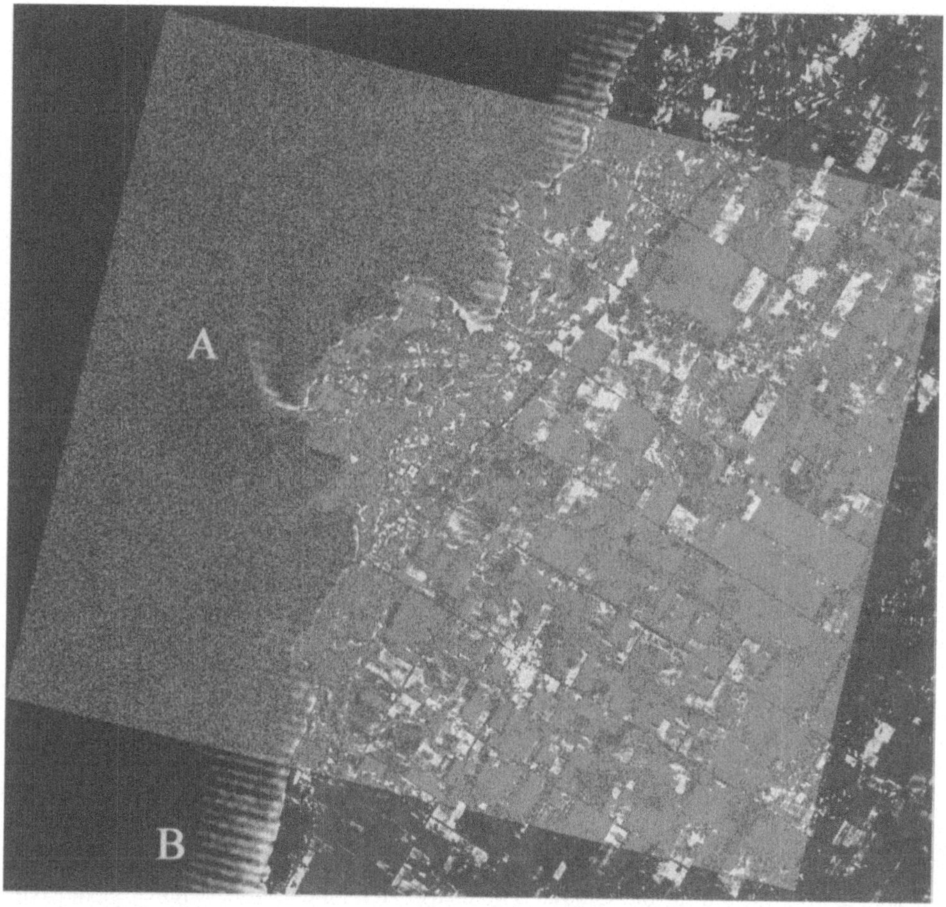

Figure 7.22. A combination of the co-registered ERS-2 image (as red) and the principal components C4–C2 (as green) and the component C1 (blue) as an RGB image. It can be seen that the warm water plumes coincide in all the images (see Plate 11).

principal components for every single spectral band, where the second principal components are a measure of change. The pixel values exceeding three standard deviations are shown in Figure 7.24. The use of other linear transformations, such as the iterative principal component analysis (IPCA) and the MAD technique (Figure 24c), identified significant changes in a scene. In this way should any changes be detected and the state has not declared them, the Agency can ask for explanation and even trigger an on-site inspection. The pixels that represented changes were superimposed in colour on the panchromatic KVR-1000 image (Figure 7.24b).

In another process, multispectral data taken at two different times were transformed so that the first transformed components were maximally and the last components minimally correlated. The corresponding components were then subtracted

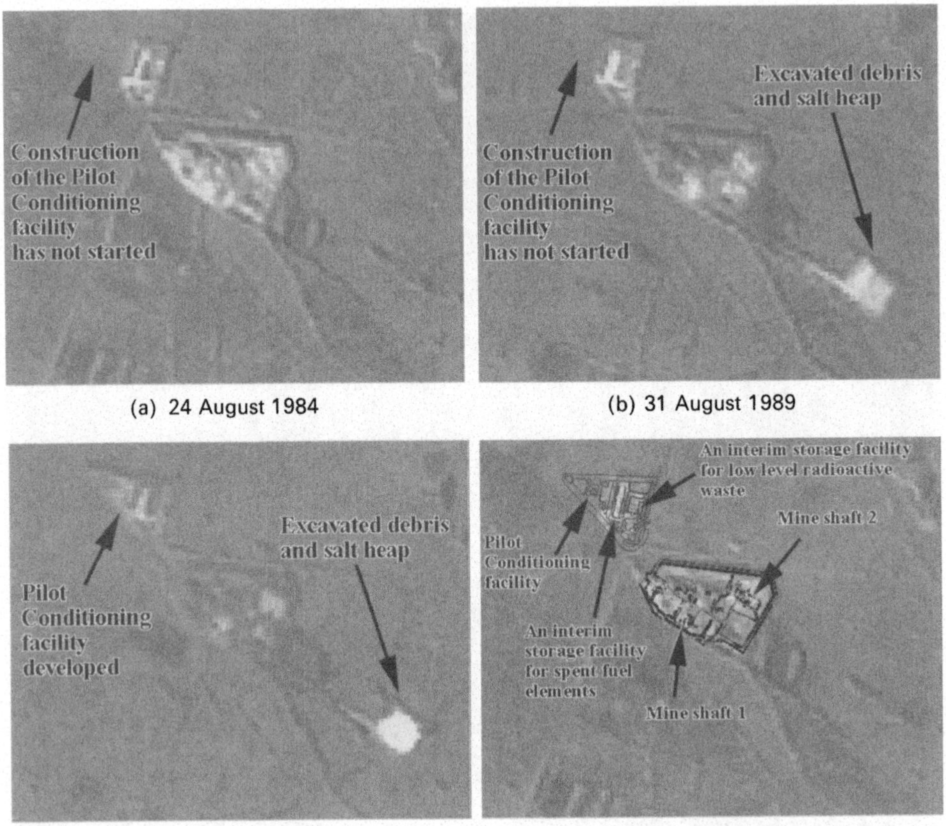

(a) 24 August 1984 (b) 31 August 1989

(c) 21 August 1997 (d) 24 August 1984

Figure 7.23. Extracts from the Landsat TM images (resolution 30 m) acquired at different times over Gorleben, shown as combinations of the principal components 1 (red), 2 (green) and 3 (blue) (a–c) and the site map superimposed over the image (d). The dates are indicated under each image (see Plate 13).

from one another in order to detect changes. Here, pixels, the values of which exceeded five standard deviations, were superimposed on the KVR-1000 image (see Figure 7.24c) so that only changes in the site were registered. It can be seen that the construction of the road is not well identified but other changes in the site are shown more positively.

A useful technique to combine images of different types is IHS. Here, it was used to superimpose on the high-resolution KVR-1000 image multispectral images containing only those pixels that represented changes in the scene. This was achieved by choosing the standard deviation of the values of the pixels such that they represented all the changes in the scene. Red, green, blue IHS transformation was carried out using the three bands of the Landsat TM analysis (three second principal components or three MAD components, for example). The intensity component was replaced by the high resolution KVR-1000 image. In Figure 7.24d, therefore, the

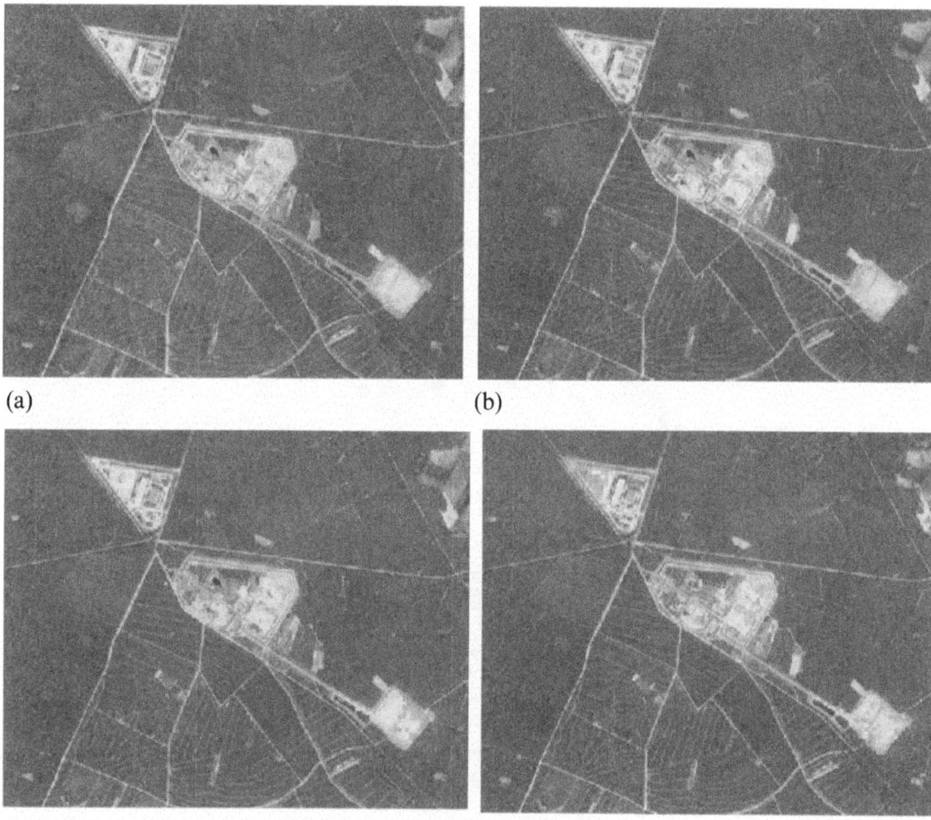

Figure 7.24. Bi-temporal change detection for the Landsat TM (resolution 30 m) extracts over Gorleben acquired on 24 August 1984 and 31 August 1989 fused with a high-resolution KVR-1000 image. (a) PCA: Shown is the second principal component of the channels 5 (red), 4 (green) and 3 (blue) exceeding three standard deviations. (b) IPCA: Shown is the second iterated principal component of the channels 5 (red), 4 (green) and 3 (blue) exceeding three standard deviations. (c) MAD: Combinations of the MAD Components 6 (red), 5 (green) and 4 (blue) exceeding five standard deviations. (d) Bi-temporal change detection for the Landsat TM extracts over Gorleben acquired on 31 August 1989 and 21 August 1997, fused with a high resolution KVR-1000 image: Combinations of the MAD components 6 (red), 5 (green) and 4 (blue) exceeding five standard deviations (see Plate 14).

coloured areas represent changes that occurred between 31 August 1989 and 21 August 1997.

In the above cases, the areas in which changes were investigated were relatively small. A much wider area was selected at the US defence site near Hanford. At this site several plutonium production reactors and reprocessing facilities were constructed. Changes between 1983 and 1987 were investigated by first co-registering the Landsat images of 1983 and 1987. The images were combined as band five (red) of 1987 with bands five (green) and four (blue) of 1983. The combination is show in

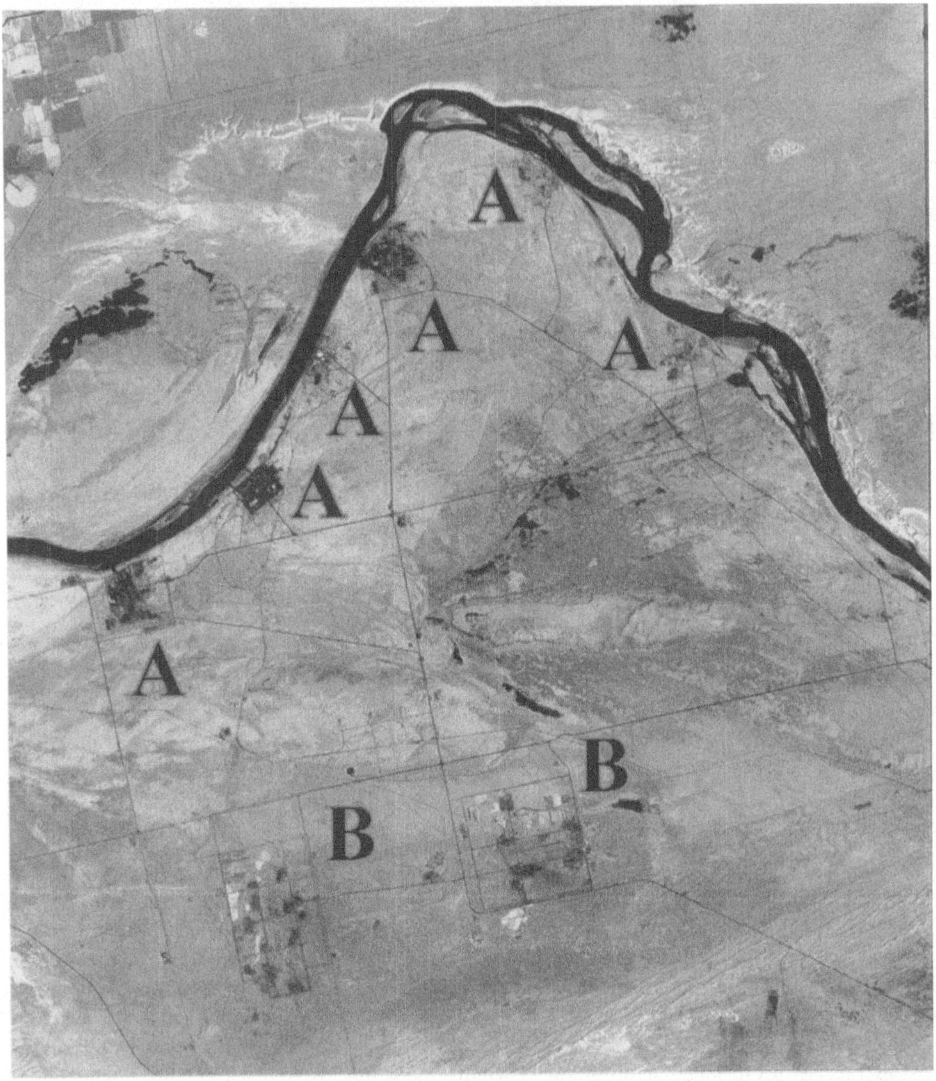

Figure 7.25. In this figure the images were combined as band five (red) of 1987 with bands five (green) and four (blue) of 1983. Sites marked **A** and **B** are production reactors and reprocessing facilities. In the image red and green colours represent changed areas. Compare this with the image in Figure 7.26 (see Plate 15).

Figure 7.25. In this image, red and green colours represent changed areas. Compare this with the image in Figure 7.26.

Using the two Landsat images, the MAD technique was used to detect changes in the Hanford site. The image processing was the same as that used in the case of the Gorleben facility. The resulting image is shown in Figure 7.26. It shows three linear

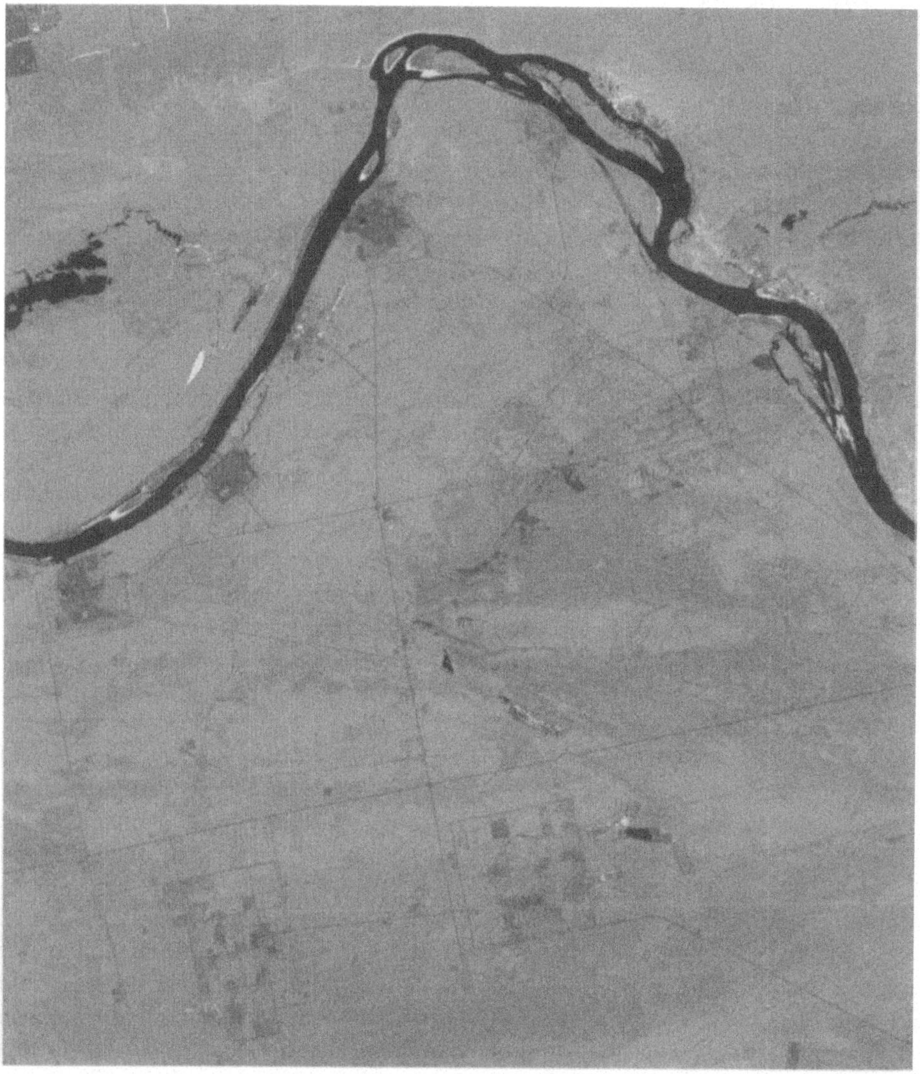

Figure 7.26. This shows a 3 linear combination of the twelve bands of the bi-temporal image in red, green and blue resulting from the analysis using MAD. Twelve infrared bands of Landsat images dated 12 August 1983, 27 July 1986 and 28 June 1987 were used. The changes represented by colours were then superimposed on a single band from the image dated 12 August 1983. Compare this with the image in Figure 7.25 (see Plate 16).

combinations of the 12 bands of the bi-temporal image in red, green and blue resulting from the analysis using MAD. Twelve infrared bands of Landsat images dated 12 August 1983, 27 July 1986 and 28 June 1987 were used. The changes represented by colours were then superimposed on a single band from the image dated 12 August 1983.

7.3 CONCLUSIONS

It must noted at the outset that information gathered from commercial observation satellites is only one among several sources of information. Recently the US launched a commercial observation satellite with sensors having resolutions of 0.6 m panchromatic and 2.44 m multispectral. While the use of such data will certainly enhance the IAEA's safeguards procedures, the work done so far using currently available optical (resolutions between 1 m and 30 m) and radar (resolution between 12 m and 25 m) images has shown that even these relatively low-resolution commercial satellite imageries have considerable potential. The following conclusions were drawn:

- It is possible to detect such large facilities as nuclear power reactors, reprocessing and enrichment plants, and research centres, new construction activities at known nuclear sites.
- Satellite images have been shown to be a useful tool to identify various facilities on sites and to correlate them with maps and details of buildings provided by a state, as well as with information collected from open sources, a task the IAEA will need to perform. The use of satellite images offers considerable scope in the planning of the Agency's on-site inspections. The Agency's safeguards procedures necessarily involves the assessment of the correctness and completeness of member states' declarations. Here again, the use of commercial satellite imagery has considerable potential.
- Even at 30 m resolutions, cooling towers associated with nuclear power reactors and enrichment facilities and, with good contrast, power lines, can be detected by optical sensors. Radar sensors can also detect pylons associated with power lines and rail tracks. They can penetrate clouds and can be used day and night under all weather conditions.
- Passive and active security features, often associated with important national strategic assets, including nuclear facilities, were also detectable in some of the sites studied. The protective structures can be multiple perimeter fences, and active defensive systems such as anti-aircraft gun and missile sites. Such features, often associated with facilities belonging to defence establishments and to covert nuclear activities, can help in distinguishing a civil facility from a military one thus alerting the Agency of any covert activity. One of the tasks of the Agency's safeguards procedures is to identify undeclared activities in a member state.
- It should be noted that generally at a defence facility, a plutonium production reactor and a reprocessing facility are in the same complex.
- It was also shown that radar and multispectral images with far infrared and thermal bands can be used to detect discharged cooling water from a nuclear reactor giving some indication of its operating status. Frequent shut down of most types of reactors would be a trigger to initiate a further investigation.
- There is considerable potential for detecting changes in a scene both by eye and with the use of mathematical techniques using computers; the latter are most

useful for detecting spectral changes in a scene. A number of existing image-processing programmes have been tried and the utility in detecting changes on some of the sites demonstrated.

From the above, it can be seen that commercial satellite imagery will have a very important role in the work of the IAEA. With the use of other open sources of information even undeclared activities could be detected and monitored.

7.4 REFERENCES

Federation of American Scientists Web Site: http://www.fas.org/nuke/guide/usa/facility/k25.htm.

Hipple, F., and Levi, B.G. (1986) Controlling nuclear weapons at the source: Verification of a cut-off in the production of plutonium and highly enriched uranium weapons. In: Tsipis, K., Hafemeister, D. W., and Janeway, P., *Arms Control Verification*, Pergamon-Brassey's International Defense Publishers, pp. 338–388.

Jasani, B. (1990) Commercial observation satellites and verification. In: Krepon, M. *et al.* (eds), *Commercial Observation Satellites and International Security*, MacMillan, Basingstoke, pp. 142–150.

Jasani, B., and Larsson, C. (1986) Security implications of remote sensing. *Space Policy*, Vol. 4, No. 1, February, pp. 46–59.

Jasani, B., Richter, B., Stein, G., Ward, M. and Killinger, M. (1996) *Enhancing IAEA Safeguards Using Commercial Satellite Imagery: A pilot study*. November, Report number SRDP-R240, JOPAG/12.95-PRG-265, Task A.16/JNT D882.

Jasani, B., Canty, M., Richter, B., Stein, G., Niemeyer, I. and Ward, M. (1998) *Commercial Satellite Imagery and Safeguards: Some case studies using multispectral and radar data.* November, Report number SRDP-R256, JOPAG/11.98-PRG-284, Task D00988.

Jasani, B., Canty, M., Richter, B., Stein, G., Niemeyer, I. and Ward, M. (1999) *Commercial Satellite Imagery and Safeguards: Some case studies using multispectral and radar data.* August, Report number SRDP-R266, JOPAG/05.99-PRG-293, Task D00988.

Jasani, B., Canty, M., Niemeyer, I., Richter, B., Stein, G. and Ward, M. (2000) *Commercial Satellite Imagery and Safeguards: Further case studies using multispectral and radar data.* January, Report number SRDP-R269, JOPAG/01.00-PRG-301, Task JNT D00988.

Kiser, E.C. (1975) Operational Planning. In the Proceedings of the Uranium Enrichment Conference, 13–14 February 1975, Oak Ridge, Tennessee, US Energy Research and Development Administration, Office of Public Affairs, Technical Information Center, pp. 35–42.

Kokoski, R. (1995) *Technology and the proliferation of nuclear weapons*, SIPRI/Oxford University Press, Oxford, p. 108.

National Imagery and Mapping Agency (NIMA) (1997), Vector Map Level 0, Ed. No. 003, January.

8

Geographical information systems in nuclear safeguards

A. Poucet, S. Contini and F. Bellezza

8.1 INTRODUCTION

The rapid increase in power of personal computers has made their use possible for applications that require resources previously available only in more expensive workstations. One of these concerns the use of Geographical Information Systems (GIS), a quite recent technology initially applied to study environmental problems, that rapidly found application in business and in many industrial fields. GIS are software tools suitable to deal with data that are characterised by their position on the earth.

As far as nuclear safeguards is concerned, the advent of the strengthened safeguards systems, as described in the model protocol INFCIRC/540, entails a new series of more qualitative verification activities, in addition to the more quantitative assessment that was inherent to the traditional nuclear material accounting (INFCIRC/153). The new requirement asks for an expanded declaration of a state's nuclear fuel cycle related activities with the objective to ensure that there are neither clandestine materials nor facilities on the state's territory and that no undeclared activities are taking place. IAEA inspectors have now the right to access any building in the site. The traditional and the additional protocols, together with open source information, will allow the Agency to get a more detailed description of a state's nuclear activities.

Both the states and the Agency have to deal with a large amount of information of different types, (i.e. texts, numerical data, images, CAD drawings, maps, etc). A relevant percentage of this information is spatially referenced (i.e. it can be characterised by its location on the earth). Therefore, the use of GIS is a natural choice for transforming this large volume of spatially related data into usable information. Different GIS-based support systems can be conceived, depending on the user needs. This chapter provides first a general introduction to GIS; it then describes an example of a GIS application developed for safeguards.

8.2 BASIC CONCEPTS OF GEOGRAPHICAL INFORMATION SYSTEMS

8.2.1 GIS definition

Scientific literature offers several definitions of a GIS. A good example is that provided by Maguire *et al.* 1991, as it is based on its main functionality. A Geographic Information System is an informatic tool for capturing, checking, storing, retrieving, manipulating, analysing and displaying spatial data (i.e. data which are spatially referenced to the earth. It is estimated that about 80% of the data used in business and government are of spatial type.

GIS is a quite recent technology that has many features in common with other previously consolidated systems: databases, remote sensing, CAD, and computer cartography.

Database systems

Data Base Management Systems (DBMS) store, retrieve, and analyse alphanumerical data, also referred to as aspatial data. These systems efficiently handle huge amounts of information. Several commercial DBMS display data on a raster map, but they have very limited or no capability at all for graphical data manipulation and analysis.

Remote sensing systems

These systems are designed for the manipulation and analysis of remotely sensed images supplied by satellites orbiting around the earth. A raw image is processed to remove noise and distortions, to improve readability and to georeference it, in order to make it clearer for subsequent interpretation. Other important operations include image rectification, change detection, supervised and unsupervised classification.

CAD systems

Developed to support the design of complex structures, computer aided design systems have a powerful graphical user interface and several functionalities to draw and manipulate graphical data. Generally, they are interfaced with simple databases and are not suitable for dealing with images.

Computer cartography systems

These systems are used for the production of high-quality maps in vector format. They assign attributes to graphical features and manage maps and use satellite images and aerial photogrammetry to produce high-quality maps.

GIS technology, being more recent than the systems mentioned above, has taken several functionalities from them, with the important additional possibility to perform *spatial data analysis*. Therefore, GIS can deal with maps and images, assign attributes to graphical features using a powerful DBMS, and produce maps as a result of spatial data analysis.

8.2.2 Data formats in a GIS

The method of representing spatial data in a GIS is through maps, a two-dimensional collection of graphical features (objects) characterised by their x,y co-ordinates. In a GIS, spatial data can be represented in *vector, raster or grid formats* (Heywood *et al.*, 1998).

Vector format

The fundamental types of entities used are points, lines and polygons. They portray real life objects. For instance, in a nuclear facility, points may represent entrance gates, or the location of surveillance equipment, whereas lines may represent roads, rivers, electric power lines. Polygons may represent buildings, site fences, facility boundaries, material balance areas, plant units, etc. Each feature can be described by means of some associated attributes (numbers, characters, Boolean). Attributes are stored into a record of an attribute table managed by a DBMS. It follows that attributes can be queried and used for calculations.

Raster format

Raster data formats record spatial data in a matrix data-structure, in which the smallest logical unit represents a pixel (i.e. a dot on the screen); these formats are typical of scanned maps, digital photos and satellite images.

The row-column numbers of the array give the coordinates of a pixel. Values associated to a pixel depend upon the number of bits used to store data. For example, with only one bit the possible values are 1 and 0 (typical of black-and-white scanned maps), whereas with 8 bits (i.e. one byte) they are $2^8 = 256$, and so on. The information stored in a pixel depends on what the picture represents (e.g. it may represent the light reflected by the object in a satellite image or the brightness value in a scanned map). A pixel does not have an associated attribute table. Attributes can only be associated to the image as a whole.

Raster data require more memory space than vector data and for this reason they are normally stored by using data compression techniques.

Grid format

Similar to raster data, grid data represent spatial data in a matrix data-structure. The difference between grid and raster is related to the information associated to each (square) grid cell, whose dimension is set by the user: the smaller the grid cell the higher the resolution. The content of a cell may be an integer or a real number. Integers are generally used to represent features as in vector format; each feature type of the grid format can be associated with an attribute table stored in the DBMS, in the same way as vector themes can. Real numbers are used to describe the spatial distribution of graduated phenomena (e.g. temperature, humidity, etc). In this case the grid format is more similar to the raster format, in the sense that there is no table of associated attributes.

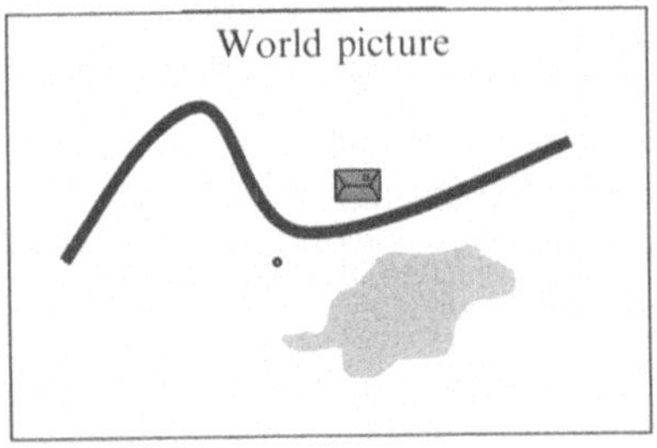

The world picture contains a road, a house, a well and a lake.

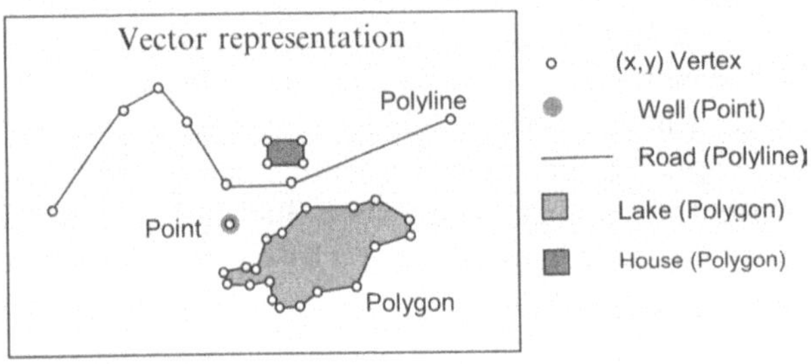

In the *vector format* each feature is represented by points, lines and polygons. Features have associated attribute tables.

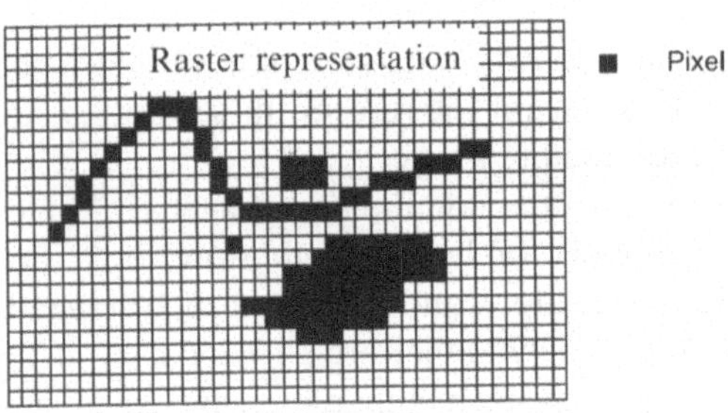

In the *raster format* each cell is a pixel. The pixel value may be stored in a bit, a byte or two bytes and represents a value in a gray or color scale. Pixels have no associated attribute tables. Suitable for representing graduated phenomena and images.

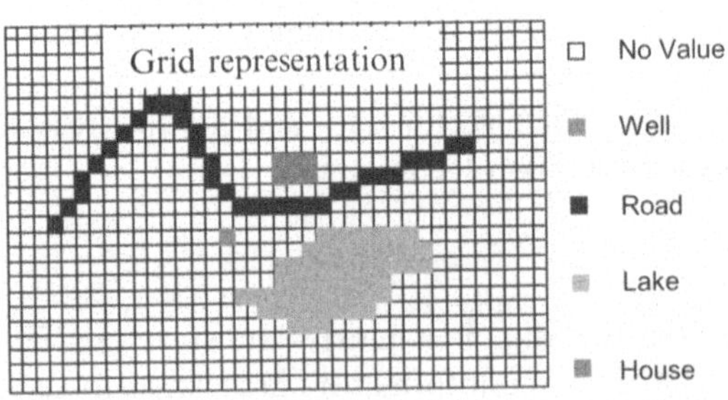

In the *grid format* each cell is square and has a user-defined dimension. Features have associated attribute tables. Also suitable for representing graduated phenomena.

Figure 8.1. Different workable data formats in a GIS.

Figure 8.1 graphically describes the differences among vector, image and grid formats with reference to a sample picture.

The combined use of these data formats increases the power of GIS in modelling and solving problems requiring spatial data representation and analysis. There is no data format type that may be considered superior. The choice of the suitable format depends on many factors: problem type, data type, accuracy needed, etc.

8.2.3 Maps

Maps represent the fundamental information in GIS applications. They can be classified according to their scale:

- Planisphere and globe (scale < 1:50,000,000)
- Geographical maps (scale from 1:50,000,000 to 1:1,000,000)
- Topographical maps (limited zones, scale from 1:1,000,000 to 1:2,000)
- Layout maps (from 1:2,000 to 1:100)

The map scale represents the proportion between the map dimensions and reality; it is expressed as the ratio between a reference distance on the map and the corresponding real distance (e.g. 1:500 means that one length unit on the map represents 500 units in the real world).

Thematic map

For vector and grid data formats, a set of objects of the same type (i.e. characterised by the same attributes,) represent a thematic map, also referred to as theme or layer. In other words, a thematic map portrays the spatial distribution of a single topic. For instance, cities, rivers, lakes, are all examples of thematic maps. The set of buildings in a site is another example. Two or more thematic maps may be separately generated and then combined to generate a new map by applying the overlay operation. Overlay is fundamental for spatial data analysis involving two or more themes.

Raster scanned maps or satellite images can be used as background to generate thematic maps or to update maps.

Topographical map

A topographical map contains data related to different themes (e.g. land use, roads, rivers, etc). A scanned map of a small area (raster format) is an example of a topographical map. A topographical map can also be seen as the result of the overlay operation applied to vector or grid thematic maps.

Geographical maps

Geographical maps represent the earth on a flat surface with reference to a two dimensional coordinate system, so that the position of any object on the earth can be univocally determined. Several types of maps are available for different applications and with different scale. The way in which geographical maps are obtained is briefly described below.

8.2.4 Ellipsoids and map projections

The earth can be represented as an ellipsoid (i.e. a three-dimensional object generated by rotating an ellipse around its minor axis). The major and minor radiuses of the earth are called respectively *equatorial* and *polar* radius (see Figure 8.2).

Generally, earth's dimensions are expressed in terms of the major radius and of the flattening ratio $f = (a - b)/a$. The difference between the two radiuses $(a - b)$ is about 21.5 km. Several ellipsoids have been proposed; they differ from the measure of the equator radius and the flattening ratio (Maling, 1992; Pearson, II, 1990). Table 8.1 contains the main parameters of some ellipsoids. The approximation of the ellipsoid with a sphere is acceptable for world maps only. Coordinates in a three-dimensional representation referred to as longitude (λ) and latitude (φ) allow determining the geographical position of any point on the earth's surface.

Latitude and longitude define a grid on the ellipsoid whose dimensions change from point to point. The length of meridians from 0° to 1° is 110,567.3 m, whereas from 89° to 90° is 111,669.3 m. The variation, very small indeed, is due to the oblateness of the earth. The length of parallels varies from 111,321 m at the equator to 1,949 m at 89°.

The relation between the length of a meridian D_m and that of a parallel D_p at a given latitude φ is given by: $D_p = D_m \cos\varphi$. For instance, at $\varphi = 60°$ latitude, $\cos\varphi = 0.5$: the length of the parallel is half that of the meridian.

A simple way to represent the earth is to map it on a globe. This mapping is obtained by simply scaling the earth. The advantage is that there is little distortion, but it has several practical disadvantages (e.g. distances are difficult to measure and it is difficult to store it in a computer). It is therefore preferable to represent the globe on a flat surface (i.e. on a 2D-plane system).

The operation that allows the representation of the earth on a Cartesian 2D-plane co-ordinate system is referred to as *map projection*. A projection is a

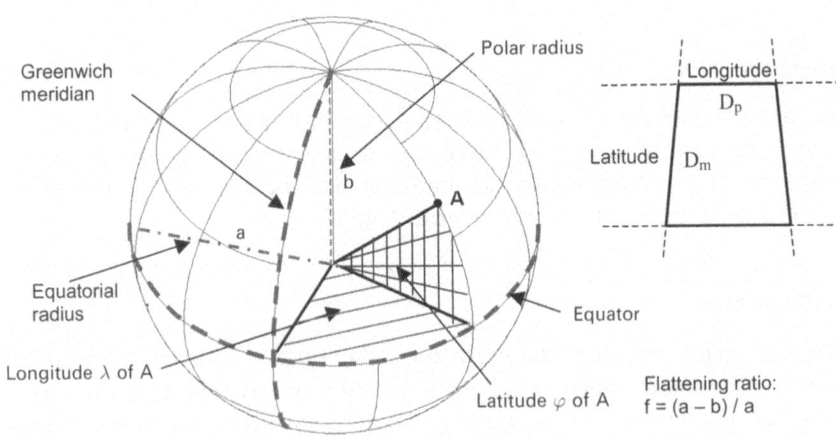

Figure 8.2. Ellipsoid as a 3D model of the earth.

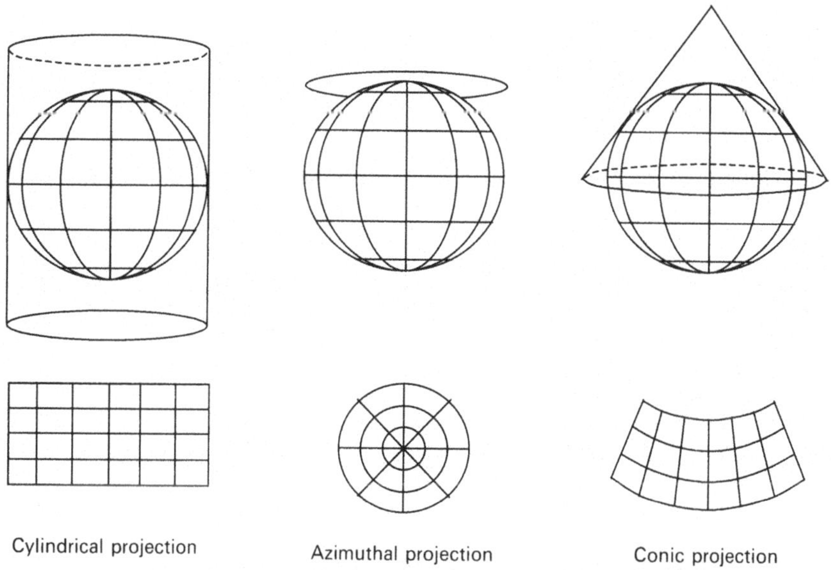

Figure 8.3. The basic types of plotting surfaces.

Table 8.1. Main data of some ellipsoids.

Ellipsoid	Equatorial radius (m)	Polar radius (m)	Flattering ratio
Bessel 1841	6,377,397.2	6,356,079.0	1/299.15
Clark 1880	6,378,249.1	6,356,514.9	1/298.46
Australian 1965	6,378,160.0	6,356,774,7	1/298.25
WGS72, 1972	6,378,135.0	6,356,750,7	1/298.26
GRS80, 1980	6,378,137.0	6,356,752.3	1/298.257
WGS84, 1984	6,378,137.0	6,356,752.3	1/298.26

mathematical transformation to determine the x,y coordinates from λ, φ (i.e.: $x = f_x(\lambda, \varphi); y = f_x(\lambda, \varphi)$). Equally important is the inverse transformation (i.e. $\lambda = g_\lambda(x, y)$; $\varphi = g_\varphi(x, y)$).

A large number of projection methods have been developed in the past and others will certainly be proposed in the future (see Maling). The three basic types of plotting surfaces are azimuthal, cylindrical, and conic along with the corresponding graticule of meridians and parallels.

Unfortunately all projections inevitably induce distortions in at least one of the geographical characteristics: area, distance, angle, and shape. This is demonstrated in Figure 8.4, as:

a A projection that does not distort area is called *equal-area* or *equivalent*.
b In the linear distortion (i.e. distortion of distances) a segment may be longer,

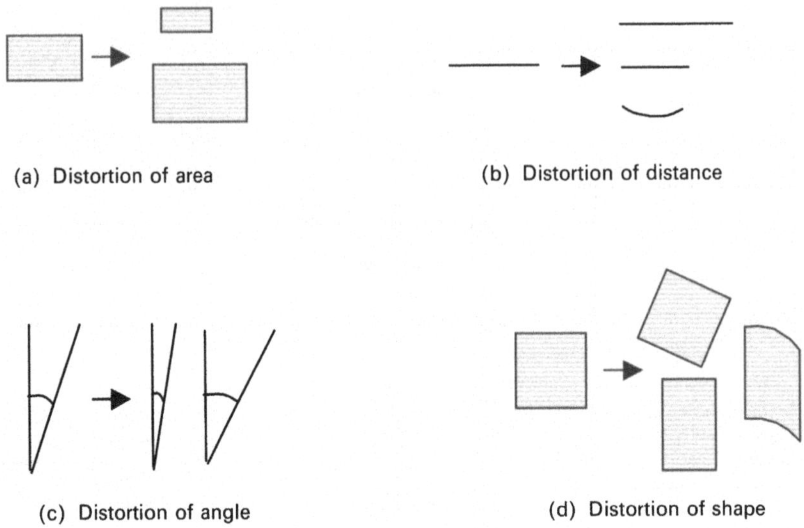

Figure 8.4. Possible distortions occurring in map projections.

shorter than the real one or even curved. A projection that does not distort distances is called *equidistant.*

c Also angles may be distorted. A projection that maintains angles is said to be *conformal.*

d The distortion of shape, maintaining the area, is due to the modification of angles and distances. The shape may be distorted in several ways (e.g. a square may be transformed into a rhombus, a rectangle or even with an equivalent surface w:th curving effects.)

Some projection systems change a point into a line. This occurs at the poles, which are represented as segments of finite length. In these cases the distortion of the area, length, angle, and shape is greater there. The degree of distortion of a projected map varies from point to point and can be calculated (see Pearson, II).

Each basic projection system represented in Figure 8.3 (see p. 217) can be further classified by reference of the point of contact between the ellipsoid and of the plane or the developable surfaces. For instance, the azimuthal projection, which represents one hemisphere, can be classified as Polar, Equatorial, or Oblique according to the position of the plane.

Moreover, a further classification can be done by reference to the projection point. Figure 8.5(a) shows some common azimuthal projection: 1 = *Gnomonic*, 2 = *Sterographic*, 3 = *Equivalent*, and 4 = *Orthographic*.

The plane can be tangent or secant to the sphere. It can easily be realised that meridians are equally spaced lines radiating from the centre, whereas the parallels are spaced differently, as shown in Figures 8.5(b, c) where two examples of the earth maps are shown using: (b) the sterographic polar projection, and (c) the orthographic polar projection.

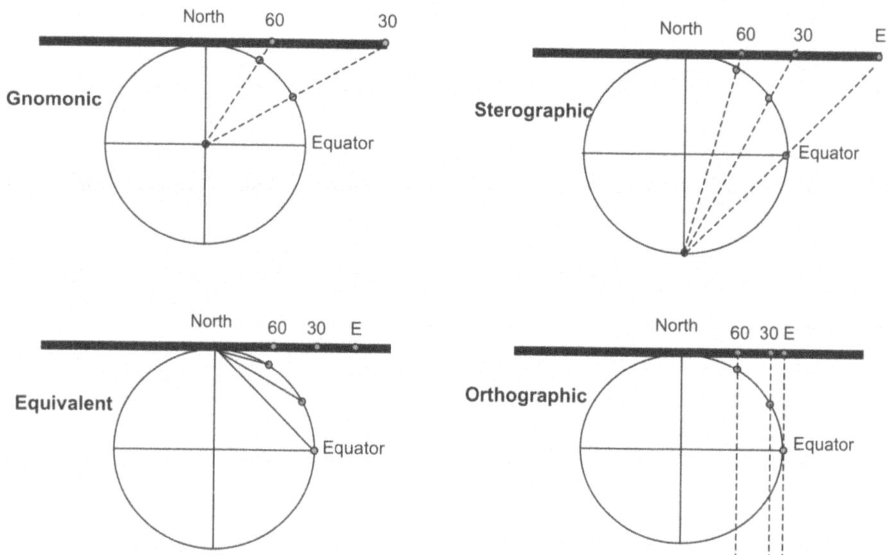

(a) Some common azimuthal polar projections

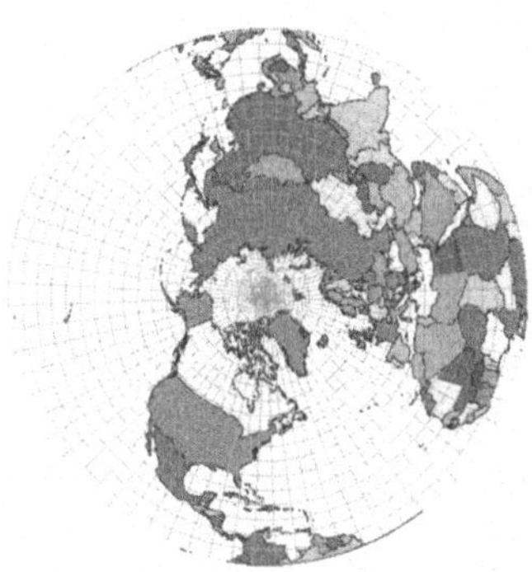

(b) Sterographic Polar Projection

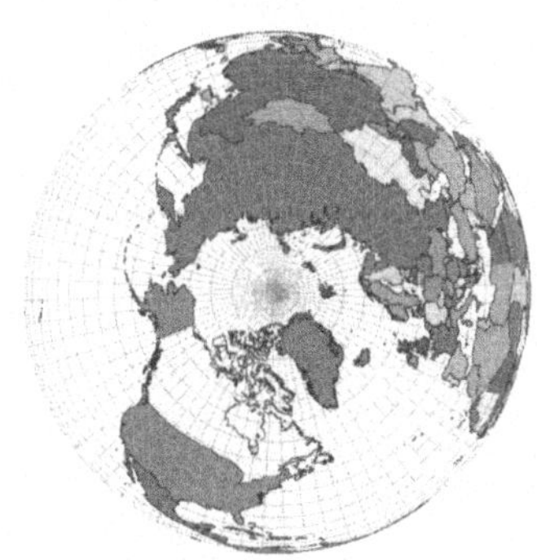

(c) Orthographic Polar Projection

Figure 8.5. Some commonly azimuthal polar projections.

With cylindrical and conic projections the earth is projected on a solid surface, tangent or secant, giving maps with different characteristics. One of the most important cylindrical projections is the Mercator, direct and transverse.

Direct Mercator projection

This is obtained by considering the cylinder tangent to the equator on which all points of the ellipsoid are projected, as shown in Figure 8.6(a).

In this projection, shape and area are distorted when moving away from the central meridian, as can be noticed by looking at the distortion of the grid (meridians are parallel lines!). For instance, Brazil and Alaska are shown as having comparable

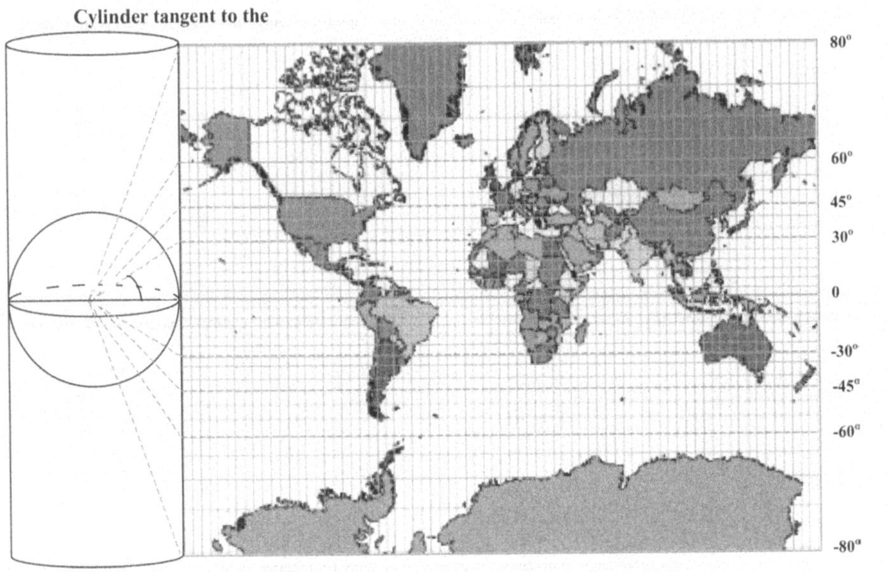

Figure 8.6(a). Direct Mercator (cylindrical) projection.

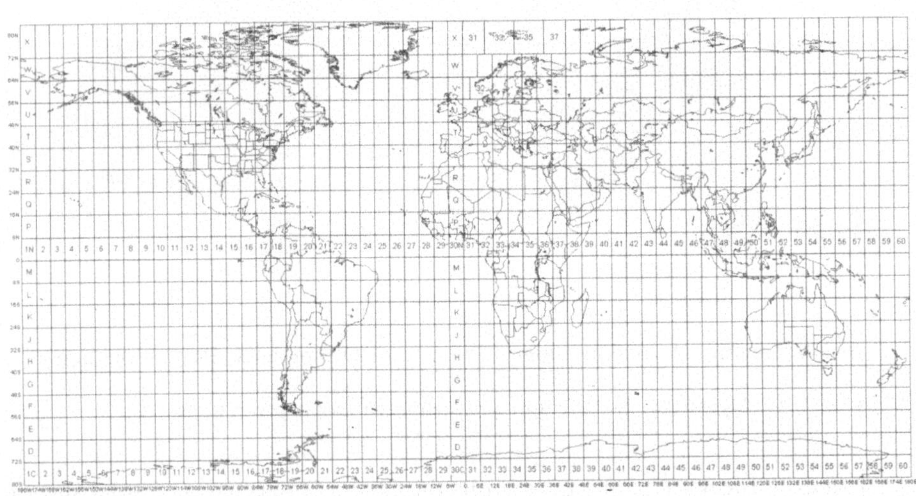

Figure 8.6(b). The Universal Transverse Mercator (UTM) projection zones [source: www.dmap.co.uk].

areas, whereas Brazil is about five times larger than Alaska. The deformation of the South Pole, of Greenland and Canada is similar. Accurate scaling for distances is only true along the equator (i.e. the tangency point) where the map is conformal and equidistant. The linear distortion module m, which is the ratio between the length of a small segment on the map and the real distance, is very low at $\pm 3^\circ$ ($m = 1.0007$), but rapidly increases as φ increases (e.g. $m = 1.064$ at 20°, $m = 1.304$ at 40°, and $m = 1.995$ at 60°).

Transverse Mercator projection

This is obtained when the cylinder is tangent to a meridian. As with the direct projection within ±3° from the meridian of contact the deformation is less then 0.07%.

Universal Transverse Mercator

The projection known as the Universal Transverse Mercator (UTM) is internationally adopted to map the whole globe. The earth is subdivided into 60 zones of 6° each; the meridian in the centre of a generic zone is called the central meridian. In the UTM the cylinder is tangent to the central meridian. Consequently, the maximum distortion is always less than 0.07%. The origin of each zone is the intersection between the central meridian and the equator. Zones are numbered starting from the anti prime meridian towards east (see Figure 8.6(b)). Latitude values range from 80°S to 84°N. Rows are identified with letters from C to X (I and O are omitted) starting from 80°S.

To eliminate negative values the coordinates of the origin are altered. Values assigned to the central meridian and to the equator are respectively referred to as *false easting* and *false northing*. For locations in the Northern Hemisphere the false easting is 500,000 m and the false northing 0.0. For locations in the Southern Hemisphere the false easting is 500,000 m and the false northing 10,000,000.

Within each zone this projection is conformal, the scale is constant along the central meridian and local angles are correct. For these reasons the UTM projection is widely adopted for topographical maps, satellite images and other applications requiring precise position.

The choice of the suitable projection depends on the use of the map. Some criteria for the selection of a suitable projection can be found in Pearson, II (and Bugayevskiy *et al.*, 1995). These criteria are based on the type of distortion that needs to be minimised.

In a GIS, thematic maps and remote sensing images are frequently overlapped to solve specific problems. To do that it is necessary that all themes be based on the same coordinate system. Map projection is one of the operations applied to this purpose.

The application described later in this chapter (see Section 8.3) uses three different levels of maps: world/continent, country and site. The first two levels are not projected (i.e. their co-ordinates are in latitude-longitude), whereas for the third level, which represents a site layout, the UTM projection system has been chosen.

8.2.5 Maps rectification and georeferencing

In order to correctly overlap different maps it is necessary that they have the same coordinate system, (e.g. the real-world system or a projected coordinate system).

Rectification

This operation assigns a raster map the same coordinate system of another image or vector map, so that the two images, or the image and the vector map, are correctly aligned and can be overlapped. The operation is based on the selection of control points in both maps. At the end of the procedure the first map is assigned the same coordinate system of the second one. For instance, if the second map is stored in real-world coordinates (i.e. latitude, longitude), the rectification operation establishes an image-to-world-transformation that converts the image coordinates to real-world coordinates. Figure 8.7 represents the rectification of the image on the left with the vector file on the right. The result is the correct alignment of the two layers.

Georeferencing

This operation assigns geographical coordinates (e.g. latitude and longitude, or x,y in the UTM system, or whatever) to each pixel of a raster map, or to each cell of a grid, or to each vertex of a vector theme. Operatively, the georeference operation entails the assignment of known coordinates to some points of the map (i.e. GCPs). The coordinates of the GCPs can be obtained from another georeferenced map, a satellite image or from the Global Positioning System (GPS). Figure 8.8 shows an example of a map georeferenced using four GCP; the coordinates are given in the table.

8.2.6 Satellite images

One of the most interesting data sources in GIS applications is represented by digital images supplied every day by several satellites orbiting around the earth (Moik, 1980). The remote sensing theory is covered extensively in other parts of the book, and thus it will not be considered here. However, it is interesting to show, by examples, some uses of remote sensing images in a GIS system.

Figure 8.9 shows an IRS-1C panchromatic (one-band) image with spatial resolution of 5.8 m per pixel which has been rectified to a map coordinate system in order

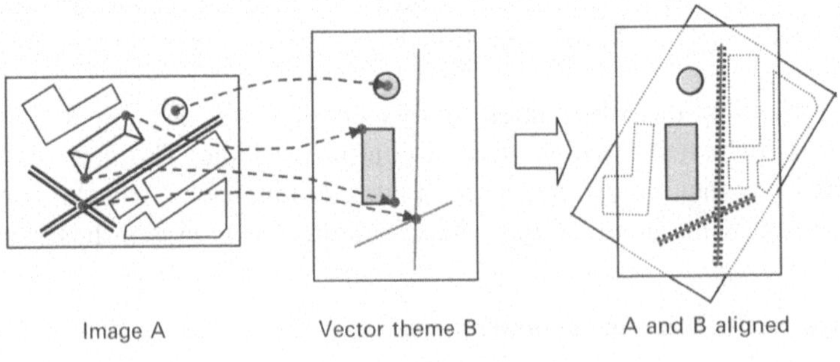

Figure 8.7. Example of the rectification operation applied to the image A and the vector map B.

GCP#	X	Y
1	10	190
2	365	173
3	421	5
4	8	20

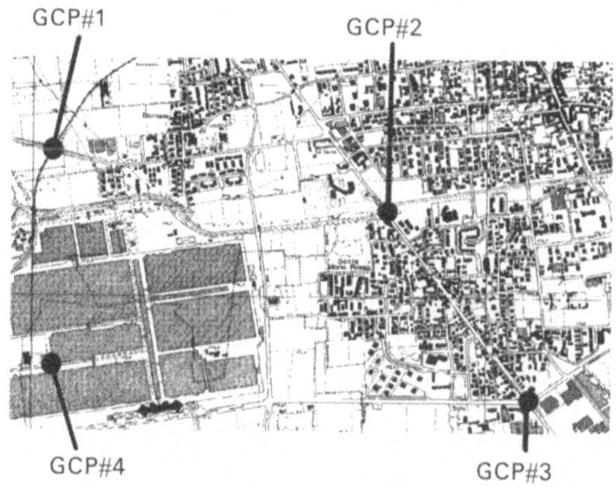

Figure 8.8. Example of the application of the georeferencing operation to a raster map.

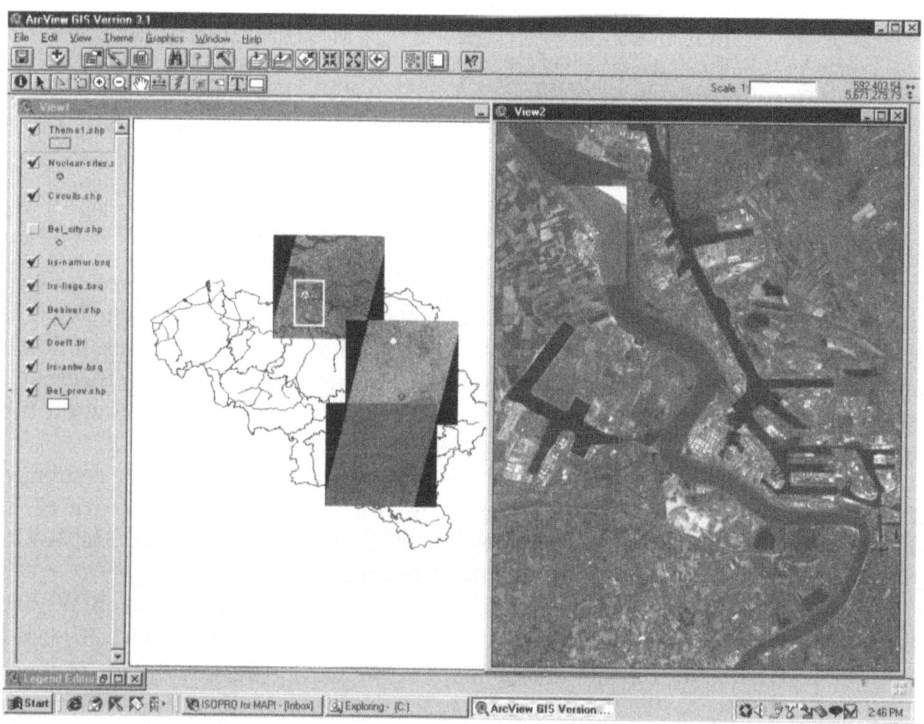

Figure 8.9. IRS-1C images and zoom in of a part of an image showing the wider area around the Doel Nuclear Power Plant.

to allow the use of the image data in conjunction with other data organised in thematic layers. At the left side the map of Belgium is shown with, superimposed, the three geo-referenced IRS-1C images. In this view some geographical themes are also shown (shapes of provinces, cities, rivers, nuclear plants). At the right side a smaller zone north of Antwerp is shown. This image has further been aligned with

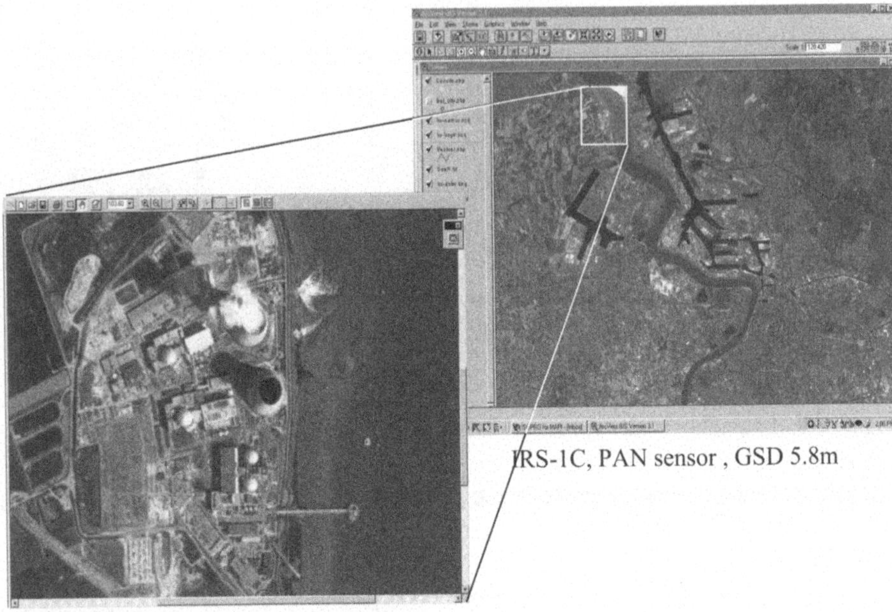

Figure 8.10. The Doel Nuclear Power Plant taken by the Russian KVR-1000 camera.

the KVR-1000 image of the Doel Nuclear Power Plant (the small rectangle). The KVR instrument produces images on photographic film, whose spatial resolution is around 2 m (the declared spatial resolution is 1.56 m/pixel but this is a theoretical value calculated from digitalisation of the photographic images). Zooming on this area one can see the nuclear plant quite clearly (see Figure 8.10).

This example shows how images of different resolutions can easily be combined in a GIS. In this case, the IRS-1C image shows the wider context and provides the information about the industrial area where the reactor site is located. The KVR-1000 on the other hand shows the details of the site itself.

Images taken by the IKONOS satellite have 1 m spatial resolution (B/W) and 4 m multispectral. Due to their high geometrical resolution these images represent a valuable tool to verify the site layout maps requested by the additional protocol. Figure 8.11 shows the IKONOS image obtained combining the 1 m panchromatic with the 4 m multispectral of a research centre. Zooming in on part of the image it is possible to clearly identify buildings at a scale of 1:2,000.

Within a GIS, satellite images can be used for several other purposes. For example, to:

- georeference plant layout maps, using GCP taken from the satellite image;
- provide a background to generate vector thematic maps of features of interest;
- provide a background picture to verify declared site infrastructures;
- detect changes in infrastructures;

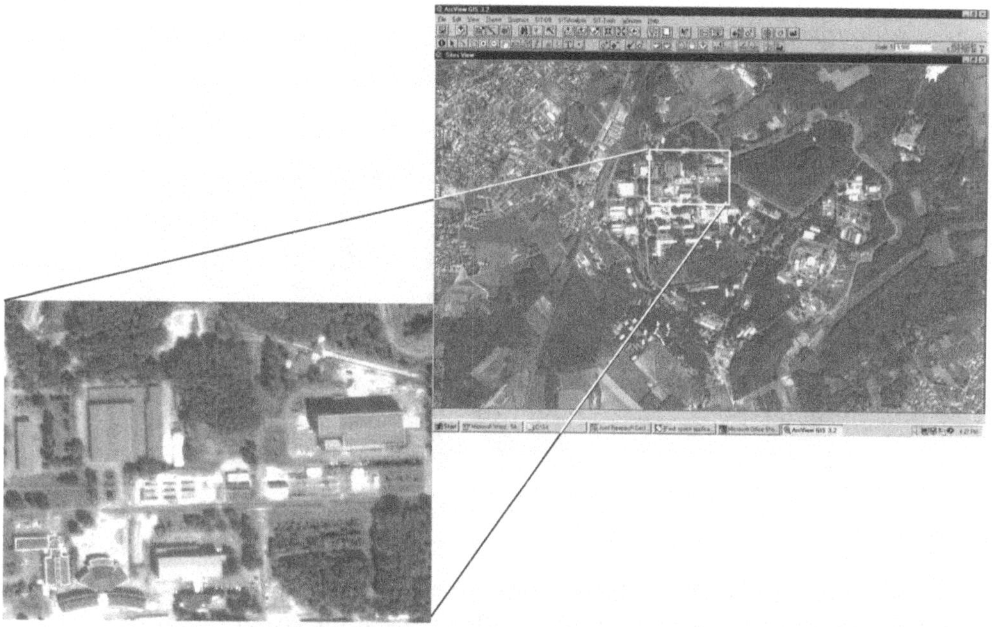

Figure 8.11. IKONOS image and zoom in of part of it showing some buildings.

Figure 8.12. Road net update.

- update old maps;
- generate the digital terrain model using a stereo pair of images;
- represent a site in 3D after estimating the high of buildings.

Figure 8.12 shows some examples of the use of a SPOT pan image – and provides an example of the use of SPOT to update the road network of the area for emergency

Figure 8.13. Hillshade.

planning purposes, since available topographical maps were not up to date. In the figure other thematic maps are visible: hydrology and the layout of a plant.

GIS offers a set of complex functions for rapidly solving several problems (e.g. minimum cost path between two points in complex terrain, slope, visibility).

Figure 8.13 represents the hillshade from the digital elevation model of part of the area represented in Figure 8.12. From the elevation model it is also possible to derive the elevation contours map.

All these applications, together with those extensively described elsewhere in this book are particularly interesting in the declaration and verification processes (described in Section 8.3).

8.2.7 Spatial analysis

Spatial analysis represents one of the most important features of a GIS. It is fundamental in supporting decision making in several fields: business, environment, hydrology, agriculture, social, etc. Examples of complex problems that can be faced with spatial analysis are *What is the best location for a new nuclear power plant?* or *Which is the optimum path for a new pipeline?*

A GIS contains *spatial data* represented on thematic maps along with the associated attributes, also referred to as *aspatial data.* Attribute databases can be queried to get data that satisfy certain conditions (e.g. *How much nuclear material is held in a given site?*). Similarly, spatial queries can be used to get results involving spatial data (e.g. *Where are PWR nuclear plants located in the European Union? or how many plants are there within 10 km of my town?*).

The former type of query gives results in tabular form, whereas the latter is presented in a more understandable way: on a map. Both types can be combined to answer more complex questions (e.g. *Where are PWR NPP with power greater than*

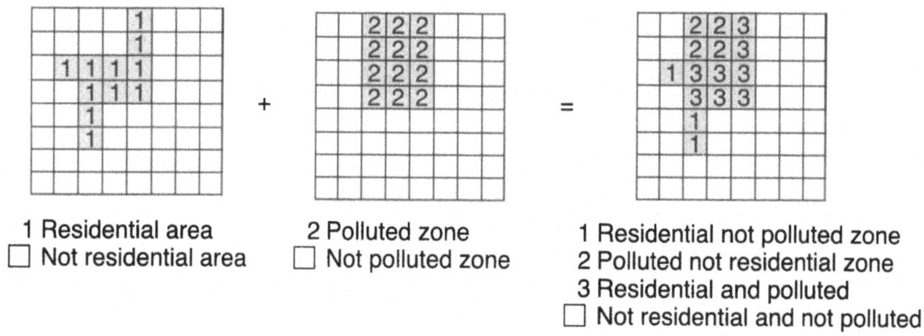

Figure 8.14. Example of overlay applied to features in grid data format.

1,000 MWe and built after 1990 are located in the EU?). Results are displayed on a new map.

Spatial aanalysis is applicable to any of the three fundamental data formats previously described, even though the efficiency may be different. Generally, these operations are easier to perform on raster/grid data than on vector data.

Spatial analysis functions range in complexity from simple operations, such as the calculation of the length of a segment to more sophisticated ones (e.g. buffering, slope and visibility). A rich set of operators applicable to single or multiple layers are available. For multiple layer operations the necessary condition is that maps are referenced to the same coordinate system (i.e. they can be overlaid). Overlay and operators allow making thematic compositions. For example, the sum operator applied to themes 1 and 2 representing respectively a residential area and a polluted zone produce (Figure 8.14) a new map showing the residential polluted (3) and not polluted zones.

Map algebra (Tomlin, 1990) uses math-like expressions, which are applied to cells of one or more layers and return numerical values that are assigned to the output grid layer. Operators are arithmetic, logarithmic, trigonometric, Boolean, Relational, etc. Some examples are given in Table 8.2.

As an example of application of spatial analysis, consider the problem of determining the spatial distribution of the population at risk in case of accident in a hypothetical chemical plant. The accident may involve the release of toxic substances, the fire/explosion of flammable substances, and their like. The following data are used:

- land-use of the area of interest, derived from the SPOT image (represented in Figure 8.11);
- population distribution, taken from a census database;
- accident damage zones, obtained from an accident consequence analysis model; the result of these models is generally represented by zones with different effects on the population, the environment and the plant.

The map of the residential area (top right in Figure 8.15) is extracted from the land-use map. To each resulting polygon the population data is assigned, thus

Table 8.2. Example of some operator of map analysis applied to grid data.

<table>
<tr><th>Operator</th><th>Example</th></tr>
<tr><td>Euclidean Distance (points A and B are taken at the centre of their cells)</td><td>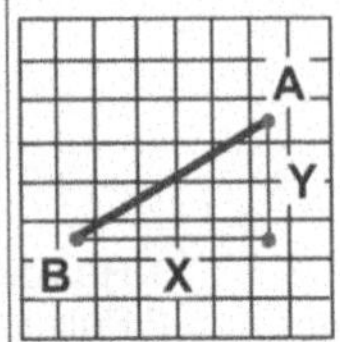

$d_{AB} = \sqrt{X^2 + Y^2}$ in cells units
X, Y = number of cells</td></tr>
<tr><td>Perimeter</td><td>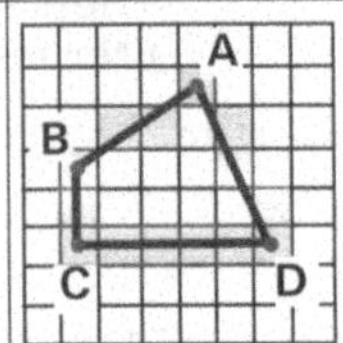

$p_{ABCD} = d_{AB} + d_{BC} + d_{CD} + d_{DA}$</td></tr>
<tr><td>Area</td><td>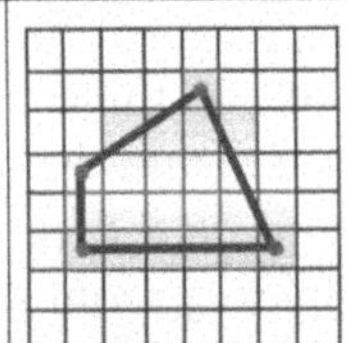
Area = Total number of cells</td></tr>
<tr><td>Buffering, the generation of a zone (buffer) around a point, a line, a polygon</td><td>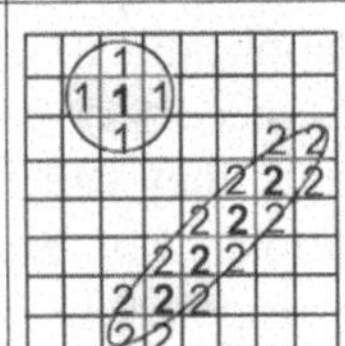</td></tr>
<tr><td>Mathematical operators The example shows the calculation of the temperature variation at time T1 and T2.
□ = no value</td><td>
Temper. at T2
<table>
<tr><td>13.5</td><td>14</td><td>15</td><td>18</td></tr>
<tr><td>13</td><td>10</td><td>17</td><td>22</td></tr>
<tr><td>14</td><td>17.5</td><td>22</td><td>25</td></tr>
<tr><td>16</td><td>18</td><td>24</td><td>28</td></tr>
</table>
-
Temper. at T1
<table>
<tr><td>14.5</td><td>14</td><td>16.5</td><td>20</td></tr>
<tr><td>14</td><td>12</td><td>15.5</td><td>18.5</td></tr>
<tr><td>14</td><td>16.5</td><td>20</td><td>22</td></tr>
<tr><td></td><td>17</td><td>21</td><td>24.6</td></tr>
</table>
=
Difference of temper.
<table>
<tr><td>-1</td><td>0</td><td>-1.5</td><td>-2</td></tr>
<tr><td>-1</td><td>-2</td><td>1.5</td><td>3.5</td></tr>
<tr><td>0</td><td>1</td><td>2</td><td>3</td></tr>
<tr><td></td><td>1</td><td>3</td><td>3.4</td></tr>
</table>
</td></tr>
<tr><td>Boolean operators (AND, OR, NOT) on overlaid layers. Result is 1, 0 or no value.</td><td>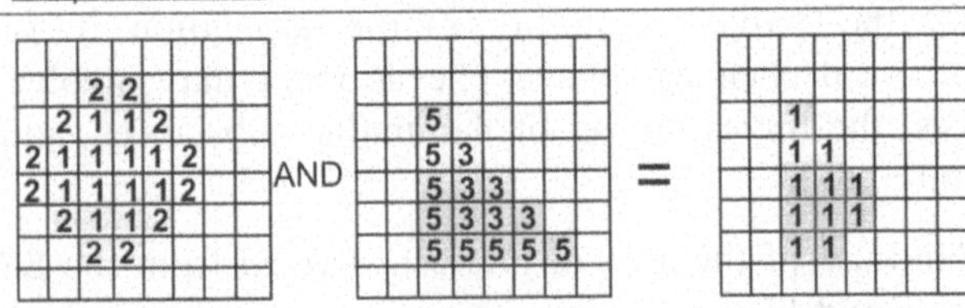
</td></tr>
<tr><td>Relational, comparison between two or more layers. Result 1, 0 or no value</td><td>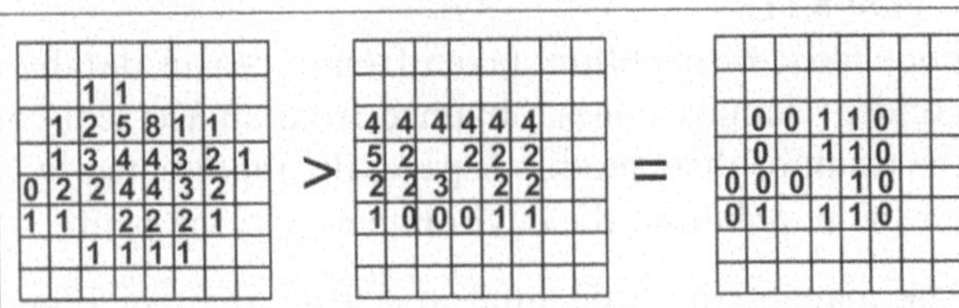
</td></tr>
</table>

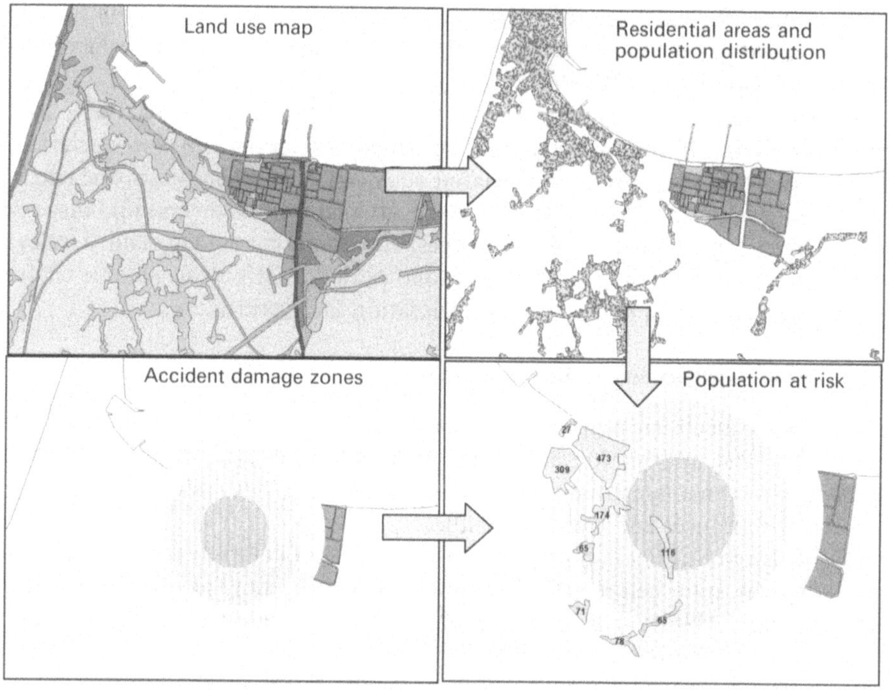

Figure 8.15. Example of calculation of the number of people at risk.

giving the spatial distribution of the population. Next, the map with accident damage zones is considered (bottom left in Figure 8.15). The intersection between the population distribution and the map of accident damage zones gives the number of people at risk. A similar analysis can be performed for calculating the environmental damage map.

8.3 AN EXAMPLE OF A GIS APPLICATION IN NUCLEAR SAFEGUARDS

8.3.1 Problem description

Nuclear material in the civil nuclear industry is safeguarded to avoid its diversion for the construction of nuclear weapons, or other explosive devices. Control is exercised by the IAEA in some 180 countries, all of which have voluntarily signed the NPT. These countries have to periodically declare all their activities according to the protocols signed with IAEA, by means of INFCIRC/153 (IAEA, 1972), which focuses mostly on nuclear material accountancy, and INFCIRC/540 (IAEA, 1997), which asks for an expanded declaration of a state's nuclear fuel cycle related activities.

Procedures for verifying the correctness and completeness of a state's declaration at site level are the responsibility of the IAEA, using information from different sources (Chitumbo and Nicholas, 2000).

At *site level* correctness and completeness are intended, respectively, as the verification that all installations and activities correspond to the declarations provided and that there are no undeclared activities or installations at the site.

The huge amount of information to be examined is quite diverse and dispersed, ranging from text, numerical data, images, drawings, maps, etc, and a relevant percentage of it is spatially referenced. It follows that a GIS is the suitable environment for developing support tools for declaration and verification.

8.3.2 The declaration and verification process

An idealised conceptual scheme for declaration and verification is represented in Figure 8.16. The information submitted under the safeguards agreements can be organised into three interconnected models: *topological, functional*, and *operational* (Poucet, 1998; Poucet and Contini, 2000).

The declared information may be subject to a set of completeness and coherence controls by the state before being submitted to the Agency, possibly in electronic form. On the declaration as a whole, the Agency's quality control procedure is applied to verify the conformity with the agreed protocols. In case of incoherence or insufficient details, additional requests are made to the state concerned.

The contents of the declaration can then be verified by considering previously submitted declarations, open source information and in-house knowledge. Satellite images and the cartography of the area around a site may be used to verify the declared topological model, information on the nuclear fuel cycle is used to check the functional model, and so on.

The subdivision into topological, functional and operational models to represent the data is useful for data management purposes.

Topological model

Comprising all geographically referenced data, the topological model contains at site level thematic maps on:

- boundary of facilities and related units;
- boundaries of material balance areas, and location of the associated key measurement points (i.e. key points where the nuclear material is measured according to specific procedures);
- buildings and the associated layouts.

Other thematic layers of interest (e.g. site entrance gates, site fence if different from the site boundary, land use outside the site fence, etc.) can be added by the user.

Functional model

In the functional, or technical, model the information concerns the description of processes, technical characteristics of the installation, materials used and procedures,

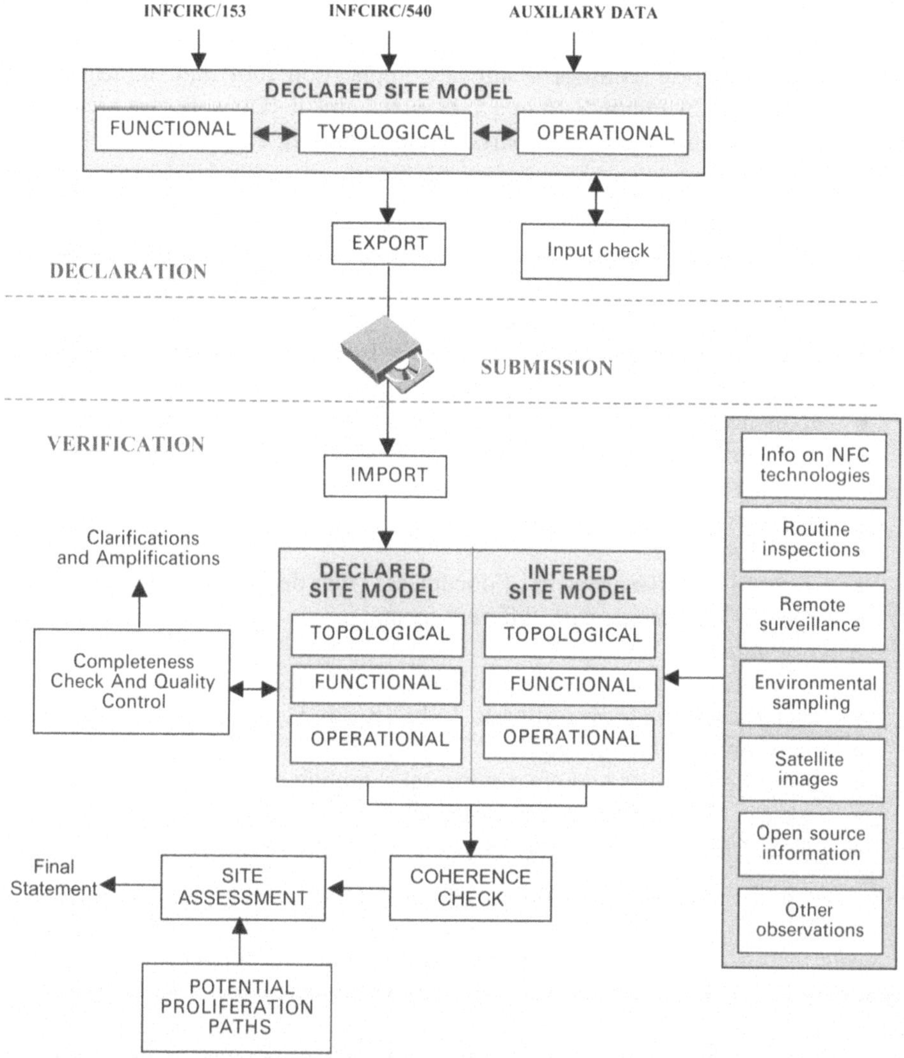

Figure 8.16. Conceptual scheme for declaration and verification.

material flows and indication of material balance areas and points where materials are measured, procedures, etc.

Operational model

Here, the model contains information about the operational programme at site level (e.g. overall production numbers, operating schedules, modification or construction activities) and at facility level (e.g. outages, refuelling schedules).

8.3.3 The SIT support system

SIT (Site Investigation Tool) is a software application tool that implements the conceptual scheme of Figure 8.16 in a GIS environment. The software architecture can be subdivided into three main parts, namely (Poucet, *et al.*, 2001):

1. The *graphical user interface* (GUI) based on a commercial desktop GIS, in which data can be organised according to the topological, functional and operational models. ESRI ArcView for Windows has been adopted with some extensions to deal with satellite images, CAD files and to perform spatial analysis.
2. A *set of tools* embedded into the GIS for:
   - importing/exporting declarations (made by e.g. the protocol reporter tool of the IAEA) and performing quality controls;
   - accessing, retrieving and managing data from local and remote databases;
   - performing data analysis, completeness and consistency checks;
   - georeferencing raster and vector files;
   - image analysis;
   - spatial analysis;
   - retrieval and management of documents from declaration and open source;
   - opening documents with different formats.
3. A *set of databases* on:
   - declarations made according to INFCIRC/540;
   - declarations made according to INFCIRC/153;
   - maps and satellite images;
   - inspection reports and environmental sampling data;
   - nuclear fuel cycle generic knowledge;
   - documents from open sources (e.g. via the internet).

Thematic maps can be drawn using the site map as background as represented in Figure 8.17 (a screen dump of the two levels of the topological model, showing the distribution of sites on the country map and different themes of a site located in Italy). The user can add any themes considered useful to better describe the site.

The need to deal with a large amount and variety of documents calls for the use of a document management system embedded in the GIS. In fact the integration of such a system with a GIS makes data storage and retrieval particularly simple and flexible. For example, it is possible to select an object in a theme and then to find all documents related to it, or to select all documents that are associated to objects resulting from a spatial query (e.g. located within a specified zone). Any graphical object in any thematic map can be described by means of any type of document, and/or linked to external databases and models as shown in Figure 8.18.

In the verification process, image analysis tools can be used (e.g. to compare the site layout map with a satellite image) in order to identify possible discrepancies between the declaration and the real situation. Figure 8.19 is a screen shot of SIT showing the declared map of a site superimposed with the IKONOS image. The site map has been georeferenced with GCP taken from the image, and then its

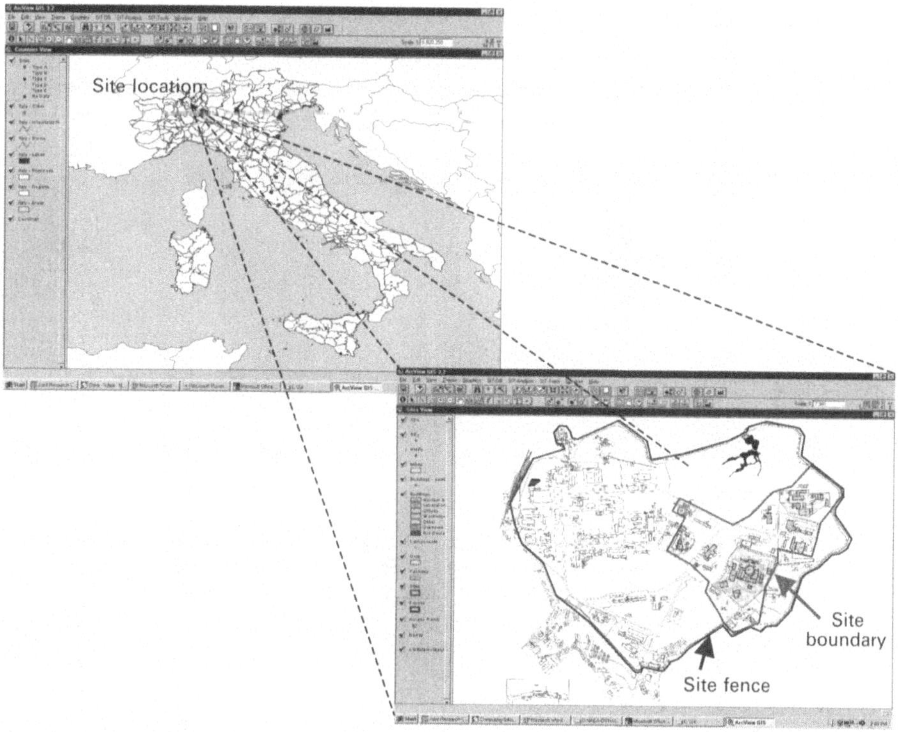

Figure 8.17. Screen shots of SIT showing the location of a site and its vector thematic maps superimposed to the scanned and scaled raster site map.

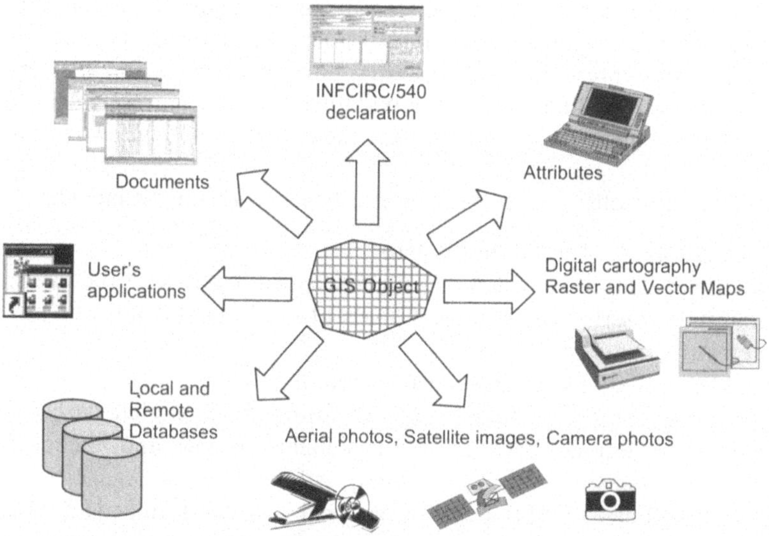

Figure 8.18. Schematic representation of information that can be associated to any graphical object.

Figure 8.19. Example of verification of a site map with a high-resolution satellite image.

background colour has been made transparent. In this figure the declared buildings are represented with semi-transparent polygons. Mismatches between the map and the image are represented with coloured polygons (yellow for unclear objects, red for undeclared buildings, green for buildings that are represented in the map but not in the satellite image, and blue for those that are in the image but not in the map).

The SIT system offers to the user several benefits. Examples include:

- representation of information on maps;
- linking the declarations to maps, with consequent queries on the maps;
- possibility to describe any graphical object at any level of detail with any type of document;
- easy access to data stored into local/remote databases;
- a lower probability of mistakes or of submitting an incomplete declaration (due to the cross-checking between the topological model and the declaration) resulting in a reduction of clarification questions from the Agency;
- total compatibility of SIT with the protocol reporter format [14], means that declarations made with the latter can be imported and associated to the topological model;
- declarations made with SIT can be automatically imported for verification,

including maps, themes and other associated documents, thus avoiding the digitalisation phase;

- since the whole declaration is in digital form it can be submitted on CD or via the internet using appropriate data security techniques;
- possibility to perform spatial analysis in addition to the conventional data analysis;
- representation of the results of data analysis in graphical form on maps or as charts;
- easy and fast procedure to georeference raster and vector maps;
- comparison of satellite images with declared site layout maps;
- comparison of images/maps taken at different times to detect changes.

The open architecture of SIT offers the possibility to easily add new modules for data and image analysis to suit further user needs.

8.4 REFERENCES

Bugayevskiy, M., Snyder, P. (1995) *Map Projection: a Reference Manual*, Taylor & Francis, London.

Chitumbo, K. and Nicholas, M. (2000) *The Role of Diverse Sources of Information in Safeguards*. Proceedings of the ESARDA seminar on Strengthening of Safeguards: integrating the new and the old. Dresden 2000, EUR Report 19587.

Heywood, I., Cornelius, S. and Carver, S. (1998) *An Introduction to Geographical Information Systems*, Addison Wesley Longman.

IAEA (1972) *The Structure and Content of Agreements between the Agency and States required in connection with the Treaty on the Non-Proliferation of Nuclear Weapons*, INFCIRC 153, June.

IAEA (1997a) *Model Protocol Additional to the Agreement(s) between State(s) and the International Atomic Energy Agency for the Application of Safeguards*, INFCIRC/540, Vienna.

IAEA (1997b). Department of Safeguards, *Guidelines and Format for Preparation and Submission of Declarations Pursuant to Art. 2&3 of the Model Protocol Additional to Safeguards Agreements*, Vienna.

Maguire, M. F., Goodchild, D. W. and Rhind, D. J. (eds) (1991). *Geographical Information Systems*, Longman, Harlow.

Maling, D. H. (1992). *Coordinate Systems and Map Projections*, Pergamon Press, Oxford.

Moik, J. G. (1980) *Digital Processing of Remotely Sensed Images*, NASA SP 431, Washington, D.C.

Pearson, II, F. (1990) *Map Projections: Theory and Applications*, CRC Press, Florida.

Poucet, A. (1998) *Integration of Information in the Content of the Strengthening of the Safeguards System*. ESARDA Seminar on Modern Verification Regimes. Helsinki 12-14 May 1998.

Poucet, A. and Contini, S. (2000) *Use of Geographical Information Systems and Satellite Images for Safeguards*. Proceedings of the ESARDA Seminar on Strengthening of Safeguards: integrating the new and the old. Dresden 2000. EUR Report 19587.

Poucet, A., Contini, S., Bellezza, F. (2001) *A GIS-based System for Declaration and Verfication*. Symposium on International Safeguards: verification and nuclear material security. Vienna, 2001.

Tomlin, C. Dana (1990) *Geographic Information Systems and Cartographic Modeling*, Prentice Hall, Englewood Cliff, New Jersey.

9

Satellite imagery and GIS: the Canadian experience

Q. S. Bob Truong and C. Vincent Tao

In the handling of spatial data and information, GIS technology has proven to be a powerful tool. As a foundation technology that is an integral part of mainstream IT, GIS has been used extensively in non-safeguards areas and has recently attracted considerable interest in the international safeguards community. This chapter presents the Canadian experience with the application of GIS to facilitate management, analysis, and reporting of various safeguards data including satellite imagery; the emphasis is on the integration of GIS and remote sensing technologies from the Canadian perspective.

9.1 INTRODUCTION

A collection of information technology, data, and procedures for collecting, storing, manipulating, analysing, and presenting maps and spatial information about features on earth, GIS is becoming common in many different organisations that use geographic information and maps to support their activities.

Organisations around the world have embarked on the development of GIS systems and they include local and state government, national and international agencies, transportation companies, marketing firms – just to name a few. GIS systems are becoming integration platforms for various sets of corporate database information and GIS programs are used to incorporate maps with demographics and various other types of numerical information usually figured in tabular form. People are combining the information from the maps and numbers in different ways in order to make sense of 'numbers' and conducting 'what if' analyses. Many relationships may also be discovered that were not apparent when the information was only in tabular form. The increasing power of software and hardware systems as well as the increasing availability of spatial data (maps and remote sensing images) further stimulate the acceptance of GIS technology in organisations.

In early 1998, the CNSC (formerly the Atomic Energy Control Board), in conjunction with the University of Calgary and the GeoTango International Corporation launched a collaborative research project to investigate the potential applications of GIS technology to safeguards; this being an early attempt to perform research and development on geospatially enabled processing of safeguards information using GIS. The research was successfully carried out.

A prototype system, GeoATOMS, has been developed. GeoATOMS is a GIS-based component system that is designed to manage, analyse and report nuclear material information. GeoATOMS has been demonstrated many times in the above conferences and workshops, and has been presented to several IAEA members. A GIS-based system, Safeguards Imagery Archiving and Management System (SIAM), has also been developed to manage, archive and browse large amounts of satellite imagery data. The CNSC's efforts in developing GIS technology for safeguards applications have been well recognised. Due to the significant efforts by the Canadian groups as well as other groups at both member state and Agency levels, awareness of GIS technology and its use for safeguards applications has been greatly improved in the nuclear safeguards community.

9.2 INTEGRATION OF GIS AND REMOTE SENSING

Major developments in the integration of GIS and remote sensing are now taking place. In many areas the mutual needs are evident. This convergence is not unexpected, since as the demand for spatial data grows so GIS practitioners are looking at remote sensing technology as a means to collect the required information. As the satellite imagery holding becomes substantial, it would be appropriate to use tools such as GIS to manage the information holdings. Satellite imagery is just another source of data and hence can readily be integrated with non-imagery data in a GIS, which can also incorporate environmental analysis such as consequence analysis, and 3-D simulation using remote sensing technology for creating DEMs and 'fly-throughs'.

9.2.1 GIS fundamentals

During the course of our investigation of potential applications of GIS in safeguards it is realised that more than 50% of the data required by the Additional Protocol, INFCIRC540, is geographically related. Information such as site maps and diagrams, photos, plans, drawings, and satellite imagery is extremely important for safeguards reporting and data management.

A GIS stores information about the world as a collection of *layers* that can be linked together by geography or geospatial coordinates. This simple but extremely powerful concept has proven valuable for solving many real-world problems (Figure 9.1). The layer approach allows us to organise the complexity of the real

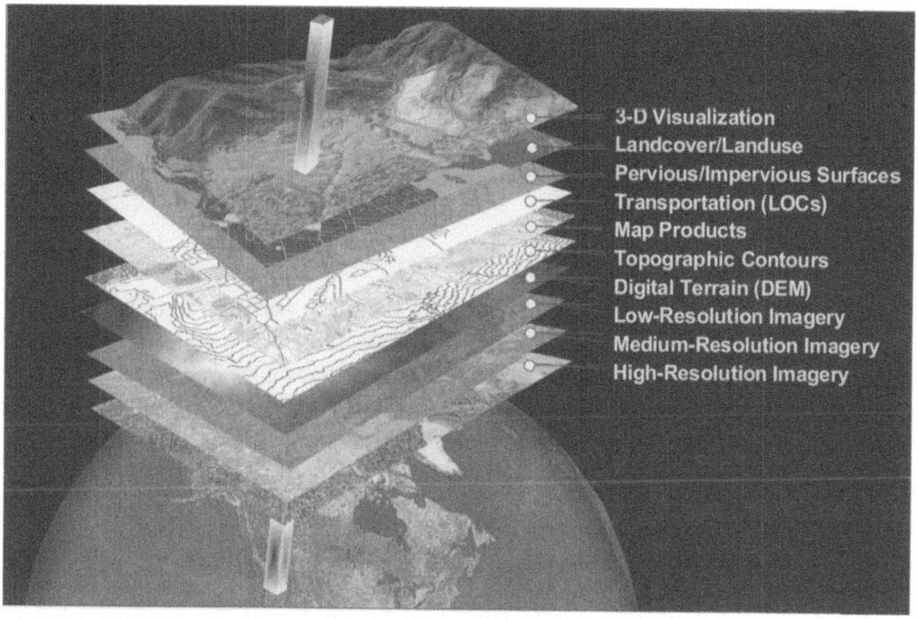

Figure 9.1. GIS layers [courtesy of Space Imaging Inc].

world into a simple representation to help facilitate our understanding of natural relationships. The basic data types in a GIS are *spatial data*, and *attribute data* (or *non-spatial* data). The spatial data are encoded as either *vector data* or *raster data*. Image data and DEM data are often considered as raster due to the similar data characteristics and data handling methodology.

Vector data encodes spatial features by using geometric elements (points, lines and polygons). Each geometric element is defined using one vertex or sequential *vertices* in some geospatial coordinate system (such as WGS84). The spatial relationship between vectors is represented using an explicit or implicit topological data structure, a key to the GIS functionality and performance. Many of the complex data analysis functions cannot effectively be undertaken without a topologic vector data structure.

Raster data models incorporate the use of a grid-cell data structure where the geographic area is divided into cells identified by row and column. The size of cells is selected on the basis of the data accuracy and the resolution required. A raster data structure is in fact a matrix where any coordinate can be quickly calculated if the origin point is known, and the size of the grid cells is known. Since grid-cells can be handled as two-dimensional arrays in a computer, encoding many analytical operations are easy to program. Image data is a type of raster. However, it is typically stored in a variety of *de facto* industry standard proprietary formats such as TIFF, GIF, JPG, etc. These often reflect the most popular image processing systems.

The selection of a particular data model, vector or raster, is dependent on the source and type of data, as well as the intended use of the data. Certain analytical procedures require raster data while others are better suited to vector data. For instance, the vector data model does not handle continuous data (e.g. elevation or gravity distribution) very well while the raster data model is more ideally suited for this type of analysis. Accordingly, the raster structure does not handle linear data analysis (e.g. shortest path) while vector does. For detailed information about the GIS data models, see Buckley (1997) and Bernhardsen (1999).

9.2.2 Complementarity between GIS and remote sensing

Over the next ten years the complexity of remotely sensed datasets will grow significantly through use of multi-sensor, multi-time, hyperspectral, and multi-view angle approaches. As the demand for spatial information grows there is an ever-increasing synergy between remote sensing and GIS. There are three main ways in which remote sensing and GIS technologies are complementary to each other (Wilkinson, 1996):

1 Remote sensing can be used as a tool to gather datasets for use in GIS.
2 GIS datasets can be used as ancillary information to improve products derived from remote sensing.
3 Remote sensing data and GIS data can be used together in environmental modelling and analysis.

There is a close interrelationship between these three types of complementarity. Both the academic research and commercial development communities have realised that the synergy of these two technologies can result in much more powerful tools and methods for effective spatial data handling, analysis and mining. Figure 9.2 shows an example of image-based GIS investigation and analysis of a nuclear facility site (Tao *et al.*, 1998), where vector data such as roads, buildings and waste areas were overlaid on a satellite image to facilitate the investigation. In the following sections, we will present an overview of some of work performed by the Canadian groups on GIS and its integration with remote sensing technology.

9.3 GEO-ATOMS: A HYBRID DATA MANAGEMENT SYSTEM

9.3.1 Spatially enabled information management: a GIS-based approach

Canada has been involved in nuclear activities for 50 years including nuclear generation stations, research reactor licenses to universities, nuclear research and test establishments, active uranium mines, fuel fabrication plants, and waste management licenses. Today, Canada has the richest uranium deposits in the world and is the world's largest producer and exporter of uranium. Nuclear activity is distributed across the country. It has been a challenging task to manage vast amounts of safeguards data sets in an efficient manner. They are often heterogeneous in terms of formats and sources, can be reports, drafting papers, maps, tables, charts, figures, and even video clips. However, these data sets are extremely

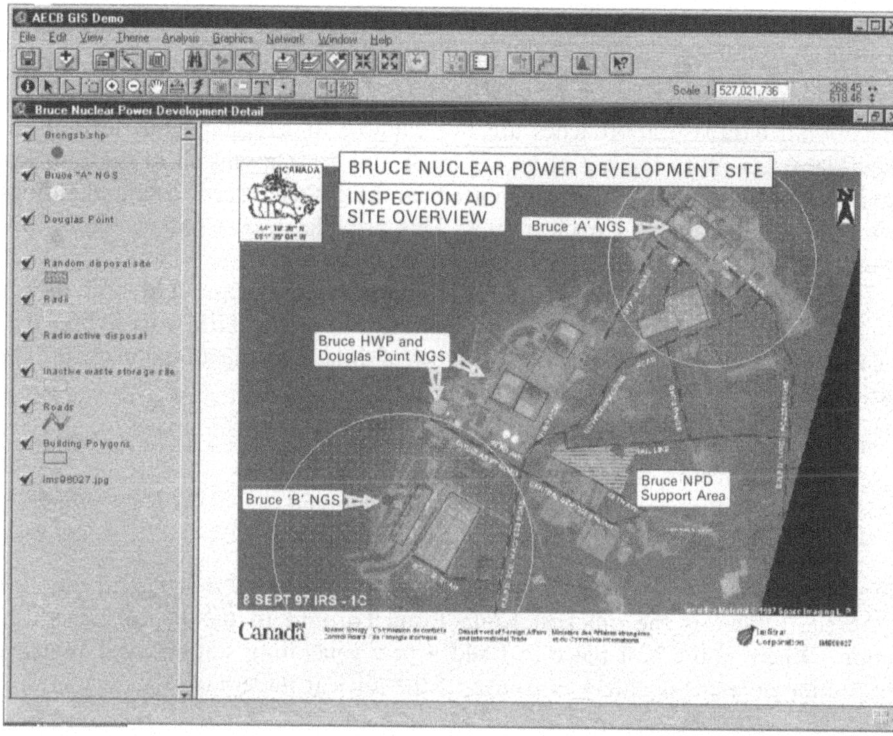

Figure 9.2. Image-based GIS investigation and analysis of a nuclear facility site.

valuable for safeguards management and operations and must be managed and achieved appropriately.

The CNSC is an agency that manages the use of nuclear materials in Canada. It exercises control through a regulatory system that establishes the health, safety, security and environmental standards for uses of nuclear energy. With the growing nuclear activities in Canada, the volume of reports and the declarations of materials movements and stocks variations are dramatically increased. The current system becomes outdated and unable to manage the vast amount of data and various types of data. CNSC is now investigating the development of a new system. There are several requirements to address. The system should:

- be capable of managing various types of multimedia data, such as textual data, maps, images, video clips as well as reports and documents;
- be capable of managing these data sets in an integrated manner so that a sophisticated query can be developed to retrieve data efficiently;
- be capable of handling both vector based maps and raster based remotely sensed images; and

- be capable of viewing, editing and query of data through the internet so that inspectors can use the system during on-site inspections.

It is realised that 70%–80% of the information in the system is spatially related. The conventional database approaches are not capable of handling both textual and spatial data sets. With the use of vector maps and remote sensing images, nuclear materials stored, processed and transported can be georeferenced to spatial entities, such as buildings, roads or nuclear sites. This provides a transparent control and monitoring of the movements of nuclear materials.

In order to link the spatial and non-spatial components, a hybrid GIS is required. The key to the design and implementation of this GIS is its hybrid data management model that handles various types of data in an integrated manner. This hybrid data model should be able to support:

- integrated spatial and non-spatial data query;
- handling of vector, raster and multi-media data types;
- integrated spatial analysis and decision making; and
- visualised presentation and rapid map generation.

For example, it can answer questions like: Where are the closest emergency services to the site? What is the quickest route from Toronto to the Bruce Generating Station? Where is the best place to build a new generating station? and Which is best routine in terms of safety to transport the nuclear materials?

9.3.2 Design and implementation of GeoATOMS

The prototype system mentioned earlier in this chapter (see 9.1), GeoATOMS – Geospatial Atomic Tracking and Organisation Manager System, has been developed for organising and tracking vast amounts of nuclear materials (Tao *et al.*, 1999 and 2000). Multimedia information such as imagery, various types of reports and documents as well as video clips can be managed integrally. The prototype was implemented using the component-ware technology in which software ActiveX controls, ESRI MapObjects and Microsoft Jet Engine were used. An example of the system interface and some functions is shown in Figure 9.3. The user is able to edit and query the sites, facilities and equipment status through the GUI. In the example, the user clicked a building at a nuclear site. The system brought up a big window showing this building along with all the equipment information related to this building. The link to all the documents related to this building is also available. The seamless link between the GIS graphic engine and the database has been constructed in GeoATOMS.

Organisation of spatial data is crucial to maintaining an efficient program when querying for spatial data. A new hyperspatial media data model was built in GeoATOMS. The advantage of this model is its hierarchical structure and scale-dependent data cataloging. There are three different types of spatial data in the system. A small-scale data set of Canada including themes such as provinces,

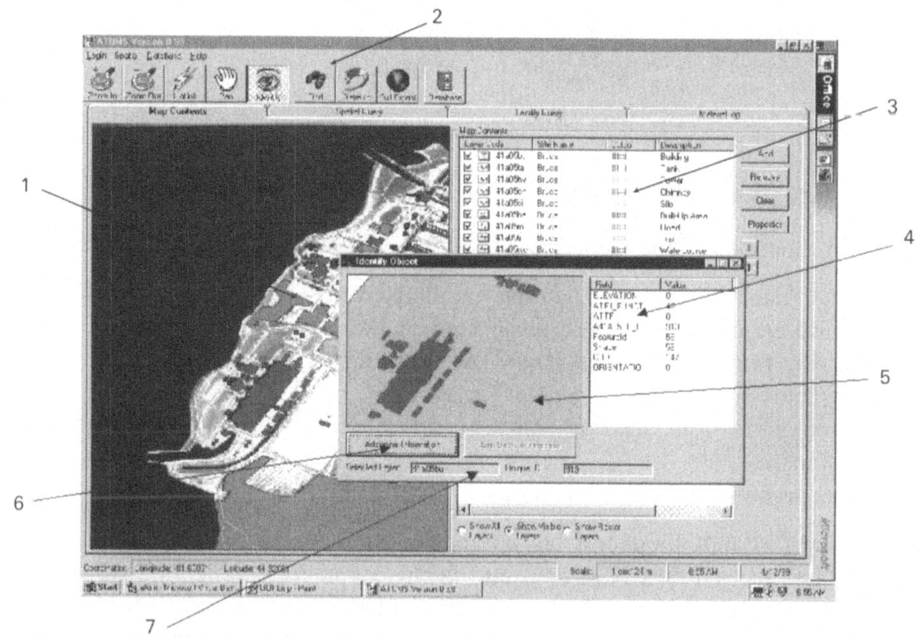

Figure 9.3. GeoATOMS system interface (1 – map window, 2 – toolbar, 3 – data control, 4 – facility information, 5 – query facility, 6 – links to documents, 7 – facility ID).

cities, lakes and primary road networks was generated for the initial map display and indexing. As the purpose of the system is to manage nuclear facilities within the main sites, large-scale data sets of the areas that are administered by the CNSC were populated into the system. These data sets come from the Canadian National Topographic Data Base (NTDB). In total, 39 layers at scales of 1:50,000 and 1:250,000 pertaining to the CNSC administered sites were available. To assist in the monitoring of nuclear facilities and to detect any potentially important changes, high-resolution aerial photos and satellite imagery of the selected sites were also acquired, geometrically rectified and integrated to its related layers. The non-spatial data in GeoATOMS is hierarchical as well: member states, sites and facilities. Within the member states, there are sites. Facilities exist at these sites. Materials, equipment and documents exist in facilities. The site, equipment and facilities tables will have tables that will track the changes to these tables, while the materials table will log each time the material is traveling between facilities.

GeoATOMS has been implemented using component concept to ensure its extensibility and flexibility. The system is composed of many 'plug and play' software components, such as database component, spatial component, visual graphics component, spatial/non-spatial data link component, query component, data management and data analysis components, etc. All of these components work independently of each other, and each of these components can be used in another program without any complications. ESRI MapObjects was used as a development toolkit in GeoATOMS.

9.3.3 Applications of GeoATOMS

The main functions of the GeoATOMS can be grouped into four categories:

1 Spatial functions such as zooming, panning, digitising, identity, spatial finder, distance measure, Properties, Layer control, etc.:
2 Querying functions such as spatial query, facility query, database query, SQL query, etc.:
3 Database functions such as viewing, editing and logging, as well as security functions:
4 Image archiving and manipulations.

Due to the fact that there will be different levels of users using the system (i.e., international inspectors, CNSC staff and facility managers) three levels of security controls have been developed. Users with different security levels are only allowed to access certain functions. GeoATOMS can be used for the following applications:

Integration with imagery for inspection

GeoATOMS supports the incorporation of satellite imagery with vector layers. This helps to determine whether an inspection is escalating. Furthermore, there is a possibility of digitising spatial entities within the system.

Linking the Protocol Reporter and other databases

The Protocol Reporter is another database that requires a GIS link. This system has a capability to link other database structures. It allows spatially enabled data reporting and presentation.

Navigation tool for inspectors

The system can be used for inspectors to create a plan of navigation, even when the inspector is not familiar with the site. It helps to determine the inspection route as well.

Nuclear material accounting and tracking

GeoATOMS can be used as a front-end viewer to extract information from accounting databases and perform visualised tracking analysis of nuclear material.

Figure 9.4 shows the function of the spatial hyperlink. Once a site is clicked by a user, a detailed site map will be displayed and the relevant information about this site can be queried and accessed. Figure 9.5 shows the tracking record of nuclear

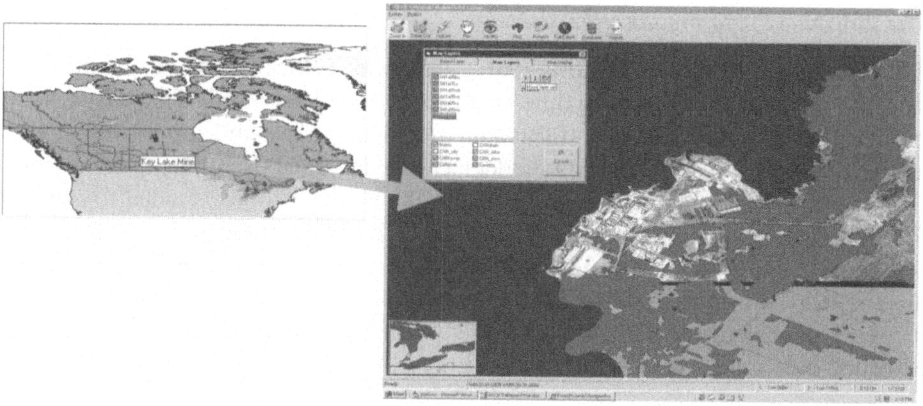

Figure 9.4. Hyperspatial data link for information query.

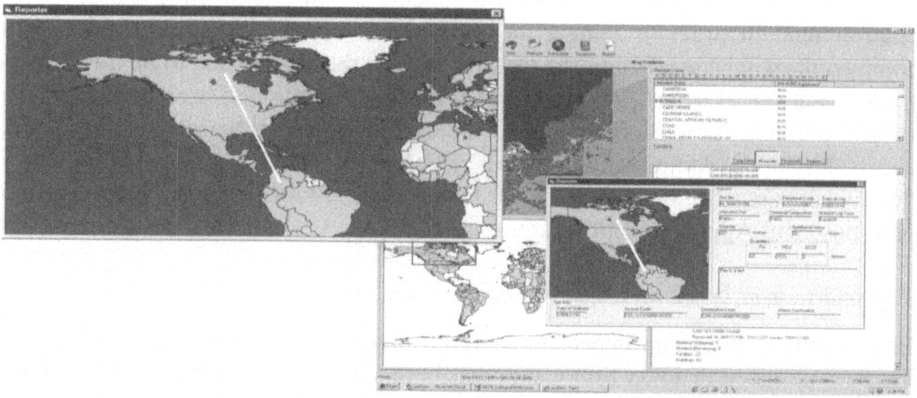

Figure 9.5. Dynamic tracking of nuclear materials in GeoATOMS.

materials. The system was built using a dynamic data model and therefore all the historical data can be kept in the system for material tracking and accounting.

9.3.4 Web-based GeoATOMS

It is a natural extension to make GeoATOMS accessible via the internet. There are many advantages on offering an internet-based system. The most important one is that both managers and inspectors will be able to query the information and monitor/track the status of the nuclear materials via the internet. It enables an efficient field-to-office solution to managing nuclear materials. Although the data model and the system architecture have been developed for GeoATOMS, they cannot be employed directly to the internet environment. The emphasis of the project is placed on restructuring the system architecture.

The distributed data organisation strategy

The data model of GeoATOMS is spatial-temporal. It means that all the information entered into the system is time-stamped and georeferenced. This design is of particular importance in terms of tracking nuclear materials and supporting of time and spatial-based query. The data model has been designed using the object orientation concept. With the internet environment, the model will be extended to account for issues that relate to *distributed transactions*. Therefore, information such as who is accessing the data or where the data is being accessed needs to be included in the data model. Transaction control is about support for concurrency. Most commercial database management systems offer tools to handle this issue. In the project, an extended data model has been investigated.

The client/server system architecture

GeoATOMS was developed using the component approach. It essentially consists of several ActiveX components. These components can be employed in a client/server environment. In order to achieve an optimised computation balance, some of the components will be located on the server site while the others need to be transformed to the client site. A smart client distributed GIS architecture has been developed. The current implementation utilised the Autodesk Mapguide software. With Mapguide, many developed components can be reusable or extensible since Mapguide is developed based on the COM/DCOM standards. Mapguide also provides users with powerful Application Programming Interface (API) for further customisation.

The smart client based user access model

Bandwidth has been a bottleneck issue to the internet-based applications. The system has been designed in a way that it supports both users with large bandwidth and ones with limited bandwidth. This system can be developed using the smart client architecture (Tao, 2000), allowing users to determine the physical locations (client site or server site) of the computing and the data resources.

Several examples of functions in the web-based GeoATOM are shown in Figure 9.6. The system can be accessed via the popular web browsers (e.g. Internet Explorer and Netscape).

9.4 SIAM: GIS-BASED IMAGERY ARCHIVING SYSTEM

9.4.1 SIAM system development

The increasing use of satellite imagery for safeguards demands a data management and archiving system. The development of SIAM, was motivated by two overriding concerns:

1 The need to archive and manage efficiently the enormous volumes of data generated by commercial high-resolution satellite imagery,

2 The major challenge of making such data known to and accessible by inspectors, analysts, and managers.

Better making use of these data sources depends very much on the capabilities of data and information systems to enable access to extensive data holdings, facilitate the analysis and stimulate exchange and transfer of data sets.

In 1999, GeoTango International were awarded a contract from the CNSC to develop SIAM, based on the commercial product, Carterra Analyst, developed by Space Imaging. Carterra Analyst is a hybrid system which integrates GIS, remote sensing, image processing, photogrammetry and cartography tools into a single workstation enabling the user to search, browse images and data from a database; perform multispectral analysis, create and combine vector and raster files; generate reports and products. The software incorporates third party tools, such as ESRI ArcView, ERDAS Imagine, Sensor Systems RemoteView, Netscape Navigator, Core TerraSoar, GDE SocetSet, Applix Graphics and Microsoft Office. The system is powerful yet is not designed for non-professionals. Moreover, the system does not have capabilities of image archiving as well as friendly tools for safeguards applications.

Significant efforts have been made to develop SIAM. An intelligent search model has been researched. This is the so-called content-based image query. This query technique is extremely complex in that it requires that several disparate image query techniques be combined. First, the feature query component requires the assessment of the feature similarities of regions. Second, the spatial query component requires the assessment of the similarities in spatial locations and sizes of regions. Third, the system requires the comparison of images consisting of multiple regions. Last, the system requires that the spatial relationships, such as 'above,' 'below,' 'near,' and so forth, be resolved. As depicted in Figure 9.7, the overall comparison of images utilises both the feature and spatial attributes of the regions in computing their similarity (Smith, 1997).

9.4.2 SIAM system functions

SIAM supports three groups of functionalities that are most popular to safeguards users. For detailed description of the SIAM system, refer to the SIAM user manual (GeoTango, 1999).

Image metadata management

This group of functions was developed based on the ESRI ArcView GIS module as well as Microsoft Access database engine. It allows users to archive images based on

- image type (aerial photos, satellite images);
- imaging information (spatial resolution, spectrum resolution);
- temporal information (date acquired);
- geographical information (coordinates, region or country information);

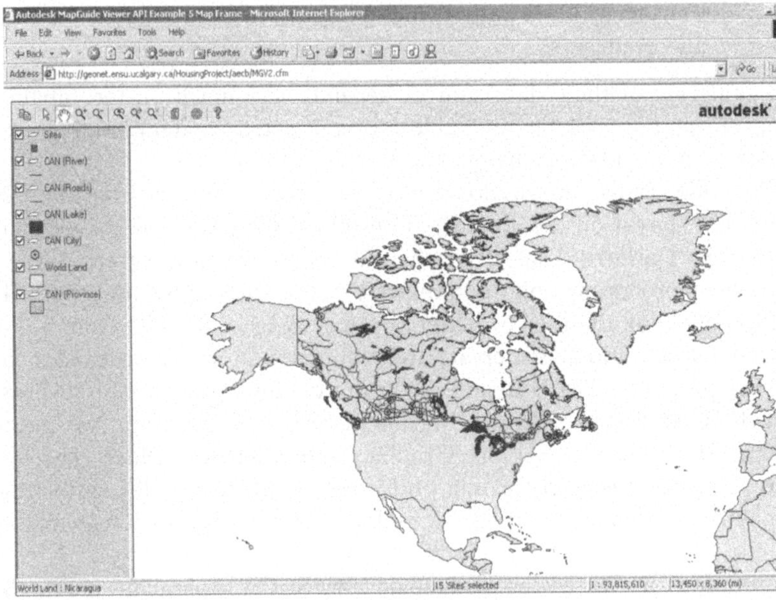

(a)

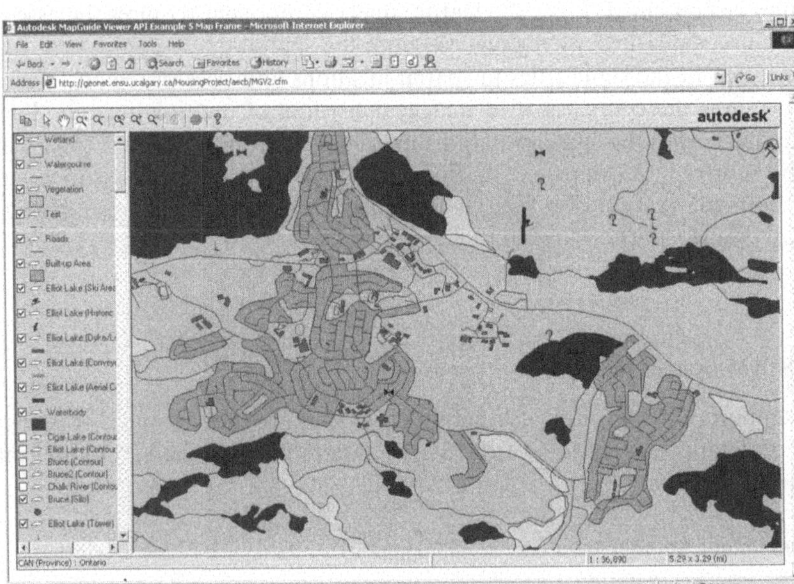

(b)

Figure 9.6. Functionality of GeoATOMS online (web-based access and retrieval): (a) Overview of Canadian nuclear activities; (b) information retrieval of a Canadian nuclear site; (c) display and visualisation of nuclear facilities at that site; (d) detailed query of information at the building level.

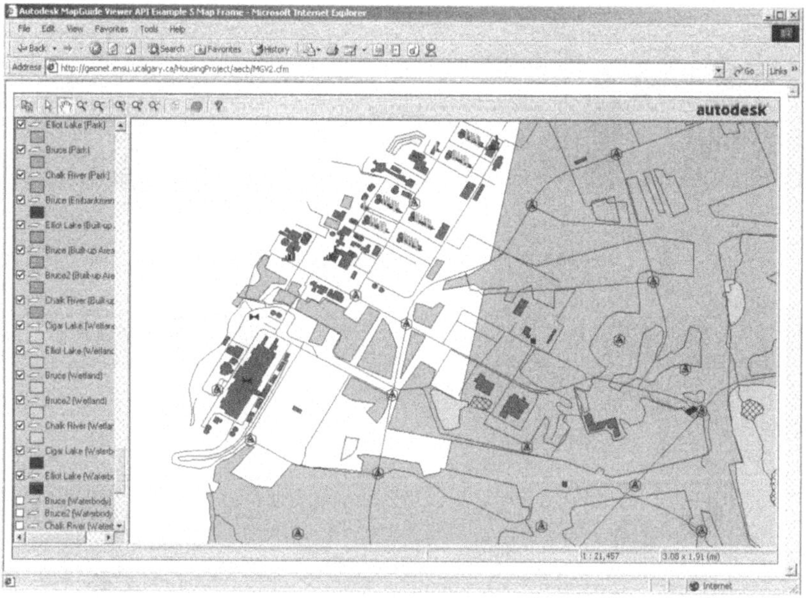
Autodesk MapGuide Viewer API Example 5 Map Frame - Microsoft Internet Explorer
File Edit View Favorites Tools Help
Back Search Favorites History
Address http://geonet.ensu.ucalgary.ca/HousingProject/aecb/MGV2.cfm
autodesk
Elliot Lake (Park)
Bruce (Park)
Chalk River (Park)
1 : 21,457
Internet

(c)

Autodesk MapGuide Viewer API Example 5 Map Frame - Microsoft Internet Explorer
File Edit View Favorites Tools Help
Address http://geonet.ensu.ucalgary.ca/HousingProject/aecb/MGV2.cfm
autodesk
123
Internet

(d)

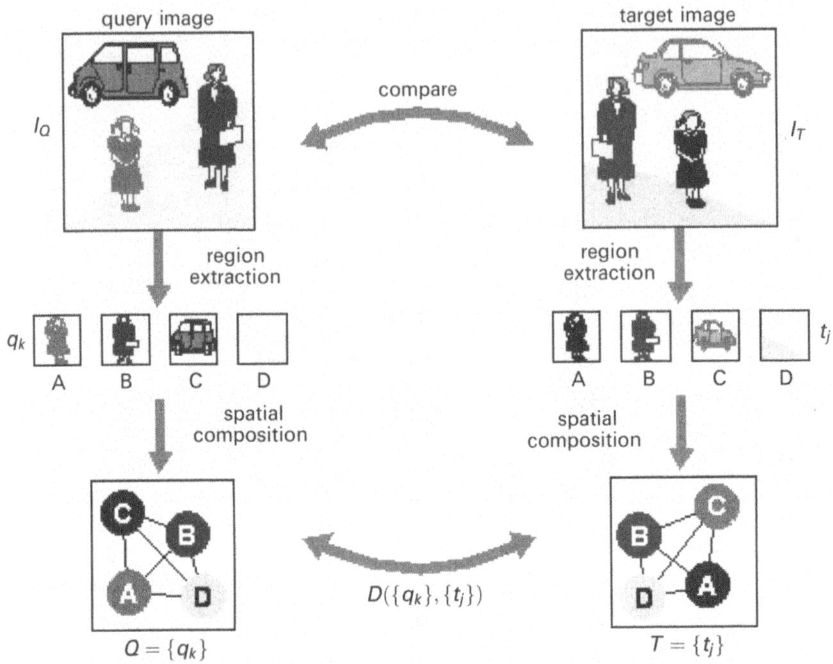

Figure 9.7. Integrated spatial and feature query. Images are compared by comparing the spatial arrangements of regions.

- other information (source, vendor); and
- general content description (contents, comments).

There are issues regarding the metadata standard. ISO (International Standard Organisation) is now working with the imaging industry to develop an internationally recognised metadata standard to facilitate data exchange and transfer. The image metadata management component in the SIAM has been designed in a way that it can be extended to support any standards proposed by the standard committees.

Map-based query and index

Use of map for query is probably the most frequent tool users and analysts apply. It simplifies the search and query process, as it is intuitive and a very user-friendly way for non-professionals. Currently, SIAM supports the following query by:

- geographical information (coordinates, country or regions);
- direct map manipulation (drill down and zoom in);
- image type;
- image content;
- image date;

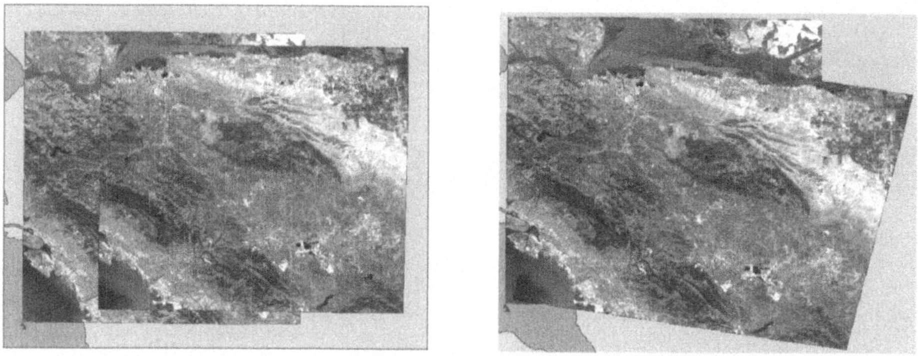

(a) (b)

Figure 9.8. Image registration: (left) images not lined up; (right) rectified image.

- image temporal information; and
- input polygon.

Image browsing, annotation and analysis

One of main advantages of SIAM is its full capabilities to integrate various tools, GIS or image processing, remote sensing analysis into a single interface. Therefore, users would not need to switch different systems to perform their work. SIAM supports functions in:

- image manipulation;
- image registration (image to image) and image rectification (image to map);
- change detection;
- feature annotation; and
- image format transformation.

An image registration example in SIAM, where one image is rectified with respect to anther image (maser image) to form a seamless mosaic is shown in Figure 9.8. In Figure 9.9 an example is shown on change detection in SIAM; the difference image was generated by a change detection algorithm (right).

9.5 GIS GENERATION USING SATELLITE IMAGERY

Mapping from space-based sensor systems is one of the main uses of remote sensing. The data sets produced can be used as additional layers in a GIS. The core components of mapping from satellites are photogrammetric reconstruction, classification and spatial generation. It is a fact that the majority of small-scale maps and DEMs were generated using aerial and satellite photogrammetry. Photogrammetry is a technology that utilises stereovision to generate 3-D information from stereo

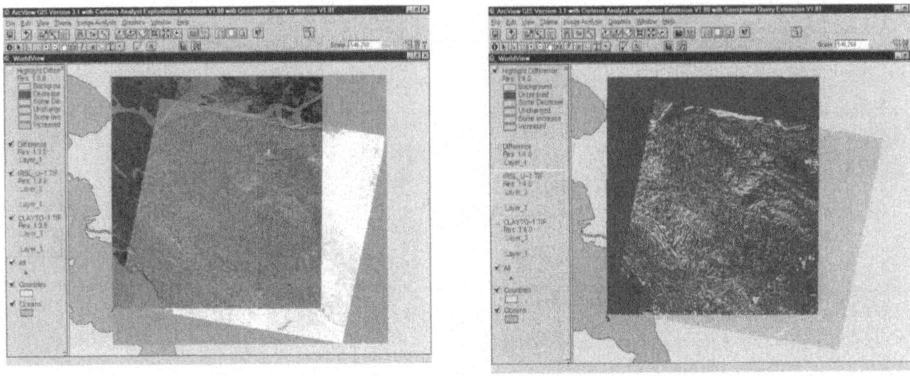

(a) (b)

Figure 9.9. Change detection: (left) two raw images; (right) difference image after the change detection applied.

Figure 9.10. A pair of stereo IKONOS images of a nuclear site.

image pairs. In principle, any ground objects that are visible from images taken at different positions can be reconstructed in 3-D. Figure 9.10 shows a pair of stereo IKONOS images covering a nuclear facility site. One can tell that these two images are taken at different positions by examining the directions of the shadows. This pair of images can be used to extract 3-D coordinate information of objects such as footprints and heights of buildings and tanks as well as DEMs (Tao and Hu, 2001). GeoTango International is working with CNSC on the investigation of the capabilities of IKONOS stereo reconstruction.

Figure 9.11. Automatic road extraction (black lines in the right image) from IKONOS imagery (courtesy of CCRS).

High-resolution imagery from satellites such as IKONOS, QuickBird and EROS can provide detailed representations of discrete objects such as roads, buildings, and large vehicles. There have been many research activities on developing automatic ways to extract features by using various spatial/spectral/contextual attribute combinations. the CCRS has developed a framework for accessing the feasibility of integrated feature extraction and recognition (Guindon, 2000). In this approach, images are first segmented into homogenous regions where most segments constitute part or whole examples of thematic objects. A series of recognition modules are then sequentially executed to label a subset of the image segments, thereby generating a partial interpretation of the scene that includes only those objects classed of interest. Labelling decisions are based on accumulated evidence. Contextual cues, such as logical extension and perceptual grouping, have been found to play pivotal roles in road tracking and the recognition of residential housing units respectively. Figure 9.11 shows partial results of road extraction from IKONOS imagery.

Image draping on a DEM is the most efficient and convincing method of introducing details into a landscape. 'Fly-throughs' is essentially generated by rendering draped images from a variety of viewpoints. This is an extremely powerful tool for investigation of nuclear sites and unfamiliar regions. Many GIS applications provide draping capabilities. Most include rectification procedures to properly orient and align a digital image to the ground surface. Simple applications utilise orthophoto images that have already been registered to the ground surface and corrected for elevation, or relief, displacement. There are three ingredients required to perform satellite image draping:

- a digital terrain model (DTM);
- a satellite image of the same area as the DTM; and on the DTM
- at least two, preferably three, coordination points that can be precisely located in the image.

Figure 9.12. An IKONOS image draped on a DE.

Figure 9.12 shows an example of IKONOS image draped on a DEM generated by Intermap Technologies Ltd, Canada.

9.6 CONCLUDING REMARKS

The rapid developments now being made in remote sensing have many implications for the integration of satellite imagery and GIS data. The new remote sensing datasets, especially with many spectral channels and with frequent repeat coverage, will contain large numbers of spatial coincident data layers for GIS. Utilisation and exploration of such layers with pre-existing GIS data would offer more powerful outcomes.

Geospatial technology including GIS and remote sensing is playing a significant role in the nuclear safeguards community. It has been demonstrated that GeoATOMS and SIAM offer a unique solution on spatially enabled data management, analysis and archiving. The common characteristics of these tools are that they both support vector and raster data types, and both GIS and imagery data. With the increasing availability of GIS and image data, and with the improved visibility of this technology, such systems can be further developed to provide more advanced GIS and remote sensing capabilities and to perform more effective and efficient analysis, management and decision making.

Besides improving our ability to acquire spatial data, it should not be overlooked that most data sets captured from space have a temporal aspect. Therefore, it requires that a GIS model be capable of handling time and temporal changes in the data. This trend needs to be appreciated by GIS developers and users (see Wilkinson).

9.7 ACKNOWLEDGMENTS

The authors would like to thank Jon Ruttle, Rafael Lucero and Suen Lee for their contribution to the system development and programming of GeoATOMS and

SIAM systems. Technical support and advice provided by Robert Benjamin and Richard Keeffe of the CNSC are gratefully appreciated.

9.8 REFERENCES

Bernhardsen T. (1999) *Geographic Information Systems: An Introduction*, 2nd edition, John Wiley & Sons, Chichester.

Buckley, D. J. (1977) The GIS Primer. http://www.innovativegis.com/education/primer/primer.html

GeoTango International Corp. (1999.) User Manual of SIAM – Safeguards Image Archiving and Management System.

Guindon, B. (2000) A framework for the development and assessment of object recognition modules for high-resolution satellite images, *Canadian Journal of Remote Sensing*, Vol. 26, pp. 1–15.

International Atomic Energy Agency INFCIRC/540, 1997. Model Protocol Additional to the Agreement(s) between State(s) and the International Atomic Energy Agency for the Applications of Safeguards. Vienna, Austria.

Smith, J. R., (1997) Integrated spatial and feature image systems: retrieval, analysis and compression. PhD thesis, Columbia University, New York, February.

Tao, C. V., Truong, B. and Keeffe, R. (1998) GIS-based Nuclear Data Management and Safeguards Information Reporting, Proc. of INMM/ESRDA Workshop, Sept. 21–23, Albuquerque, USA.

Tao, C. V. *et al.*, (1999) GIS-Based Nuclear Data Management and Safeguards Information Reporting, proceedings of 40th Annual Meeting of the Institute of Nuclear Materials Management (CD-ROM), Phoenix, Arizona, USA, July 26–28, 1999.

Tao, C. V., (2000) Development of Internet-based GIService, Proceedings of GIS 2000 conference, Toronto, March 14–16, 2000.

Tao, C. V. and Hu, Y. (2001) The Rational Function Model: A Tool for Processing High-Resolution Imagery, Earth Observation Magazine (EOM), Vol. 8, No. 1, pp 13–16, 2001

Wilkinson, G. G., 1996. A review of current issues in the integration of GIS and remote sensing. *International Journal of Geographical Information Systems*, Vol. 10, No. 1, pp. 85–101.

10

Application of satellite imagery – a country case: Canada

Q. S. Bob Truong, R. Keeffe, P. Baines, J. P. Paquette

10.1 INTRODUCTION

The applications of satellite imagery have been widespread in non-safeguards domains (e.g., agriculture and forestry management, environment assessment, and urban planning) for at least 25 years. When the potential use of commercial satellite imagery was initially discussed in the international safeguards forum, several issues were identified. These included the imagery costs, the costs of setting up the infrastructure, the potential benefits, and legality of the use of this new tool. At the request of the IAEA, several member state support programmes were initiated to address the above issues. This chapter presents the results of the work carried out by CSSP in support of the IAEA with specific focus on the applications of satellite imagery.

During the past several years the CSSP in collaboration with its partners, notably Canada's Department of Foreign Affairs and International Trade, has undertaken a comprehensive project to investigate the potential applications of commercial satellite imagery for international safeguards. The main objective was to provide technical input to the IAEA internal discussion on this topic, with emphasis on the practical applications that could benefit IAEA inspectors. Significant results have been presented at various international conferences and symposia and published in various reports and papers; see for example Truong et al., 1998 and 1999. Complete details of this country case are contained in a report to the IAEA, CSSP-122 (as referenced in Section 10.7). Some of these results have been used in recommendations to management to proceed with the use of satellite imagery for safeguards by IAEA staff.

The Canadian work made use of commercial satellite imagery from various sensors including panchromatic, multispectral, and SAR, with resolution ranging from 120 m to 1 m. The aims of the study were to investigate, using concrete examples, the following tasks:

- evaluation of imagery spatial resolution;
- comparison between multispectral and panchromatic imagery;
- comparison between SAR and panchromatic imagery;
- monitoring of reactor operation using thermal infrared imagery;
- evaluation of a member state declaration;
- inspection aid;
- change detection; and
- detecting undeclared activities using satellite imagery and other open-source information.

The results of the study are discussed in the following sections, starting with the sensors and sites used in the study to be followed by the general findings in Sections 10.2 and 10.3. The case studies are discussed in detail in Section 10.4. Some conclusions are presented in Section 10.5.

10.2 IMAGERY AND SITES USED IN THIS STUDY

There are two SAR and several electro-optical/multispectral commercial satellites available today with spatial resolution ranging from 30 m (120 m for infrared) to 1 m. The sensors provide imagery of sufficient quality to provide accurate assessment of infrastructure change detection and activity monitoring. By the year 2002, there will be additional imaging systems deployed, with accuracy of 5 m or better. Four of these will provide 1 m resolution imagery.

For brevity, only a small selection of imagery is presented in this chapter. More detailed information is contained in the CSSP report for the IAEA, and Baines 1998. It should be noted that, due to printing degradation, the graphics presented in this chapter are not as clear as the original imagery. In the case of multispectral imagery, black and white reproduction of a colour image does significantly reduce the clarity of the images. The commercial systems used in this study included:

Landsat-5 and 7 TM	(USA)
SPOT -2 and 4	(France)
RADARSAT-1 SAR	(Canada)
IRS-1C/D Pan and LISS-3MS	(India)
KVR-1000	(Russia)
IKONOS	(USA)

As appropriate, high-resolution archived (1 m) and open skies aerial photographs (0.3 m) were also used to supplement satellite imagery. Imagery for major nuclear facilities in Canada were purchased. The majority of imagery and photographs, including an IKONOS 1 m stereo pair, were for the Bruce Nuclear Power Development site. This is a large Canadian nuclear facility, where the multi-unit reactors at Bruce A and B, the decommissioned Douglas Point reactor, the Bruce Heavy Water plants and a spent fuel dry storage facility are located. Other images were for Gentilly-2, Point Lepreau, Pickering and Darlington reactors, the Chalk River Nuclear Laboratories, several uranium mines in the provinces of Ontario and Saskatchewan, a uranium refinery and a processing facility in Ontario, an abandoned uranium mine in a remote area of northern Saskatchewan, and a diamond mine in the Canadian Northwest Territories.

10.3 GENERAL FINDINGS

General findings are discussed in this section.

10.3.1 Costs

The average cost per image is about US$2,500 for electro-optical imagery. The current cost of SAR imagery is approximately US$3,500.

A high-quality personal computer with a large monitor and peripherals with remote sensing software applications are required to perform imagery analysis and produce a high-quality print-out of the results. The total cost of such a system is in the range US$10,000–20,000, depending on the configuration of the system. It is expected that the cost of imagery and computer hardware will continue to decrease due to market competition.

For projects involving a large collection of images, the costs for archiving images are additional to the above and should be considered in the overall budget.

10.3.2 Imagery resolution

Imagery of all resolution (ranging from 120 m to 1 m in this study) would be useful, depending on the application. High-resolution sensors are required for identifying the details, but it is more advantageous to use low-resolution sensors for locating a facility. In addition, the use of available aerial photographs, and archived satellite imagery help in reducing the cost of imagery acquisition.

Using Landsat-5 (30 m), SPOT-2 (10 m), IRS-1C (5 m), and simulated IKONOS (1 m) air photo imagery, a comparison of imagery resolution was carried out. At 30 m resolution, major features such as the four-unit Bruce A and Bruce B reactors, the single unit (decommissioned) Douglas Point reactor, and the Bruce Heavy Water plants can be discerned. At 10 m resolution large buildings and structures, such as the vacuum building, the four reactor buildings, powerhouse, and the switchyard, can be seen. Additionally, the 10 m resolution permits the interpreter to observe, with an acceptable degree of clarity, objects such as fuel storage tanks. At 5 m resolution, smaller objects, such as the water treatment buildings and the pump houses can be discerned. Figure 10.1 shows a comparison between the IRS-1C (5 m) and an airborne image simulating the quality of the 1m IKONOS sensor. Prior to the launch of IKONOS an aerial photograph was used to show the potential clarity

Figure 10.1. Imagery resolution comparison.

and the amount of detail this new sensor would provide. Even without magnification, objects as small as the standby generators can be clearly identified.

As discussed above, the initial work on resolution comparison was based on simulated IKONOS imagery using 1 m aerial photography. Since then we have acquired several IKONOS images and have found that our expectations based on aerial photography were below what has actually become available from the IKONOS system. The IKONOS images actually provide much better resolution quality, as shown in Figure 10.2 for the Bruce site.

10.3.3 SAR imagery

SAR imagery is useful for detecting small structures and roadways and for determining whether an electro-optical image has been altered. A good feature worth mentioning is that SAR sensors are capable of imaging during day/night and in all-weather conditions. SAR sensors are therefore a valuable tool for detecting sites located near the polar regions or in the areas which are often under cloud cover or affected by inclement weather, or when timeliness is of utmost concern.

Under some conditions, SAR imagery could be used to detect cooling water discharges, hence determining the operating status of the facility (as discussed in more detail in Section 10.4.2).

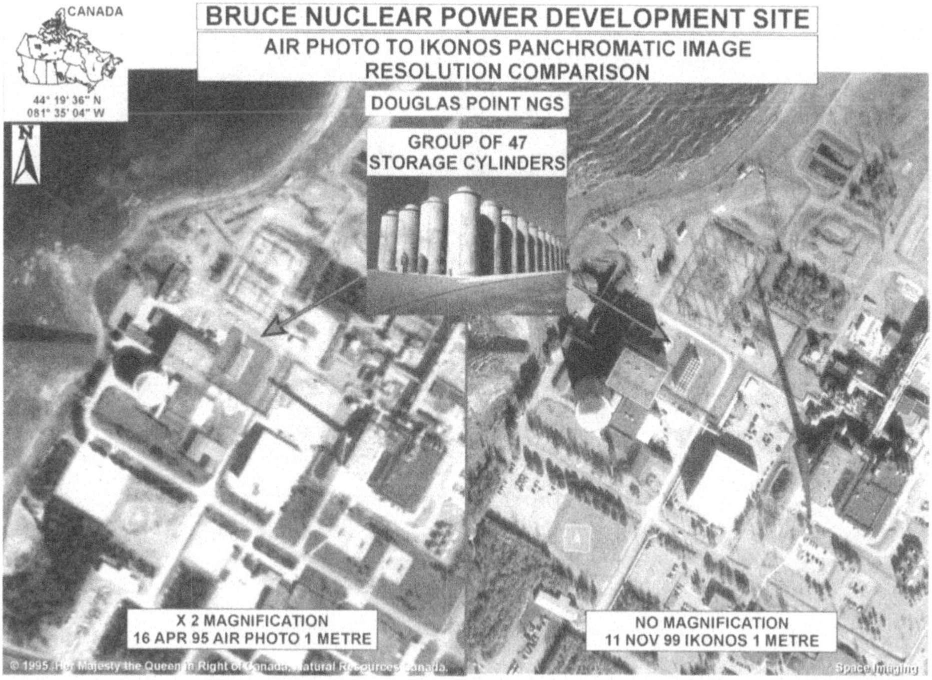

Figure 10.2. Comparison between air photo and IKONOS imagery.

Using the Bruce site, SAR imagery was analysed to investigate its potential capabilities in detecting structures and objects that have height and depth. In addition, RADARSAT-1 fine beam mode imagery (8 m) was compared with a digitally altered IRS-1C image for the Bruce site to demonstrate that SAR imagery could be used to verify the authenticity of panchromatic imagery. It should be noted that SAR imagery is the most difficult to alter and that a diversity of commercial sources deters image manipulation.

10.3.4 Stereo pair and advanced processing/analysis techniques

Useful information can be extracted from the imagery by using stereo pairs to give the perception of depth and height, or digital enhancement processing techniques.

In certain situations, advanced processing techniques have the potential for expanding the information extraction capability of SAR imagery over some types of nuclear facilities. The current RADARSAT-1 and ERS-2 sensor have spatial resolutions in the range of 8 m to 25 m. This level of resolution is not sufficient to identify details of small objects. SAR resolution enhancement has the potential to contribute in three basic areas of surveillance: object detection and false target discrimination, multiple object resolution, and object classification. This advanced

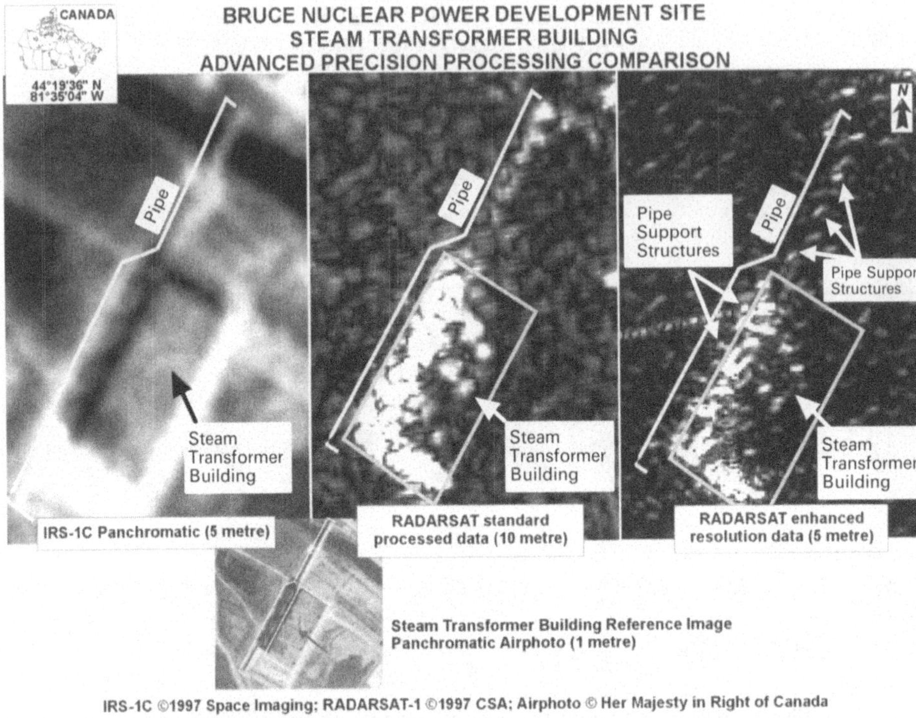

Figure 10.3. Comparison between advanced precision and standard processing techniques. The pipe adjacent to the steam transformer building is better delineated in the resolution-enhanced RADARSAT-1 image.

precision processing technique could provide information on the number of objects present as well as provide greater clues as to what the object might be (e.g. ten cylinders as opposed to a small building). Figure 10.3 illustrates how resolution enhancements of SAR data improve the identification of structures when compared with standard processing techniques.

10.4 CASE STUDIES

Several case studies were created using imagery of Canadian nuclear facilities to illustrate the potential applications of commercial imagery benefits to IAEA inspectors. These applications are listed below (and will be discussed in detail in Sections 10.4.1 to 10.4.6):

- Confirmation of member state declaration.
- Monitoring reactor operation.
- Detecting potential undeclared activities.
- Monitoring changes at a declared facility.

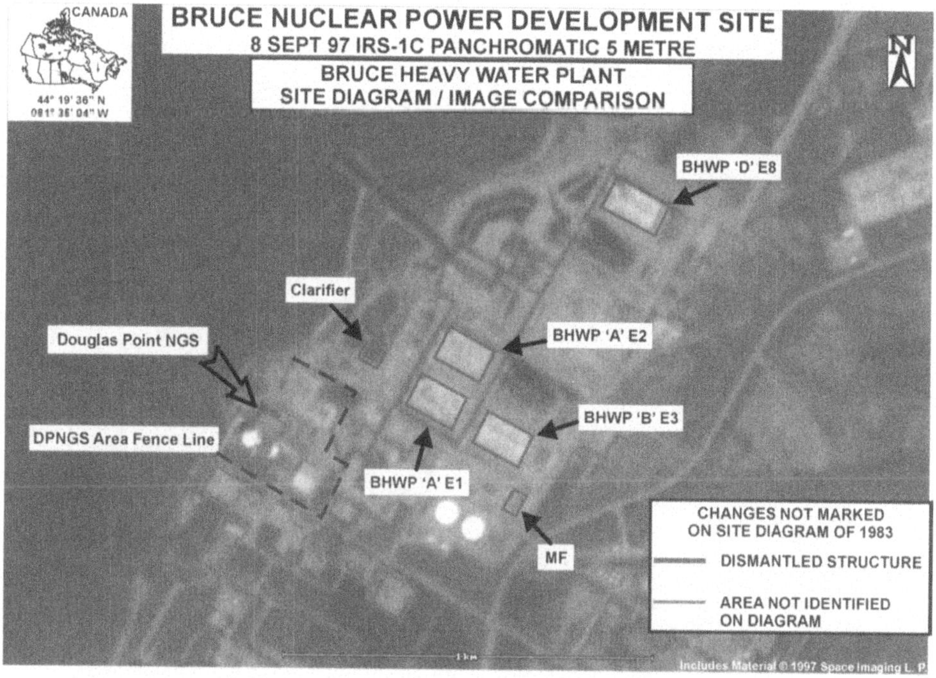

Figure 10.4. Confirmation of a member state declaration.

- Monitoring activity at a decommissioned facility; and
- Assisting in inspection planning and reporting.

10.4.1 Confirmation of member state declaration

A declaration submission often includes a site diagram. These diagrams may be out of date by the time they are printed and distributed. Also, they may not contain all relevant information such as the date, the geodetic coordinates, the distance scale, or the north arrow for orientation. A satellite image provides a date-stamped data set that includes the proper orientation and a measurement scaling capability, when processed with a geographic datum. The site line diagram can be superimposed on the satellite imagery to identify any anomalies between them.

IRS-1C imagery was compared with Bruce site diagrams, which were used as examples of the information, which might be submitted under a 'Member State Declaration'. It was evident from this comparison that numerous changes had taken place from several years ago when the site diagrams were prepared to the more recent situation when the IRS-1C imagery was taken in September 1997. In addition to the above verification aspect, IRS-1C imagery was used to produce a more up-to-date site diagram complete with roadway and building identification. This serves as an example of potential use of satellite imagery for pre-inspection preparation.

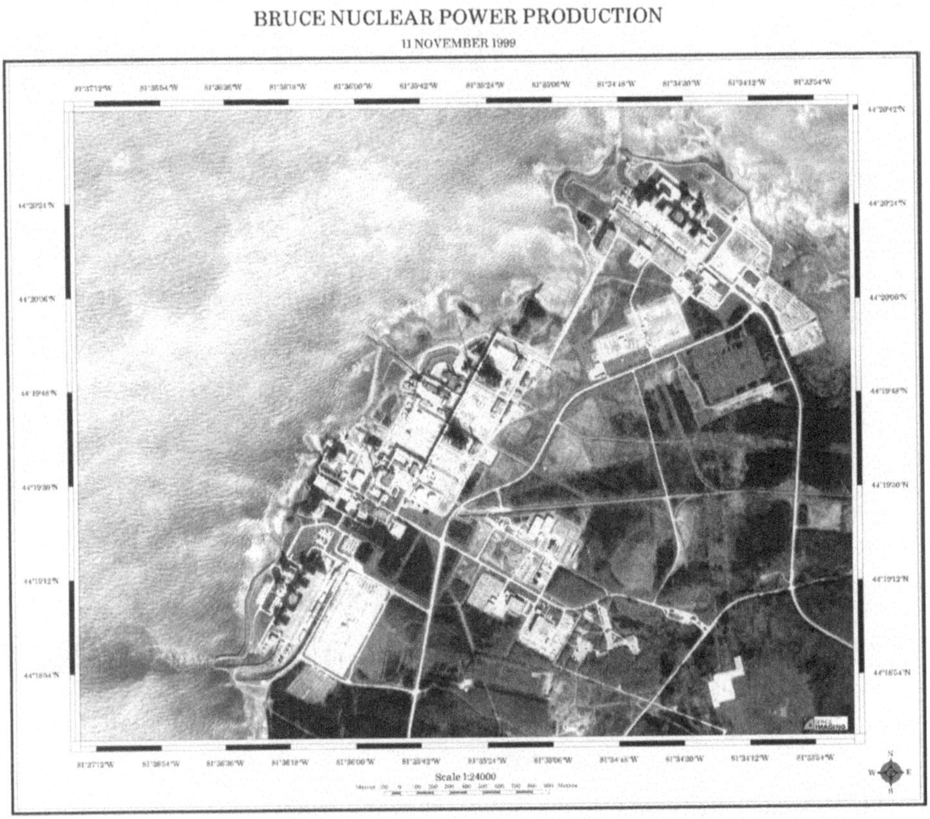

Figure 10.5. Image map for Bruce Nuclear Facilities.

To date we have acquired IKONOS imagery over several major nuclear facilities in Canada. Additional imagery of uranium mine sites and other facilities will be acquired when environmental conditions permit. The high-resolution imagery would be very useful for creating baseline reports which include image maps and information such as: name of the facility; date of image acquisition; facility location and boundary; feature details of building and road infrastructure; and terrain features in a map format with proper orientation and geodetic coordinates grid. An example of such an image map is shown in Figure 10.5.

Comparison between declared site diagrams with recent IKONOS imagery for several major Canadian nuclear facilities has resulted in more accurate site diagrams for submission under the Additional Protocol.

10.4.2 Monitoring reactor operation using thermal infrared and SAR imagery

Several Landsat images were obtained for Bruce, Gentilly-2 and Point Lepreau sites to investigate the potential use of thermal infrared imagery to monitor reactor

Figure 10.6. Detecting thermal signatures.

operation. Figure 10.6 shows clear evidence of the thermal signatures indicating reactor operation at Bruce B reactors, and no thermal signatures (indicating shut down condition) for Bruce A reactors, Douglas Point reactor, and the Bruce Heavy Water plants. Similarly, strong thermal signatures were also observed for the Gentilly-2 reactor indicating a reactor in operation. However, we have not yet been successful in detecting the thermal signatures for the Point Lepreau reactor, which is located in the Bay of Fundy where the seawater is cold and the tides are very strong. In addition, the cooling water discharge pipe is deeply submerged and designed to lower the discharge temperature. A detailed discussion on the effects of ambient temperatures and tidal conditions on the thermal signatures of the cooling water discharges from several power plants can be found in Carney, 1997.

Inspired by the work of Jasani (1998) SAR images were acquired and analysed for their potential for detection of reactor operation (see also Chapter 7). Figure 10.7 shows a comparison of the ERS-2 SAR image of 13 May 1997 with the Landsat-5 TM image (bands 6,2,3) of 7 May 1997 over the Bruce Nuclear Facilities. The thermal plume is clearly shown in the Landsat image. Also, it is quite obvious from the ERS-2 image (left side of the figure) that the cooling water discharge was detected by the ERS-2 sensor, as indicated by the semi-circular line beginning at the discharge outlet of the Bruce-B station and curving northward with approximately

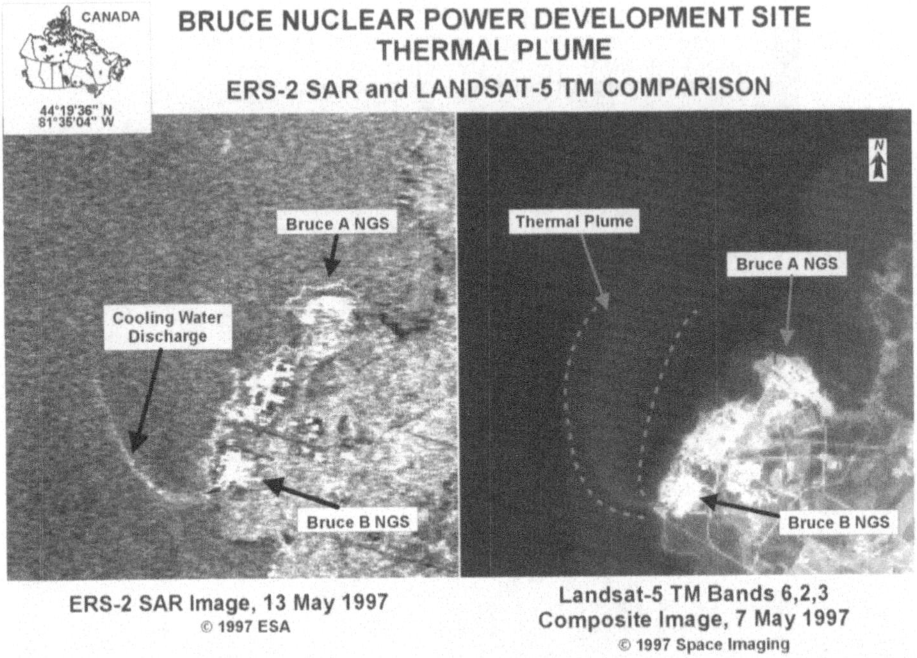

Figure 10.7. Detection of reactor operation with SAR and thermal infrared imagery.

the same curvature as the left edge of the thermal plume shown in the Landsat image. This curvature is the result of a broadside collision between the prevailing lake current and the cooling water discharge and has been observed and photographed from the roof of the reactor building.

10.4.3 Detect potential undeclared activities

According to open source information there was some mining activity in a site in the Canadian Northwest Territories (NWT) during the CSSP study. This site was extremely remote and very difficult to access by ground or by air transportation. Newspaper articles indicated that the activity was related to a diamond discovery. Since this region was also known for uranium deposits, it was decided to use this situation as an example of how verification of a potential undeclared activity could be carried out. In this exercise the analysts were supplied with open source information, the newspaper articles on the alleged activities (*Ottawa Citizen*, 1997), and publicly available maps. The analysts' task was to locate the mining site using satellite imagery and to ascertain the claim that production would commence in the autumn of 1998. A search of archived images was carried out using the supplied information. Due to the low light and environmental conditions in the region it was not possible to acquire an optical image of the area of interest. It would have taken more than six months before we could have acquired an optical

Figure 10.8. SAR picture of EKATI mine.

image. In order to get the information as soon as possible for assessing this potential situation it was necessary to acquire a RADARSAT-1 image. This image confirmed that there was mining activity in the region. It also confirmed the location and its key infrastructures such as the open pit, the main complex, and the airstrip, as illustrated in Figure 10.8. With this initial information one could proceed with a preparation for inspection, if necessary.

Follow-up information, including an aerial photograph of the site (see Figure 10.9) provided by the company managing the mine, confirmed the information gleaned from the SAR imagery and allowed us to determine that this was in fact a diamond operation and not a clandestine operation involving uranium.

Further confirmation, by acquiring an electro-optical image of the area of interest, was investigated. Three attempts to acquire an IRS-1C 5 m panchromatic image were made during the summer months of 1998. However, due to adverse environmental conditions in the area, all three attempts were unsuccessful. The failure of the electro-optical sensor in acquiring an image of a mine site further reinforces the need to employ alternative sensors such as SAR. Employing a SAR sensor permits nearly guaranteed information acquisition over a large search area. Not only did the RADARSAT-1 satellite acquire the image during a period of darkness in the NWT, it also provided the coverage in a timely manner.

Figure 10.9. BHP colour picture of Ekati mine.

10.4.4 Monitoring changes at a declared facility

Satellite imagery can successfully be used to monitor and assess changes to a facility due to construction. By using several images during a period of time the location and extent of the construction of new buildings within a small area of a site could be observed. Although an inspector can visit the facility during the period, high walls or piles of construction materials may hamper the inspector's view. With the view provided by satellite imagery the progress of construction could be periodically monitored. The information gained from this monitoring exercise would be helpful in deciding if an inspection of a new building or an update of the site diagram is necessary.

The example chosen for the change detection exercise was to detect construction activities around the area where two new radioisotope production reactors (Maple-1 and 2) were being built at Chalk River Laboratories. The air photo (1 m resolution) on the left half of Figure 10.10 (taken before the construction) shows clearly the details of buildings 100, 110, 124, and the smoke stack in June 1989. The IRS-1C (5 m) taken during the construction (winter of 1998) indicates the disappearance of the buildings and the smoke stack mentioned above. The Maple-1 reactor building and a dark area where the hole for the Maple-2 reactor and the foundation for the common processing facility for the two reactors are located now occupy that area.

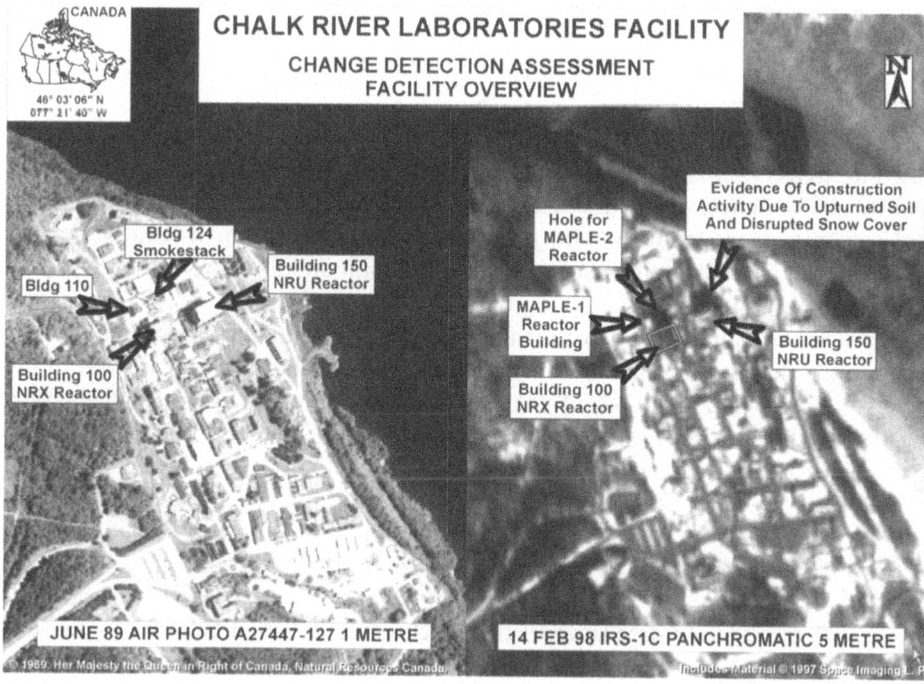

Figure 10.10. Change detection.

The site was covered with snow as indicated by the white patches. With a 5 m resolution, it is not possible to determine the depth of the dark area next to the Maple-1 building. However, it should be possible to verify if it was actually a hole with stereo imagery at higher resolution (e.g. 1 m) or with SAR imagery for its ability to detect depth (since the hole would act as a reflector). RADARSAT-1 had been tasked to obtain an image for verification of the above effect, and the results indicated that construction activities could be detected within the construction zone by the intense Radar energy returns from objects and features that have height and depth. The brightness and pattern of the energy returns indicated the presence of concrete walls, piles of construction materials, and the open lattice design of a typical construction boom type crane.

10.4.5 Monitoring activity at a decommissioned facility

Satellite imagery would be very useful in assessing if there were any activities performed at a site. Evidence of activities such as ground scarring, or the presence of vehicles and equipment could be detected using various kinds of sensors. Sites that are remote or difficult to access may be monitored with satellite imagery on a routine basis. On-site inspection could then be carried out when the monitoring programme using satellite imagery has identified evidence of suspicious activities at a decommissioned facility.

10.4.6 Assisting in inspection planning and reporting

Satellite imagery can also be used as an efficient tool for on-site inspection and reporting. The imagery provides an excellent visual aid of building sizes, dimensions, and terrain features for inspection planning purposes. Image maps can be created with annotations that correspond to the site diagram and any discrepancies can be identified.

In our initial studies the detection and identification capabilities were determined using aerial photographs with a 1 m resolution. This was done in anticipation of the successful deployment of one of the new high-resolution imagery satellites. The inspection planning benefits were later confirmed with the acquisition and assessment of the IKONOS imagery.

10.5 CONCLUSIONS

The results available to date support the conclusions discussed below.

In the next few years, there will be more than a dozen imaging satellites, including those carrying the multi-polarized RADARSAT-2 SAR sensor and Orbview-4 hyperspectral sensor, operated by various countries to provide high-resolution imagery on a commercial basis. This market availability is expected to reduce the cost of imagery and ensure the integrity of the imagery.

Each level of the commercial imagery's resolution may be useful, depending on the application. A 30 m resolution image would be sufficient to identify large objects or features, which may be useful for locating or identifying a site. However, clear identification of small objects such as stand by generators would require resolution in the 1 m range.

The use of satellite-borne thermal band infrared imagery has been used to detect the thermal emissions of operating nuclear power plants, but has some restrictions in application due to facility design, environmental conditions and sensor limitations.

SAR imagery is useful for detecting small structures, vehicles and roadways and for determining whether an electro-optical image has been altered. SAR sensors are also capable of imaging during day/night and in all-weather conditions. This makes them a valuable tool for detecting sites located near the polar regions or in areas which are often under cloud cover or affected by bad weather, or when timeliness is of utmost concern. Under certain conditions, SAR imagery can be used to detect cooling water discharge and hence the operational status of the facility.

Imagery can also be used as an effective tool for change detection, verification of member state declarations, and for providing up-to-date inspection aids such as image maps.

Combined with open source information, satellite imagery can be used in an assessment of allegations of potential clandestine operations. This could then lead to an on-site inspection of the suspected area.

Satellite imagery and open source information, including airphotos, could be collected and managed by GIS software for on-demand information retrieval. This would provide baseline information on nuclear facilities for future reference, or for submission by member states to the IAEA under the Additional Protocol for safeguards.

10.6 ACKNOWLEDGEMENTS

This study is financially supported by the CSSP of the CNSC and the International Security Research and Outreach Program of the Department of Foreign Affairs and International Trade (DFAIT), Canada.

10.7 REFERENCES

Baines, P. 1998 Exploitation of Commercial Satellite Imagery for IAEA Safeguards Analysis. Proceedings of IAEA Seminar, Vienna, Austria.

Canadian Safeguards Support Program CSSP-122 for the IAEA 1998 Potential Applications of Commercial Satellite Imagery in International Safeguards.

Carney, M. (1997) Surface temperature in Narragansett Bay: seasonal dynamics and anthropogenic effects. B.Sc. thesis, Center for Environmental Studies, Brown University, Providence, Rhode Island.

Jasani, B. (1998) Commercial satellite imagery: an important open source for the IAEA. Proceedings of IAEA Seminar, Vienna.

Ottawa Citizen (1997) Diamond discovery in the Northwest Territories. 26 December.

Truong, Q. S. Bob, Keeffe, R., Baines, P. and Paquette, J. P. (1998) Potential applications of commercial imagery in international safeguards. Proceedings of the INMM 39th Annual Meeting, Naples, Florida, USA.

Truong, Q. S. Bob, Keeffe, R., Baines, P. and Paquette, J. P. (1999) Potential application of commercial imagery in international safeguards. Journal of Nuclear Material Management. Volume XXVII, Number 2, pp. 13–18, USA.

Truong, Q. S. Bob, Keeffe, R. Baines, P. and Paquette, J. P. (1999) Potential Applications of Commercial Satellite Imagery in International Safeguards – A Progress Report. Proceedings of the INMM 40th Annual Meeting, Phoenix, Arizona, USA.

11

Acceptability of the use of satellite imagery for agency safeguards purposes

R. Loosch

11.1 INTRODUCTION

The task assigned to the IAEA on the basis of the 1993 recommendation by SAGSI is 'an assessment of the usefulness, technical feasibility, associated costs and acceptability of the Agency's obtaining satellite photographs from commercial sources.'[1] The following comments focus on 'acceptability', comprising both the – fairly objective – legal admissibility and the – more subjective – acceptance by policy makers, in particular those of the Agency.

For the sake of clarity, some preliminary remarks are appropriate. First, these comments apply equally to 'photographs' in a strictly optical sense and to other forms of imagery such as thermal infrared and SAR images. Secondly, wherever, for simplicity's sake, the term 'satellites' is used to refer to space objects that obtain remote-sensing data, it should be interpreted as including other orbital structures (e.g. space stations) which are capable of earth observation. Thirdly, they cover any purchase of remote-sensing data or derived imagery regardless of whether the provider is a 'commercial source' (or entity under commercial law) or any other institution such as a governmental or inter-governmental organisation.

11.2 ADMISSIBILITY

As regards admissibility, two questions have to be distinguished:

1 Does international or national law pertaining to earth observation from space

1 See p. 4 of IAEA working document on 'Sources and Applications of Commercial Satellite Imagery' (draft dated 24 August 1998).

permit the procurement and use of commercially available remote-sensing data and images for the Agency safeguards purposes?

2 Do the Agency's statute and other regulations permit such procurement or use by the Agency?

11.2.1 International space law

International space law (treaties and other agreements, national legislation as well as customary law) regulate only activities in space. Thus, it might be argued that it does not at all concern the Agency's possible purchase or use of satellite imagery for safeguards purposes because the Agency would, at least for the foreseeable future, not operate in outer space but exclusively on the ground. It may have been for this reason that potential implications of space law were not even discussed within the Agency when, in December 1992 and February 1993, its Board of Governors considered and accepted a proposal that the Director General may, when implementing safeguards, make use of all relevant information available to him including 'that collected by Member States.'[2] This was despite a general, though mostly tacit, awareness of, and also at least one explicit reference[3] to, satellite images as being possibly included in such information.

However, any conclusion that the Agency would remain unconcerned by space law when proceeding to – terrestrial – procurement and subsequent use of satellite imagery would be hasty. On the one hand, certain provisions of international and national space law relate to earthbound activities that are closely connected with operations in space. One particular case with regard to 'remote sensing activities' will be dealt with later (see Section 11.2.1.3). Furthermore, other than just passively accepting information (i.e. as in the case considered by the Board of Governors in 1992/93) voluntarily and perhaps gratuitously offered by governments or other outside sources, the Agency would now be directly and actively involved with earth observation. This would be achieved by seeking to obtain such images from satellite operators and might even induce them to direct their remote sensing instruments at particular spots or regions identified by the Agency for safeguards purposes. Given this more active role of the Agency – and it being not only an intergovernmental organisation but also a member of the United Nations family which is the main source, and guardian, of international space law – the Agency would be, at least morally, bound and politically well advised to satisfy itself that the satellite data and images it orders are obtained and handled in full conformity with international space law.

2 Cf. IAEA Documents GOV/2554/Attachment 1, notably para. 31, and GOV/OR.768–770 and 776–778, in particular GOV/OR. 776, para.s 48 *et seqq.* and 84.

3 IAEA Document GOV/OR. 770, para. 63.

11.2.1.1 Outer space treaty

The Outer Space Treaty[4] is the only one of the five international instruments, usually referred to as the body of primary international space law, that have a direct bearing on the acquisition, processing and utilisation of data generated through earth observation satellites.[5] There is a certain general but secondary aspect such as the obligation under Article IV of the Registration Convention[6] to register objects launched into outer space with the United Nations. As of 1999, 95 countries have ratified the Outer Space Treaty and a further 27 have signed it. This includes all those who are directly involved in operating, alone or together with other states, earth observation satellites.

For the purposes of this chapter, the most important provision of the Outer Space Treaty is its Article I, paragraph 2, which stipulates that 'Outer Space ... shall be free for exploration and use by all States without discrimination of any kind, on a basis of equality and in accordance with international law.' The term 'all States' is further defined by its Articles VI and XIII which make clear that this freedom of space exploration and use is not only enjoyed by individual states but is, or can be, extended also to non-governmental entities and international organisations. Article VI, concerned with international responsibility of states, puts forth that 'states ... shall bear international responsibility for national activities in outer space ... whether ... carried on by governmental agencies or by non-governmental entities, and for assuring that [they] ... are carried out in conformity with the provisions set forth in the present Treaty.' Furthermore, the Article XIII adds that 'The provisions of this Treaty shall apply to the activities of States Parties to this Treaty ... carried on by a single State ... or jointly with other States, including ... within the framework of international inter-governmental organisations.'

The Outer Space Treaty has only two limitations regarding the freedom of exploring and using space: banning national appropriation of space including celestial bodies and certain actions in or from space undertaken for non-peaceful purposes. Since the use of satellite images for determining or verifying compliance with, or breach of, nuclear non-proliferation or safeguards obligations is a peaceful activity and it does not involve appropriation of space, these limitations should not affect the Agency unless the notion of peaceful (or non-peaceful) purposes were defined differently. It could be defined in such a way that, for example, the operation of espionage or reconnaissance satellites which serve the objective of preparing or supporting military or strategic decisions were considered as violating

4 Treaty on Principles Governing the Activities of States in the Exploration and Use of Outer Space, Including the Moon and other Celestial Bodies, signed on 27 January 1967, entered into force on 10 October 1967, UNTS Vol. 610, p. 2058.

5 Cf. the list in footnote 1 on p. 2 on the IAEA Legal Division's Discussion Paper on 'Legal Issues Associated with IAEA Use of Commercially-Available Satellite Imagery' (draft dated 24 August 1998).

6 Convention on Registration of Objects Launched into Outer Space of 14 January 1975, entered into force on 15 September 1976 UN Treaty Series (UNTS) Vol. 1023, p. 15.

the Outer Space Treaty's peaceful intent and that this tarnish also 'infected' all data and images obtained with the help of such satellites.

The Treaty does not define 'peaceful' or 'non-peaceful' and uses, in its various preamble and operative sections, different language for expressing its pacific thrust.[7] Several attempts to develop, through analysis of the Outer Space Treaty's text and negotiating history, a conclusive definition or description of this peaceful intent have not produced an unambiguous and easily usable result.[8] Therefore, the only way to resolve the issue to the extent relevant for the possible use of satellite imagery for Agency safeguards purposes remains to rely on the explicit text of the Treaty. On this basis, only certain strictly military uses of space such as deploying nuclear weapons or other means of mass destruction in space or installing military bases or conducting military manoeuvres on celestial bodies are prohibited.[9] A liberal interpretation of 'peaceful use' permitted under the Outer Space Treaty is further indicated by its provision that even for activities on celestial bodies, for which the Treaty's language is most restrictive ('for exclusively peaceful purposes' as against simply 'for peaceful purposes' in other sections), the use of military personnel or of any (i.e. including military) equipment or facility is permitted if the activities themselves are carried out for peaceful purposes.[10]

Therefore, the acquisition, processing and use of earth observation data and imagery for Agency safeguards purposes is allowed under the Outer Space Treaty. This conclusion is supported by the – admittedly more exhortative than operative – Treaty provision that activities in outer space shall be carried out 'in the interest of maintaining international peace and security',[11] the language that could even be interpreted as a positive approval of using space for non-proliferation and safeguards purposes.

11.2.1.2 Specific international agreements on remote sensing

On a second level of international space law, two instruments specifically address earth observation by satellites, namely the 1978 Moscow Convention[12] and the 1986

[7] See, *inter alia*, its second and fourth preambular paragraphs which read, respectively, 'Recognizing the common interest of all mankind in the progress of the exploration and use of outer space for peaceful purposes' and 'Desiring to contribute to broad international cooperation in . . . the exploration and use of outer space for peaceful purpose.'

[8] Cf., *inter alia*, Benkö, M., de Graaff, W., Reijnen, G. C. M., *Space Law in the United Nations*, Dordrecht, Boston, Lancaster, 1985, pp. 162 *et seqq*, and M. G. Marcoff, *Traité de Droit international public de l'espace*, Fribourg,Genève, Paris, New York, 1973, pp. 357 *et seqq*.

[9] Cf. Article IV of the Outer Space Treaty (footnote 5).

[10] Article IV, second para., last two sentences.

[11] Article III, *ibidem*.

[12] Convention on the Transfer and Use of Data of the Remote Sensing of the Earth from Outer Space, signed on 19 May 1978 by Cuba, [former] Czechoslovakia, [the former] GDR, Hungary, Mongolia, Poland, Romania and [the former] Soviet Union, printed as UN Document A/33/162 of 29 June 1978 and published, *inter alia*, in Zeitschrift für Luft- und Weltraumrecht (ZLW) 1987, p. 37.

United Nations Principles on earth observation.[13] Both of them have a lesser status than the Outer Space Treaty. This is because they are not legally binding international conventions like the Outer Space Treaty. They were, after some argument, adopted as a mere General Assembly Resolution of a political rather than legal character. The Moscow Convention has a relatively narrow geographical scope which never went beyond the countries of the former Soviet bloc (and did not even extend to all the states that belonged to it) and therefore cannot claim any general, let alone global validity.

11.2.1.3 The United Nations Principles on Remote Sensing

The UN Principles explicitly cover only 'the sensing of the Earth's surface from space...for the purposes of improving natural resources management, land use and the protection of the environment'.[14] This raises doubts whether the Principles are at all relevant to earth observation for such purposes as, *inter alia*, Agency safeguards.

Certainly, the Principles cannot apply directly to remote sensing exclusively devoted to purposes other than the three which are positively enumerated, and not just mentioned, as illustrative examples in the clause just quoted; otherwise, one would interpret the Principles against their very and, in this respect, unambiguous text. On the other hand, neither their text nor their legislative purpose nor their negotiating history suggest that they are not applicable to cases in which an earth observation satellite serves both one or more of the objectives positively listed in Principle I (a) and, additionally, some other purpose, for instance, the verification of non-proliferation commitments. It would be strange, indeed, if by simply adding a purpose not mentioned in the Principles' definition of 'remote sensing' to one mentioned there, the operator could escape from the Principles altogether.

This conclusion is further supported by a textual analysis of the various definitions in Principle I. Only that of 'remote sensing' (as the operation carried out in and from space and referred to in later sections of the Principles as 'the operation of remote sensing space systems'[15]) contains the limitation to the three purposes mentioned above. By contrast, the definitions of 'data' and 'information' as well as of 'remote sensing activities' (which include terrestrial processing of data beamed down from satellites) do not differentiate between the uses to which such data might be put eventually or for which they might be processed. In other words, once the satellite operations in space correspond to the definition of 'remote sensing' in Principle I, the Principles apply to the use and processing of all data obtained thereby, regardless of the nature or motivation of such use or processing.

13 United Nations General Assembly Resolution 41/65 of 3 December 1986, Principles Relating to Remote Sensing of the Earth from Outer Space, Official Records of the Forty-first Session of the General Assembly, published, *inter alia*, in ZLW 1987, p. 34.

14 *Ibidem*, para. (a) of Principle I.

15 *Ibidem*, para. (e) of Principle I.

Therefore, and since practically all remote sensing satellites contemplated by the Agency's Secretariat as possible sources of imagery to be used for safeguards purposes[16] are primarily oriented towards earth observation for the management of natural resources or land-use or for environmental protection, it is to be assumed that the UN Principles apply to the remote sensing activities which the Agency may conduct for safeguards purposes.[17] It is also worth noting that, after adopting the Principles Relating to Remote Sensing, the UN General Assembly and its Committee on the Peaceful Uses of Outer Space (UNCOPUOS) have stopped work on the elaboration of further rules pertaining to earth observation from space, despite their awareness that a number of questions – including the use of satellites for the verification of arms control, disarmament and non-proliferation commitments – are not directly addressed by the Principles.[18] This is likely to strengthen the tendency to refer, for lack of any better legal basis, as much as possible to the Principles when dealing with problems which might arise in connection with other kinds of remote sensing but which are not explicitly dealt with by the Principles or other instruments of international law. This goes particularly for some general concepts which are relevant for any type of remote sensing from space, for whatever purpose. A typical example is the conflict between the nearly unlimited freedom of using space for earth observation as established by the Outer Space Treaty[19] and the conventional concept of national sovereignty which makes earth observation by other means (terrestrial or airborne) subject to some authorisation or permission by the observed state. After a lengthy debate in UNCOPUOS and other UN bodies between, on the one hand, the – primarily Western – countries defending freedom in space and, on the other hand, the former Soviet bloc and several developing countries traditionally jealous of their sovereignty in order to be able to fend off any 'intrusion' from abroad, this conflict was eventually resolved by a compromise formula which leaves the freedom of remote sensing from space unscathed but obliges the sensing state to act 'on the basis of respect for the principle of full and permanent

[16] Cf. the IAEA Working Document quoted in footnote 1 above, pp. 5 *et seqq.*

[17] In this respect, the author does not share the opinion of the IAEA's Legal Division expressed in paragraph 8 on p. 3 of the paper cited in footnote 6 above, that: 'the Principles do not pertain to certain applications, such as . . . remote sensing carried out for purposes of verification of arms control measure'. It seems to be based on too broad a conclusion from the statement by N. Jasentuliyana (also quoted there) that 'the Principles are basically an agreement on economic relations and do not really address the question of the use of satellite observations to influence military or political relations'. Though this statement is historically correct as, indeed, UNCOPUOS eventually did not address earth observation for arms control or non-proliferation verification as a distinct subject, it should not be interpreted so extensively as to deny that the Principles are agnostic with regard to the purposes for which satellite-supplied information might be used on earth.

[18] See M. Benkö/G. Gruber, The UN Committee on the Peaceful Uses of Outer Space: Adoption of Principles on Remote Sensing of the Earth from Outer Space and Other Recent Developments, ZLW 1987, pp. 18 and 24.

[19] See section 12.2.1.1.

sovereignty of all States and peoples over their own wealth and natural resources...'[20] and to provide the sensed state with access to data gained by remote sensing from space.[21] Though it is not a practical proposition at this stage, it may well happen that, through international legal action or national legislation implementing international space law or through evolving customary international law, that compromise might be extended to cover types of remote sensing from space not explicitly mentioned in the Principles.

In conclusion, it is prudent though perhaps not legally cogent that the Agency behaves in conformity with the Principles to the extent that it conducts activities of the same nature as those directed at purposes specifically enumerated in the basic definitions of the Principles.

This conclusion entails the implications discussed below for the Agency, always assuming that the Agency will not, at least for the time being, operate remote sensing satellites by itself or together with governments or other entities.

As already pointed out in connection with the Outer Space Treaty,[22] only 'States' are explicitly addressed by the relevant Principles.[23] However, like the Outer Space Treaty and by express reference to its Article VI, the Principles make clear that 'States operating remote sensing satellites shall bear international responsibility for their activities ..., irrespective of whether [these] are carried out by ... or through international organizations to which such States are parties.'[24]

The question then is which obligations or conditions under the Principles are to be met by the Agency when it procures, processes or analyses satellite-generated data for its own – verification – purposes. The Principles' key term in this respect is 'remote sensing activities' which are defined[25] as comprising both 'the operation of remote sensing space systems, primary data collection and storage stations' and 'activities in processing, interpreting and disseminating the processed data.' Under the assumption made above, the first part of this definition may be disregarded as not relevant for the Agency in the foreseeable future. The definition's second part,

20 Principle IV.

21 See, in particular, Principle XII.

22 See Section 12.2.1.1.

23 The only obligation which the Principles place directly upon international organisations, that 'the United Nations and the relevant agencies within the United Nations system shall promote international co-operation, including technical assistance and co-ordination in the area of remote sensing' (Principle VIII), has nothing to do with the objectives of this paper and, therefore, does not require any detailed consideration.

24 Principle XIV; the clause in Article VI of the Outer Space Treaty that 'when activities are carried out in outer space ... by an international organization, responsibility for compliance with this Treaty shall be borne both by the international organization and by the States Parties to the Treaty participating in such organization' is not echoed in the Principles. Whether this clause also applies to activities under the Principles may remain unresolved because it does not make any practical difference whether an international organisation is directly co-responsible together with its member states or indirectly tied in through them.

25 In Principle I, para. (e).

however, is clearly relevant as the Agency would process or interpret data obtained through the operation of remote sensing space systems, regardless of whether it would procure 'primary data', i.e. 'raw data that are acquired by remote sensors borne by a space object and that are transmitted or delivered to the ground from space ... by any ... means',[26] or would step in only at the stage of 'processed data' (i.e. 'the products resulting from the processing of the primary data, needed in order to make such data usable').[27]

Under the Principles, 'the sensed State shall, as soon as the primary data and the processed data concerning the territory under its jurisdiction are produced, have access to them on a non-discriminatory basis and on reasonable cost terms.'[28] The Principles do not specify who shall have to grant such access. The reference to a 'non-discriminatory basis' makes, however, sense only if it is the original producer of the data concerned (i.e. the operator of the remote sensing space system as regards primary data and the processor of such data with regard to processed data) who is obliged to provide that access.

The Agency would most likely avoid the considerable expenditure and technical problems involved in setting up an own capability for processing primary data to the level of usability and, therefore, probably not be approached by a sensed state with regard to access to such data. However, even when procuring only processed data, it should be aware that the sensed state might already have requested, and obtained, from the satellite operator or its marketing agency or customer the underlying primary data and, depending on the circumstances, might also have processed them to a point where their usefulness for safeguards purposes has become apparent. This would not negatively affect the Agency's objectives. The Agency would still be able to process, or have processed, those data for its own purposes without the sensed state being able to manipulate them *ex post* or to prevent the Agency of having the same area being observed again in order to detect any attempt at terrestrial camouflage by the sensed state. In the end, the only advantage a sensed state might gain on the basis of the Principles is to get an early warning that the primary data and its subsequent generations of processed data and analysed information might be available for the Agency's safeguards purposes.

In essence, the same conclusion may be drawn when a sensed state requests access to processed data; the only but ultimately irrelevant difference would be that, on the one hand, the state would probably get a clearer impression what imagery might be available to the Agency, and that, on the other hand, the Agency should provide the required access if – contrary to the assumption made above – the Agency itself has processed primary into processed data. Even then, the Agency would remain in control, in particular as regards the timing of effective access, because the Principles do not establish a time limit for concluding the negotiations on the conditions of data access 'on a non-discriminatory basis and on reasonable cost terms'.

[26] See Principle I, para. (b).
[27] Principle I, para. (c).
[28] Principle XII, first sentence.

Finally, as regards 'analysed information' (i.e. 'the information resulting from the interpretation of processed data [together with] inputs of data and knowledge from other sources'),[29] the sensed state shall have access to 'the available analysed information concerning the territory under its jurisdiction in the possession of any State participating in remote sensing activities on the same basis and terms.'[30] As pointed out by the Agency's Secretariat,[31] the Principles do not spell out where and, in particular, to whom analysed information must be available in order to trigger a sensed state's right of access, and who is obliged to provide such access. In the absence of explicit rules and in view of the wording of the respective Principle ('available analysed information . . . in the possession of any State participating in remote sensing activities'), it is to be assumed that the sensed state is entitled to access to all analysed information available to anyone performing 'remote sensing activities' as defined by the Principles,[32] so including information which is produced by a recipient of primary or processed data through further processing and interpretation of such data, with or without the help of his or her own input or of input from other sources. Therefore, the Agency might be faced, and obliged to comply, with requests by sensed states for access to its own analysed information (see above).[33]

Again, this obligation should be no reason for concern to the Agency. The objective of its analysis of remote sensing information would be to obtain additional evidence or indications of a state's compliance or non-compliance with its non-proliferation and safeguards commitments and, eventually, to present the results of the analysis to the state concerned, be it – in the case of compliance – in the form of the general Safeguards Statements in Safeguards Implementation Reports or of specific statements provided for in safeguards agreements,[34] or – in the case of suspected non-compliance or of doubts about compliance – through requests for special inspections[35] or for additional explanations or complementary access.[36] Thus the Agency would, in the normal course of its safeguards procedures, provide the sensed state with access to the analysed information available to the Agency. The only difference which the Principles might make would be with regard to the timing of the access as a sensed state might require such access already before the Agency, following its normal safeguards procedures, deems appropriate to present its analysed information to the sensed state. Even then, the need to determine through negotiations with the requesting state the terms and conditions of access

[29] Cf. the definition in Principle I, para. (d).

[30] Principle XII, second sentence.

[31] Paragraphs 15 *et seqq.* of the Working Paper cited in footnote 6.

[32] See above.

[33] When settling with the sensed state on the conditions for granting access (which must be 'on reasonable cost terms' – see footnote 30 above), the Agency could, of course, include the cost of its own analysis work and of procuring the necessary input in the overall charge to be paid by the requesting state.

[34] For example, in provisions based on Article 90 (b) of INFCIRC/153/Corr.

[35] See Article 73 of INFCIRC/153/Corr. and paragraph 53 of INFCIRC/66/Rev. 2.

[36] See, in particular, Articles 2 c. and 4 a. and d. of INFCIRC/540.

(see above) would provide the Agency with an opportunity to delay access until such time as is compatible with its safeguards objectives.

Furthermore, the Agency must be aware that the Principles require the state 'carrying out a programme of remote sensing [i.e. the state for which a remote sensing satellite is, or is to be, registered] [to] inform the Secretary-General of the United Nations.'[37] This raises the question whether that information must also include the identification of the data generated and transmitted to earth by remote sensing satellites and/or of the recipients of such data.

The Principles themselves do not specify the information to be supplied to the UN Secretary-General; they only stipulate that it be presented 'in accordance with article IV of the [Registration] Convention and article XI of the [Outer Space] Treaty.'[38] Both the text of Article IV and the purpose of the Registration Convention make it clear that the information to be registered relates only to the objects launched into space (i.e. the satellite or other space structure); details of the data obtained through operating the space system and the identity of terrestrial customers for such data are, therefore, irrelevant and, in practical terms, might even not yet be known when the registration of the space system is due. The same conclusion follows from Article XI of the Outer Space Treaty though this provision goes further than the Registration Convention by requiring 'to inform the Secretary-General of the United Nations as well as the public and the international scientific community, to the greatest extent feasible and practicable, of the nature, conduct, locations and results of such activities [in outer space]' and by instructing the UN Secretary-General 'on receiving the said information, ... to disseminate it immediately and effectively'. Literally interpreted, this provision might call for information about the data acquired and transmitted to the ground by a remote sensing satellite as 'results of such activities in outer space'. However, since the purpose of this Article is 'to promote international co-operation in the peaceful exploration and use of outer space', and since a detailed knowledge of the data generated, let alone an identification of terrestrial data recipients cannot conceivably further such co-operation, the maximum information to be provided about satellite-generated data, under the subheading 'results' in Article XI of the Outer Space Treaty, would be a description of the data acquisition and transmission capabilities of the space system itself.

Under the Principles, the state for which a remote sensing satellite has to be registered 'shall, moreover, make available any other relevant information to the greatest extent feasible and practicable to any other State... at its request.'[39] On the face of it, this is a rather sweeping requirement, and it is, therefore, difficult to exclude *a limine* that it also covers requests for information on the identity of the recipients of satellite-supplied data and information derived from them. On the other hand, it is equally difficult to establish that a state's request for such information must be fulfilled. Apparently, jurisprudence with respect to this provision does not

37 Principle IX, first sentence.

38 *Ibidem.*

39 Principle IX, second sentence.

yet exist, and requests of this kind may even not have been made so far. Taking into consideration the status of the Principles as a political resolution of the United Nations General Assembly rather than a legally binding international convention and the ensuing lack of enforceability, it will, in the end, be decided in negotiations between the state of registration and the requesting state whether the identification of data customers is 'relevant' information in the meaning of the Principles and, if so, what is 'feasible and practicable'.

Of course, nothing in the Registration Convention or the Outer Space Treaty prevents a state registering remote sensing satellites from providing the UN Secretary-General or the public or the international scientific community or any other state with more information than is strictly required under international space law. Such voluntary extra information might include details about satellite-generated data, their application potential or the identity of customers. That should, however, not jeopardise the Agency's safeguards interests. Once the Agency proceeds to procuring satellite data, this would be made known by the Secretariat to the Board of Governors and, through it, to the whole membership of the Agency (i.e. to practically all states the territories of which might be observed from satellites). Normally, the Secretariat would also inform the Board, through reports on its safeguards activities or in the course of the budget procedure, to whom it wants to award contracts and which satellites would provide the data concerned. Furthermore, if these are not defence-related reconnaissance satellites the capabilities and operational details of which are usually classified, it is generally known which territories can be sensed by them. Therefore, the possibility that the identity of the Agency as a recipient of remote sensing data is revealed in connection with the registration of satellites with the United Nations does not change the situation significantly.

If, nevertheless, the Agency wants, for whatever reason, to avoid being identified as a customer for such data, it could achieve that only by concluding, when ordering the data, a confidentiality agreement with the data supplier and, where applicable, through the data supplier with the operator of the satellite and its state of registry.

Finally, there are some other provisions of the Principles which might apply, *mutatis mutandis*, to the Agency's 'remote sensing activities'. These provisions call for making available technical assistance and for furnishing information, to the states concerned, on phenomena harmful to the Earth's natural environment or on actual or impending disasters.[40] A detailed analysis of these – ancillary – provisions does not appear necessary, especially in view of the low probability of any practical impact on the Agency's safeguards operations.

11.2.1.4 The Moscow Convention

As mentioned earlier, this Convention has been accepted only by some of the countries of the former Soviet bloc and thus cannot claim any worldwide

40 See Principles VII, X and XI.

validity,[41] neither directly nor through such wide factual acceptance that it could be considered as having evolved into customary international law.

Its scope is wider than that of the UN Principles in that it covers, in its definition of 'remote sensing of the Earth from outer space', all 'observations and measurements ... of the land, ocean and atmosphere of the Earth ... which facilitate the location, description of the nature and temporal variations of natural parameters and phenomena, natural resources of the Earth, the environment as well as anthropogenic objects and formations.'[42] The last part of this definition is of particular interest for the Agency's possible utilisation of remote sensing for safeguards purposes as safeguards focus, naturally, more on anthropogenic objects and formations than on natural parameters or the environment. At any rate, the sweeping definition just quoted leaves no doubt that the Convention addresses any remote sensing from space, no matter for what ultimate purpose, thus also 'activities in... remote sensing [for] promot[ing] peace and understanding among States.'[43] This conclusion cannot be challenged by arguing that the preamble of the Moscow Convention undeniably shows the predominant interest of the contracting parties in the economic aspects of earth observation.[44]

The Convention provides that 'a Contracting Party in possession of initial data[45] of ... remote sensing ..., with a better than 50 metres resolution on the terrain, relating to the territory of another Contracting Party, [or] that has obtained[,] as a result of the deciphering and thematic interpretation of any data of the remote sensing ...[,] information[46] about the natural resources or the economic potential of another Contracting Party, shall not disclose such information or make it available to anyone except with an explicit consent thereto of the Contracting Party to which the sensed territories and natural resources belong ...'.[47]

The Agency, not being a party to the Convention, would not itself be required to solicit such prior consent from the sensed state. However, it should bear in mind that, when seeking satellite-generated data relating to the territory of a state party to the Convention from a supplier in another Convention state, the supplier may legally provide these data to the Agency only with the explicit consent of the sensed state. In this connection, it should be underlined that the Convention only provides for rights and obligations *intra partes*, therefore, prior consent of the sensed state is not

41 See footnote 12. Upon the unification of Germany, the Convention is no longer in force for the territory of the former German Democratic Republic.

42 Article I (a) of the Convention (cf. footnote 12).

43 See also the third paragraph of the Convention's preamble.

44 See, in particular, the second and fourth preambular paragraphs.

45 Defined, in Article I (a), in a similar way as 'primary data' in the UN Principles (see Section 12.2.1.3 and footnote 26) but also including 'preprocessed data derived from the flow of data which may be used for later analysis' which, under the UN Principles, would come under 'processed data'.

46 Defined, in Article I (c), as covering a part of 'processed data' and all 'analysed information'.

47 See Articles IV and V (here editorially contracted into one phrase).

required if data obtained from that supplier relate to the territory of a state which is not a party to the Convention.

Without going into further details of the Moscow Convention, it may suffice to conclude that it might make the Agency hesitate at procuring remote sensing data from a state party to the Convention.

11.3 NATIONAL LEGISLATION

The consent requirements in states party to the Moscow Convention are not the only cases where conditions for, or restrictions of, the transfer of remote-sensing data to a third party are enacted by states without or beyond any norm of international law (in the sense of the law of nations). In fact, there is an increasing number of national laws, regulations and policies, some based on international agreements (like the Moscow Convention), others not, which restrict the transfer of remote-sensing data or information derived from them or make it subject to restrictive conditions. Such legislation or policies may provide for merely administrative arrangements such as licensing under national regulations for international trade or go all the way to an embargo of data relating to certain countries or categories of countries (see above).

Because of the prominence of American and French earth observation satellites in practical life, the relevant legislation and policy of these two countries is described briefly here.

Under US policy, the Government reserves the right to limit data collection by US satellites and/or their distribution if such activities compromise national security, respect of international obligations or other foreign policy objectives of the United States. Moreover, Congress passed, in 1996, legislation restricting the collection of images of Israel or any other area so designated by the US President with a resolution of better than 2 m.[48]

France, in principle pursuing an 'open skies, open access' policy with regard to imagery from its SPOT satellites and having maintained this policy during the US operations in Grenada and Panama, suspended, in view of the 1991 Gulf War, the sale of images of Iraq and neighbouring areas.[49]

National laws and policies of this nature are not automatically contradictory to the freedom of all states to explore and use outer space as established by the Outer Space Treaty (see Section 11.2.1.1) because they only concern trade and use of remote-sensing data on earth and not activities in outer space. If, however, such laws and policies effectively deny the peaceful use of such data to other states, whether through a general embargo[50] or through targeted injunctions against

[48] The Kyl-Bingaman Amendment to the National Defense Authorization Act for Fiscal Year 1997.

[49] See Lambakis, S. (1995) *Space Control in Desert Storm and Beyond*, Orbis, pp. 419 *et seqq.*

[50] Saudi-Arabia, for example, refuses, as a matter of general policy, to allow the transfer of remote-sensing data to neighbouring countries (see H.-J. Heintze, Rechtsfragen der Datenpolitik bei weltraumgestützter Erdfernerkundung, ZLW 1992, pp. 397, 409).

transfers to particular states or regions for which no good reason is shown, it may well be argued that this violates Outer Space Treaty obligations since the 'exploration and use of outer space...shall be carried out for the benefit and in the interests of all countries, irrespective of their degree of economic or scientific development',[51] (i.e. also for countries which themselves are not active in outer space). In this respect, the Agency should benefit from this Treaty provision to, at least, the same extent as a state not involved in remote sensing activities.

Furthermore, an Agency member's refusal to transfer, or to permit the transfer of, remote-sensing data to the Agency for safeguards purposes may be considered as non-fulfilment of obligations following from the Agency's Statute[52] or from decisions of its policy making organs, specifically from the several Board of Governors and General Conference resolutions adopted since 1991 (most of them by consensus) and calling for the strengthening of safeguards including the provision of wider access to the Director General with respect to relevant information.

There is a wealth of literature[53] on various national laws, regulations and policies which might affect the transfer of remote-sensing data to the Agency as well as its rights to use such data and follow-on information. However, no reliable, comprehensive atlas showing such legislation and policies exists, and it would go beyond the purposes of this chapter to canvas the present situation in all countries from which the Agency might seek to obtain relevant data or images. The situation might also evolve quickly as more earth observation data become available from more commercial sources which might eventually make any attempt of restraining their collection or distribution ineffective and obsolete. For the time being, however, the Agency is well advised to ascertain in each single case and before it signs a contract with a satellite operator, its marketing agent or a data processor, that the relevant national regulations do not create obstacles against the purchase and subsequent use of the data and images by the Agency.

11.4 THE AGENCY'S INTERNAL REGULATIONS

11.4.1 Safeguards procedures

At the latest, since the Board of Governors agreed that the Director General may use for safeguards purposes remote-sensing data and information resulting from their processing and analysis, the procurement of such data and information does not

[51] See Article I, first paragraph, of the Outer Space Treaty; for remote sensing, see also Principle II of the UN Principles which reads 'Remote sensing activities shall be carried out for the benefit and in the interests of all countries, irrespective of their degree of economic, social or scientific and technological development, and taking into particular consideration the needs of the developing countries.'

[52] See, in particular, Article IV C.

[53] See, e. g., para.s 21–24 on pp. 5/6 of the Secretariat's Discussion Paper cited in footnote 6, and H.-J. Heintze, Rechtsfragen der Datenpolitik. (cf. Footnote 49) as well as the numerous treatises cited there.

require any further authorisation under the safeguards regulations and agreements of the Agency. In particular, the provisions of safeguards agreements on the introduction of new safeguards tools and techniques ('any other objective measures')[54] do not apply. They were not invoked when the Board of Governors agreed with the use of such data and information in the first instance[55] and must certainly not be invoked at this stage because, technically speaking, the use of remote-sensing data and information is the same, whether these are voluntarily offered by member states to the Agency or gleaned from open literature or whether they are commercially procured by the Agency.

However, following normal practice, the Secretariat should inform the Board and, because of the ongoing discussion on strengthening safeguards effectiveness and improving their efficiency, also the General Conference about the decision to purchase remote-sensing information for broadening its safeguards database.

11.4.2 Financial regulations

As for all other expenditures, the Secretariat would need a budgetary appropriation for the purchase of remote-sensing data or images. Since the cost may turn out to be significant and, thus, to exceed the amount usually accepted under a budgetary line for sundry items, this might require a new, separate line in the budget.

Additionally, the Agency's regulations on procurement must be respected.

11.5 CONCLUSIONS AND RECOMMENDATIONS

1 Agency procurement and use for safeguards purposes of data and images obtained through remote sensing of the earth from outer space and of information including imagery generated through the processing of such data and images is in conformity with international law, in particular the Treaty on Principles Governing the Activities of States in the Exploration and Use of Outer Space, Including the Moon and Other Celestial Bodies.

2 Despite certain counterarguments, the Agency is well advised to assume that the United Nations Principles Relating to Remote Sensing of the Earth from Outer Space apply to the acquisition, processing and use of remote-sensing data for safeguards purposes.

3 This does not impede the Agency's procurement and use of such data and information. It may, however, impose on the Agency certain obligations to grant sensed states access to its data and information on a non-discriminatory basis and on reasonable cost terms. In any case, negative effects on the Agency's safeguards objectives can be avoided.

[54] Cf., *inter alia*, Article 74 (e) of INFCIRC/153/Corr. and Article 6 (a)–(d) of INFCIRC/540.

[55] Section 12.1 and footnote 2 on p. 2.

4 National laws, regulations and policies of states where the Agency seeks to acquire remote-sensing data or processed information may restrict the transfer to the Agency or make it subject to restrictive conditions with regard to procurement and/or subsequent use for safeguards purposes. What, if any, restrictions or conditions apply should be ascertained by the Agency in each single case before entering into the respective procurement contract, equally, whether they are compatible with applicable international (space) law and with obligations of the state concerned under the Statute of the Agency.

5 The procurement and use of remote-sensing data and derived information are covered by the present safeguards rules and agreements of the Agency and do not require additional authority other than a budgetary appropriation.

12

Summary and conclusions

B. Jasani and G. Stein

Under the new Model Protocol Additional to the Agreement(s) between State(s) and the IAEA for the Application of Safeguards, INFCIR/540, the Agency will receive very large volumes of information on nuclear activities of the parties to safeguards agreements, thus adding a completely new dimension to the Agency's safeguards system. For example, combined measuring and surveillance systems, remote data transmission, environmental monitoring and satellite imagery have considerable potential for making the IAEA's safeguards more effective. In this book, the technical and legal implications as well as, to some extent, the cost-effectiveness of the use of satellite imagery by the IAEA were examined in some detail. Some of the findings are summarised in the following sections.

12.1 SUMMARY

12.1.1 Usefulness of commercial observation satellites

The Agency's safeguards procedures necessarily involve assessment of the correctness and completeness of member states' declarations. Here, the use of commercial satellite imagery has considerable potential in verifying a state's declarations to the IAEA, as their capabilities are beginning to approach those of the satellites belonging to the NTMs. A number of satellites are launched and operated by several states so that there are now many independent sources of information, making it possible to check the veracity of data available to the IAEA. This can only but increase confidence in the Agency's safeguards procedures. Another advantage of several satellites in orbit is the frequent coverage of the earth's surface. It would, therefore, be difficult to avoid observations from satellites. Using multi-temporal images, changes in a site could be identified. Satellite imageries will also be used to provide up-to-date inspection aids such as image maps. They will also complement other sources of information. Based on the latter, satellites could be used to detect undeclared facilities, thus making the Agency's safeguards procedures more effective.

As the capabilities of satellites increase, their use for the Agency's safeguards procedures will be greater. Even at relatively low resolution (30 m), it was demonstrated that:

- Details – such as cooling towers associated with nuclear power reactors, enrichment facilities and, with good contrast, power lines – can be detected by optical sensors.
- It is possible to detect and even identify, with the help of maps, such large facilities as nuclear power reactors, reprocessing and enrichment plants, research centres and new construction activities at known nuclear sites.
- Passive and active security features are often associated with important national strategic assets as well as covert activities that include nuclear facilities. Such features are detectable by commercial satellites and can help in distinguishing a civil facility from a military one, thus alerting the Agency of any clandestine activities. One of the new tasks of the Agency's safeguards procedures is to identify undeclared activities in a member state. Combined with open source information, satellite imagery can be used in an assessment of allegations of potential clandestine operations. This could then lead to an on-site inspection of the suspected area.
- Multispectral images, particularly those in the far-infrared and thermal bands, are very useful for the detection of warm water discharged from nuclear facilities.
- Radar can penetrate clouds and can be used day and night under all weather conditions, making it possible to detect surface effects caused by cooling water discharge from nuclear facilities. Moreover, among other things, it can detect pylons associated with power lines and railway tracks. For example, frequent shutdown of most types of reactor would be a trigger to initiate further investigation.
- The availability of current and archived data and the use of sophisticated image processing techniques allow detection of changes in facilities.
- Satellite imagery and other open source information, including aerial photographs and maps, could be collected and managed by GIS software for on-demand information retrieval. This would provide baseline information on nuclear facilities for future reference or for submission by member states to the IAEA under the INFCIR/540.

12.1.2 Geographical information systems

GISs are software tools suitable to deal with data that are characterised by their position on the earth. Different GIS-based support systems can be conceived, depending on user needs:

- If images from satellites are geo-coded using accurate maps, sites on the ground can be located to within only a few tens of metres.
- Both states and the Agency have to deal with a large amount of information of different types (i.e. texts, numerical data, images, computer-aided drawings, maps, etc.). If these can be characterised by their location on the earth, the use of GIS is a natural choice for transforming such large volumes of spatially-related data into usable information that can enhance image interpretation.
- The rapid developments now being made in remote sensing have many implications for integration of satellite imagery and GIS data. The new remote sensing datasets, especially those with many spectral channels and frequent repeat coverage, will contain large numbers of spatially coincident data layers for GIS. Use and exploration of such layers with pre-existing GIS data would become a very powerful tool for the Agency.
- As the number of providers of satellite imagery increases, the costs of data are expected to reduce.

12.1.3 Cost/benefits and satellite imagery

Initially the cost of imagery may be high, but it can be managed, as the imagery will be acquired task by task. Moreover, as more and more countries launch and operate their own satellites and sell the data, the cost of imagery is bound to decrease:

- The benefit of satellite imagery is in adding confidence and accuracy to information available to the Agency. Imagery can act as a deterrent against undeclared activities, therefore raising the credibility of the Agency in general.
- The use of satellite images offers considerable scope in the planning of the Agency's on-site inspections.
- The duration of an inspection trip might thus be reduced, due to a better informed and prepared inspector. Visits to sites may even be avoided if the location and status of a site can be determined from satellite imagery.
- The use of commercial satellite imagery could also remove some of the political sensitivity attached to satellite monitoring.
- When ordering images, if the precise location of a site of nuclear facility is not known, then it is useful to order a low-resolution image that covers a large area. From such an image, the location is determined more precisely so that high-resolution data with smaller area coverage and more detail could be acquired. This has a considerable impact on the cost. Ordering quarter-scenes also reduces the cost of images.

12.1.4 Legal issues

- The Agency has the legal authority to acquire, from commercial suppliers, satellite imagery for its safeguards purposes. It is also in conformity with international law, in particular the Outer Space Treaty.

- However, the UN Principles Relating to Remote Sensing of the Earth from Outer Space may impose on the Agency certain obligations to make available to sensed states its data and information on a non-discriminatory basis and on reasonable cost terms.
- National laws, regulations and policies of data provider states may restrict the transfer of imagery to the Agency or make it subject to restrictive conditions regarding its procurement and/or subsequent use for safeguards purposes. The Agency should ascertain, in each single case, these restrictions or conditions before entering into any procurement contract. Equally, it should determine whether they are compatible with international space law and with obligations of the state concerned under the Statute of the Agency.
- The procurement and use of remote-sensing data and derived information are covered by the present safeguards rules and agreements of the Agency and do not require additional authority other than a budgetary appropriation.

12.1.5 Some drawbacks

- As for detection of warm water discharge plumes, their flow direction at a nuclear facility is strongly dependent on local wind conditions that could be affected by regional/global weather systems. This could make detection of plumes that can determine the status of a nuclear facility difficult.
- Even a civil remote-sensing satellite, with its relatively coarse resolution, could be used rather effectively for acquiring information about the sensed state's military capabilities. Often, such information is considered, by some states, as sensitive for wide distribution. Moreover, states are particularly concerned about the availability for sale of data on their economic resources and on activities related to their national security.
- It may be necessary to ensure that any verification method, in general, and satellite-based ones, in particular, do not become a source of instability or a basis for changing the balance of power in a region.

12.2 CONCLUSIONS

- Large parts of the electromagnetic spectrum used for remote sensing are beneficial to international safeguards applications.
- The recently launched US commercial observation satellite, QuickBird, with sensor resolutions 0.61 m panchromatic and 2.44 m multispectral, will greatly enhance the IAEA's safeguards procedures.
- As sophistication of image processing software and techniques increases and as they become available, extraction of more information from civil remote-sensing data will become possible. In this context, neural network and pattern-recognition techniques are important methods to be investigated.
- The use of GIS will no doubt increase the interpretability of commercial satellite imagery.
- With the increasing availability of GIS and image data, such systems can be further developed to provide more advanced GIS and remote-sensing capabil-

ities and to perform more effective and efficient analysis, management and decision making.

- The experiences of other users – such as the Western European Union Satellite Centre and the Agency's Action Team and UNSCOM – of satellite imagery have given many insights to the Agency, particularly from the organisational point of view.
- The Agency will not be able to conduct global searches for undeclared facilities and activities, but it will be able to acquire imagery when prompted by information based on open sources.
- Satellite imagery must be seen as an additional source of open information.
- The Agency has the legal authority to purchase commercially-available satellite imagery. However, it must be fully aware of national policy considerations, including the United Nations Principles Relating to Remote Sensing of the Earth from Outer Space, that might affect the use of satellite imagery by the Agency.
- The definitions of "sensitive" and "non-sensitive" data are yet to be resolved. One parameter, spatial resolution, has been discussed at length but no clear decision seems to have emerged. Meanwhile, the resolution of sensors on board civil remote-sensing satellites is continuously improving.
- Concerns have been expressed regarding national security measures becoming transparent to the world at large through commercial remote-sensing data. While such concerns are understandable, it is difficult to see how the dissemination of high-resolution images could be prevented: particularly, now that the very same technology could provide information on illicit nuclear activities of states.
- Finally, future work should include an examination of:
  1. the use of stereoscopic images;
  2. the usefulness of the newly-launched US Landsat 7 satellite with a much higher resolution (60 m) of the thermal band than that available from the current Landsat 5 satellite (120 m);
  3. the utility of high-resolution data (panchromatic 0.6 m and 2.44 m multi-spectral);
  4. the application of hyperspectral images for observations of, for example, mines; and
  5. further examination of the use of SARs, including those deploying multi-polarised beams. Factors such as angle of observation, the design of warm water discharge methods, local current and surface wind conditions need further study, as all or some of them may play a role in the detection of cooling water discharge from facilities.

As the capabilities of commercial observation satellites increase and as more and more countries are able to launch and operate them, their use for the IAEA safeguards application will unquestionably increase. Not only this, but also they will be used for monitoring other arms control agreements, resulting in other UN agencies such as the OPCW and the CTBTO employing such satellites.

Index